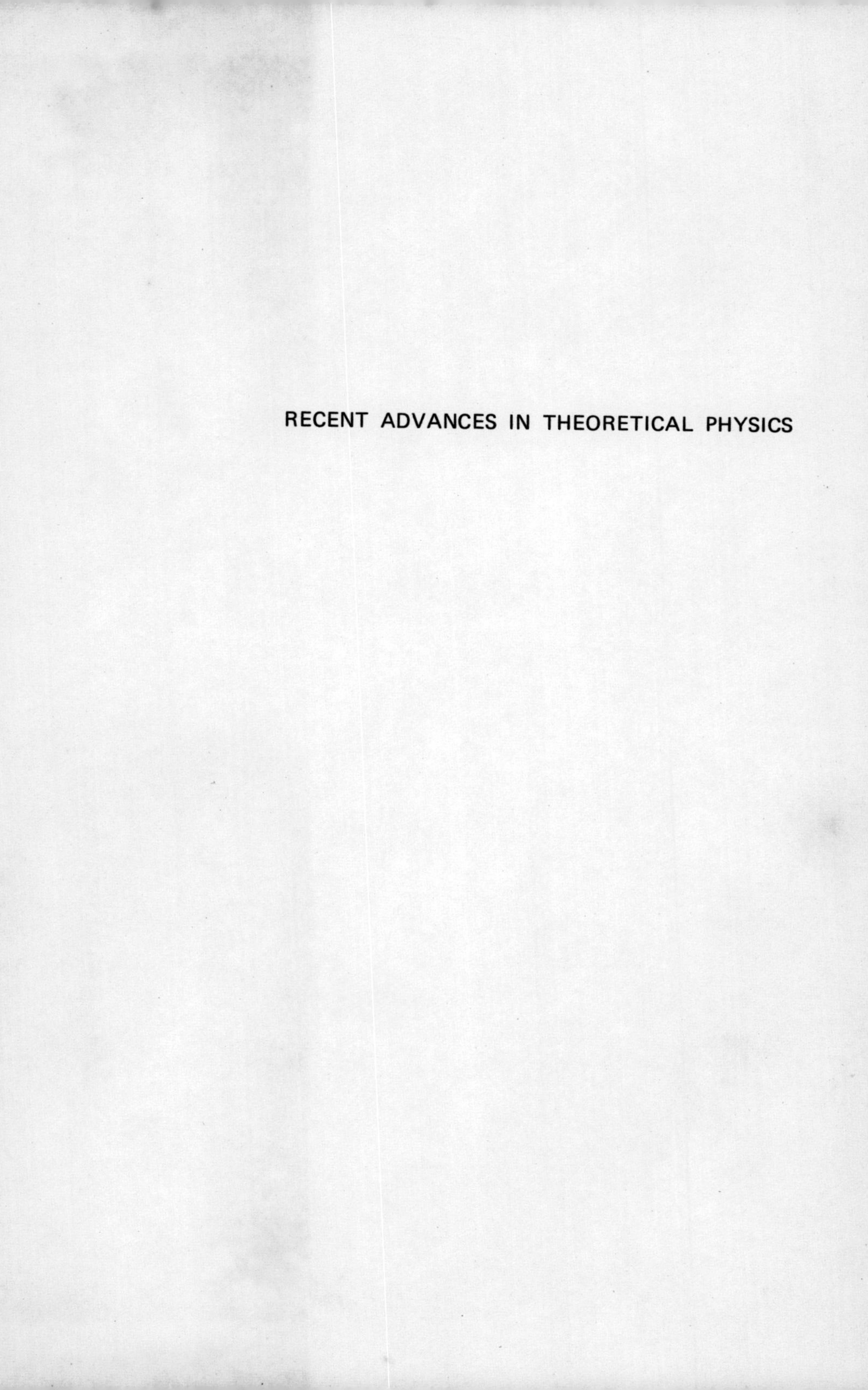

RECENT ADVANCES IN THEORETICAL PHYSICS

RECENT ADVANCES IN

THEORETICAL PHYSICS

Proceedings of the Silver Jubilee Workshop held at IIT, Kanpur, December 1984

edited by

R RAMACHANDRAN

Professor
Department of Physics
Indian Institute of Technology
Kanpur 208016
INDIA

World Scientific

Published by

World Scientific Publishing Co Pte Ltd.
P. O. Box 128, Farrer Road, Singapore 9128
242, Cherry Street, Philadelphia PA 19106-1906, USA

ISBN 9971-50-014-0

Printed in Singapore by Kim Hup Lee Printing Co. Pte. Ltd.

Editorial Remarks

The principal objective of the Silver Jubilee Workshop on "Recent Advances in Theoretical Physics" was to bring together outstanding experts as well as young scientists in diverse branches of Theoretical Physics with a view to probe the common trends and develop interdisciplinary appreciation. Recent developments in Particle Physics has established an essential link between the micro-world (basic interactions) and the super microworld (origins and composition of the universe). Ideas that had their origin in condensed matter physics are making deep inroads in the study of gauge theories and variety of phenomenon that are implied by them. Quoting Sudarshan "The miracle of theoretical physics is theoretical physics itself. In fact, it is amazing that one is able to get so much out of it".

The Workshop was designed to be both pedagogic in the main expositions by the experts and incisive in the discussions that followed. There were supplementary informal sessions that cut through the barriers between disciplines and the ever alert enthusiastic participants made the Workshop a memorable event. There were also seminars on recent work by participants.

Professor ECG Sudarshan gave the inaugural remarks to set the Silver Jubilee Workshop in action and followed it up with a guided tour, tracing the history of Particle Physics - area that was taken up for major emphasis in the Workshop. Thus begun, there was undiminished enthusiasm and excitement of hectic activities for the next twelve days that covered major front ranking topics in Particle Physics, condensed matter and Statistical Physics and Nuclear Physics. It has been a rewarding experience for me to assemble them all in this volume.

While the expository lectures and discussions are covered in reasonable detail, I have found it prudent to limit the seminar presentations to brief self contained summaries in the hope that the readers will find the missing details in the related articles that may have been published by now.

The task of organisation of the Workshop was shared enthusiastically by all members of the theoretical physics groups; I should nevertheless make a special mention of Drs. K Banerjee, JK Bhattacharjee, HS Mani, A Mukherjee, VA Singh, SK Sharma, M Yussouff and YR Waghmare.

We are very grateful for the institutional support, given to us unstintingly by Professor RM Singru, Head, Department of Physics and Professor S Sampath, Director, Indian Institute of Technology, Kanpur. In addition to IITK we received generous financial support from Department of Science and Technology (DST). Department of Atomic Energy (DAE), Council of Scientific and Industrial Research (CSIR) and Indian National Science Academy (INSA) and we are very pleased to acknowledge the same. The National Science Foundation of the U.S. provided travel support for some of the discussion leaders from the U.S. and ICTP, Trieste extended travel assistance to participants from the developing countries.

It is my pleasant duty to record my appreciation of a meticulous, accurate and neat production of several of the manuscripts in this volume by Srimathi B Rukmini and the assistance in painstaking proof reading by Dr. A Mukherjee, Messrs AK Agarwal, S Godre and Krishna Kumar.

Kanpur
June 1985 *R Ramachandran*

CONTENTS

From Left to Right –

Sitting on Chairs – D. C. Khan, R. Pandit, Y. R. Waghmare, R. Ramachandran, A. Maheshwari, I. T. Cheon, R. A. Ferrell, E. C. G. Sudarshan, B. Sakita, S. N. Biswas, G. Rajesekaran, F. Hussain, A. P. Balachandran, H. S. Mani.

Standing between 1st Row and Chairs – Rohini Godbole, Swarnamala Sirsi, S. N. Sandhya, Sumita Bhattacharya, Sutapa Sen, Neelima Gupte, Anuradha Misra, A. Mukherjee.

Standing 1st Row – D. K. Roy, V. Joshi, A. Tyagi, K. Banerjee, Deepak Kumar, C. P. Singh, R. C. Mishra, R. Parthasarathy, H. Rizvi, A. Guha, M. Yussouff; D. Subba Rao, M. C. Jain, S. K. Sharma, R. S. Kaushal, Ajit Kumar, V. K. Sharma, V. Srivastava, S. Chaturvedi, S. Maiti, B. Dey, S. D. Joglekar, D. Khandekar, M. Mathur, S. K. Paul, D. Lohiya.

Standing 2nd Row – L. S. Rathaur, H. K. Panda, P. K. Srivastava, A. K. Kapoor, V. M. Raval, P. Sharan, P. Tripathi, P. K. Raina, S. Naik, S. S. Godre, G. Mukherji, G. C. Mohanty, S. Abdullah, V. Ravi Shanker, M. S. Sriram, T. R. Govindarajan, S. Lawande, S. Panda, S. S. Manna, J. P. Singh.

(KAPOOR Studio Ph-246208. Kanpur.)

INAUGURAL REMARKS

E.C.G. Sudarshan
Institute of Mathematical Sciences, Madras

It is an honour to speak to you in this Inaugural Session of the Silver Jubilee Workshop at the Indian Institute of Technology, Kanpur. This quarter century is also the time I have been of age in professional science with its joys and disillusionments, vigorous and slow periods both in my own creative life and in particle physics.

Particle physics deals with matter into its innermost parts, the world in its making. We seek clues to the grand design of the universe and do it by humble and often tedious steps. It is the toil of many though often a few are singled out.

In its development the practising scientist is taken through stages where beliefs are questioned, "facts" are recognized to be beliefs and the design often unexpected. There is no certain knowledge, one proceeds along the path as a "radical conservative" reluctant to abandon accepted methods unless forced to do so. Yet the nature of particle theory has changed dramatically over the years.

I have been fortunate to be born where I was born and to get the education that I got; and to recognize above all that knowledge can be a source of joy, that in giving and sharing knowledge one attains greater happiness than being the "cock-of-the walk". When I went for research study this Institute was not founded and I found myself at TIFR with its unique opportunities and frustrations. From Bernard Peters I learnt hard work in science to exhaustion and from Homi Bhabha the responsibility to be a jagat-guru in science. What was lacking in not being able to get from an excellent I.I.T. graduate education was made up by having a "to pursue knowledge like a sinking star".

After my joining University of Rochester under the guidance of one of the greatest teachers I came across, Robert Marshak, I pursued the study of particle physics. It can be truly said of him:

yasya sannidhi mātrena
j̃nānam utpatyade svayaṁ
sa eva sarva sampatti
tasmāt sampūjayed guruṁ

I saw this happen not only in my case but also of Susumu Okubo. One of the visions of the teacher in the Indian tradition is that of one who does not talk at all. The interaction should be a confluence of instinct and contemplation.

My good fortune in the discovery of the universal chiral V-A interaction, the first result in broken symmetry etc. did not make it easy for a ´short brown man in search of science` to live an untroubled life. I learned painfully and gradually to distinguish between the universality of science and the partisan parochial behaviour of scientists, sometimes even great scientists. Yet the caravan must go on - one cannot let others decide your goals or your happiness.

In my second year as a young assistant professor I had to be responsible for more than half a dozen graduate students´ research. I must have driven my family to distraction since "any time was physics time". (I remember once my son telling me when I asked him who had called on the phone "It is not important, it was not physics !") One of the hardest things for me to adjust was the ingratitude of those to whom one gave without counting. But the students that I had were a continual source of stimulation. One of them who stayed on with me for a year after his Ph.D. used to greet me every morning with a few more steps of calculation than we had discussed the previous day. With such people around there is no danger of one aging.

I have traversed much ground in physics over various subdisciplines, been a member of many institutions and had the opportunity to associate with many scientists. Some of these were inspiring and I remember them with gratitude; some were painful and I am grateful they are behind me. But above all I realize that other peoples´ opinions need not be right; one needs a number of friends whom one respects who could give you their critical assessment of your work. It is in this collective endeavour that truth often reveals itself.

Now to some observations about the nature of theoretical physics itself. As its name implies it is physics that we talk about and not mathematics; and mathematical precision is to be balanced by common sense regarding the relevance of the models and equations involved. Unlike in mathematics where there are truths and falsehoods and what theoretical computer scientists like Jay Misra call "knowledge", theoretical physics deals with "beliefs" and possibly relevant models. One is never sure !

The fabric of theoretical physics has changed a great deal over the last quarter of a century. the change has been so drastic that the "star performers" of yesteryear like Arnold Sommerfeld, Max Planck etc. would not be able to comprehend many of the things we now talk about. While the fabric might have changed the methods are old like the Brahma. The miracle of theoretical physics is theoretical physics itself. In fact it is amazing that one is able to get so much out of it. The equations of theoretical physics seem to have a life of their own. They seem to give more than we put in - real things appear to come from outside, as if by grace. The discoverers are just accidental people - mere mechanisms through which the hidden truths of the nature get expressed. All that a physicist can do is to be dedicated and disciplined.

I am often asked whether theoretical physics deals with the real world or a world of concepts, mathematics and imagination. I cannot distinguish between the two - in defense of such a defenseless position I quote from Sankara´s "daksinamurti stotram":

visvaṁ darpanadruṣyamāna nagari tulyam nijāntargatam
pasyannātmani māyayā bahurivodbhutam yatha nidraya
yah sakṣāt kurute prabhodhasamaye svātmanamevādvayaṁ
tasmai´sri gurumurtaye namah iti dakṣiṇamurtaye

Every teacher should be dakṣiṇamurti, with an inside and outside (like thoeretical physics).

Like dakṣiṇamurti this Institute of Technology is in the north and benevolently looking on all who wish to learn. What is the institution if it is not the set of minds united by the search for knowledge ? We may play the role of the student or the guide: but all of you know the roles often interchange.

Institutions of higher learning should not become closed entities, otherwise they, like the black hole, do not retain memory and become irrelevant.

What is needed for encouraging and enhancing the level of scientific activity in the country ? Is it more expensive laboratories and libraries, more scholarships and salaries, more western scientific visits and visitors ? Or do we need more dhiras, more sankalpa ? A story that I learnt in my childhood was about young Solomon who had just been anointed king. Solomon performed a yaga and the Almighty was pleased with him and asked to choose one of several boons. He chose wisdom since he had to rule over the ´chosen` ; the Almighty was pleased and gave him other things too. We too have accepted the choice to this high calling. Let us choose the boon of wisdom so that we may see correctly, have insight, be able to judge without prejudice. While wealth and fame may not come immediately we will be happier.

It is my hope that this gathering here to celebrate the Silver Jubilee by dedicated work to the best of our ability is a portent of the continued and renewed dedication to knowledge in the highest by doing our best in mutual nurture and in joy.

THE PARTICLE PHYSICS KALEIDOSCOPE

E.C.G. Sudarshan
Institute of Mathematical Sciences
Madras, India

1. The Birth of Particle Physics

Particle Physics began with the turn of the century. The electron was discovered as a cathode ray and the photon was isolated from blackbody radiation. The atomic nucleus was a complex structure but electrons and photons were universal. It took us a while to understand that they were neither classical particles nor waves but new kinds of objects obeying a new physical discipline. To the extent they were to be thought of as particles the electrons obeyed Fermi statistics in accordance with the "exclusion principle" while photons obeyed Bose statistics allowing for their accumulation and eventual almost classical behaviour in aggregates. Once the laws of quantum mechanics were established the law of interaction of photons and electrons turned out to be the familiar laws of electromagnetism adapted to this quantum system. The electromagnetic field played a dual role: on the one hand it was the binding force acting on electrons holding them in the vicinity of the nucleus; on the other it was that entity whose quanta were the photons.

Later experience showed that this was true of the electrons also though the electron field had no immediate classical analogue. Field quantization could be viewed as second quantization, the classical field being viewed as a one-particle wave function sans normalization and accompanied by the conjugate wavefunctions.

With the discovery of the neutron, the proton having been identified as the hydrogen ion, the table of fundamental particles seemed to be sufficient to deal with most of the physics known then. Atomic physics involved only the electrons and photons with their known electromagnetic interaction; and nuclear physics could handle complex nuclei and gamma radioactivity. Alpha radioactivity was quantum tunneling. Only beta radioactivity created a new problem because there was a new transmutation involved. And, of course, there was the problem of the nuclear force. Why should the electric force have quanta called photons and the nuclear force no quanta ?

The mid nineteen thirties were the time when both these questions were answered, the first by Fermi in his beta decay theory and the second by Yukawa in his meson theory. The cosmic ray penetrating component was not due to a meson at all even though it usurped that name for a while: now we call it the muon. The muon was for a long time the odd particle with no function (except to serve as the decay product of pions and in turn to decay). But now it has acquired a muonic neutrino as partner and together they are seen as a recurrence of the electron, neutrino doublet.

The Yukawa meson which was responsible for the nuclear forces was identified with the pion which was seen in nuclear emulsion by Powell even as it was being predicted as the parent in a decay scheme by Bethe and Marshak. Since the pion is pseudoscalar its coupling with the nucleon should be straightforward: either directly with a pseudoscalar density of the nucleon or via its derivative to the nucleon pseudovector current. In a scalar theory Yukawa had derived an exponential cut-off for the Coulomb potential but for a pseudoscalar this would lead to a tensor force in the lowest approximation. In the meantime independent of the details of the nuclear force calculation, the "charge independence" of the nuclear-nucleon force was shown to demand the contribution of a neutral pion coupled to both proton and neutron with half the strength as the charged pion. Kemmer showed that this involved a new continuous symmetry group, the "isotopic spin" group SU(2). The nucleons were the fundamental representation of this group and the pions were the adjoint representation. Thus was born the first broken symmetry : the charge dependent electromagnetism breaking the charge independent pion-nucleon interaction. Later on with the advent of renormalization theory it became apparent that a quartic pion-pion self coupling term should be added to the "strong" interaction. The pion was the star actor on the nuclear stage.

In the meantime beta radioactivity had been formulated as a new theory by Fermi on the basis of a new fundamental interaction coupling neutron, proton, electron and neutrino. This coupling constant, unlike the coupling constants of electromagnetism and pion-nucleon interaction, had a nontrivial dimension; but in terms of the low energy reaction it engendered the effective coupling strength was much smaller than that for electromagnetism. So these couplings were "weak".

2. Towards a Quantum Field Theory

In the World War years and immediately after quantum field theory, particularly quantum electrodynamics, underwent major development. The problem of infinite number of degrees of freedom and the consequent divergences in the classical statistical mechanics of the electromagnetic field which plagued the Rayleigh-Jeans formula was overcome by the quantum hypothesis of Planck. While this made the effective degrees of freedom finite at any finite temperature as far as real photons are concerned, the same does not apply to virtual photons. Consequently the naive calculations of the higher order corrections involving virtual photon contributions gave divergent answers. There had to be new theoretical developments to handle this: and the idea of reexpressing all physical results in terms of directly observable coupling strengths and masses turned out to take care of these divergences. Mass renormalization was presumably discovered by fishermen plying their boats in sea weed and kelp beds and coupling constant renormalization by Archimedes. In the quantum field theory context the unrenormalized constants were not easily "observed" - it was as if Archimedes was permanently in his bath.

Beta decay theory would not lend itself to the renormalization methodology.

Not only was the simple world of neutrons, protons, electrons and photons complicated by the neutrino, the muon and the pions but more was in store. Rochester and Butler in the early fifties saw evidence for a new set of particles whih with subsequent systematic cosmic ray studies turned out to be the fore runners of "strange particles". Some order was restored to these in the mid-fifties (when I was a research student at TIFR), but the real advance was the discovery that isotopic spins could be integral or half integral for bosons and fermions. This could make it possible to introduce a new quantum number called "strangeness" to be introduced by Nishijima and by Gell-Mann which was again a broken symmetry. The copious production and reluctant decay could be accounted for as strangeness preserving strong interactions and strangeness violating weak decays. By this development one enlarged the symmetry scheme of strong interactions from SU(2) to SU(2) x U(1); and enlarged the scope of decay interactions from beta decay to weak interactions.

Already in the late forties people like Lee, Rosenbluth and Yang, Tiomno and Wheeler, and Klein had recognized that muon decay was not unlike beta decay and could be governed by similar coupling strengths. Thus was born the idea of "universal Fermi interaction" of which the capture of orbiting negative muons by nuclei was the third leg of a triangle. The dominant decay of the pion into the muon-neutrino pair was another expression of this leg. It required another decade for Marshak and me to discover the chiral V-A form of the universal Fermi interaction.

3. The Four Interactions

By this time there were already four different types of interactions - the strong interactions responsible for nuclear forces and meson production, the electromagnetic interaction, the weak interaction and, of course, universal gravitation. Electromagnetism had classical roots but worked well for atomic physics with natural extension to the nucleus. The photons were everywhere beginning with the experiments of Hertz and Jagadish Bose. For gravitation we only had the force but no detectable quanta. For weak interactions we had the transitions but no appreciable force. And in strong interactions, for most of the nuclear force we had no real quanta (as seen from the uncertainty relations) but the pions and strange particles could be produced. All attempts at getting detailed meson theoretic potentials of sufficient precision failed.

The development of renormalized relativistic quantum field theory in terms of a perturbation expansion, so successful for quantum electrodynamics, was not as useful for strong interactions. New methods were being sought which were nonperturbative and generally applicable. During the early fifties several authors contributed to the development of such a scheme. Some of those pioneered a mathematically more satisfactory formulation of quantum field theory while others were inclined to search for nonperturbative general physical results. The Kamefuchi-Umezawa-Källen-Lehmann representation of the exact propagator was the first example of such a result. In the meantime Schwinger and Feynman had formulated nonperturbative methods using different starting points: Schwinger from the "action principle" and Feynman from "integral over all paths". The latter has been of particular interest in recent years for the quantization of complex gauge theories. But this was still to be in the future.

The interest in general field theoretic transition amplitudes reached a peak with the discovery of "dispersion relations" introduced by Goldberger and others. The program which related real and imaginary parts of transition amplitudes via Hilbert transforms gradually became the basis of a new dynamical formulation at the hands of Chew who saw in it many virtues to calculate the effective low energy low momentum transfer strong interaction amplitudes without having to bother about Lagrangians and renormalization. Later on Regge discovered the large momentum transfer behaviour potential scattering amplitudes parametrized by complex angular momenta and interpreted it as a parametrization of high energy behaviour of scattering amplitudes. This approach dominated high energy hadron phenomenology in the early sixties.

4. The Parity Revolution

In the intervening time great new developments were taking place in weak interaction physics. Lee and Yang proposed the possibility of parity violation in weak interactions which not only was immediately confirmed by experiments but gave an excellent new avenue for more detailed study. The question of universal Fermi interaction was revived in this context. The culmination of these was in the chiral V-A interaction of Sudarshan and Marshak in which only the left chiral components were involved independent of the mass. Electromagnetism of charged Dirac particles coupled left chiral components to themselves and right chiral components to themselves, but weak interactions coupled left chiral components to themselves and the right chiral components not at all. Mass of a Dirac particle now appeared as an interloper breaking the chiral symmetry and so did the pion coupling. So what one eventually did was to tie up these two problems into one by considering a nonlinear realization of chiral symmetry.

Back to chiral fields: if both electromagnetism and weak interactions coupled left chiral components to themselves then the pion couplings were not the fundamental interactions but some other vectorial interaction. If it was mediated by mesons they had to be vector mesons. Even in such a case the effective low energy and hence long range nuclear force would be due to pion exchange.

5. The Eight-fold Way and Current Algebra

During the nineteen sixties two important developments took place. One was the experimental discovery of pion-pion resonances using the effective mass plots that Sudarshan had advocated in the X Rochester Conference; and subsequently the baryon resonances of which the first member, the (3,3) nuclear resonance, was found by Fermi and collaborators in the early nineteen fifties. The other not unrelated development was the completing of SU(2) x U(1) with SU(3) with the "eightfold way" realization of Gell-Mann and of Ne'eman - they used the same SU(3) group as Sakata but entirely different assignment of the hadrons. The hadron resonances easily fell into natural representations of SU(3) and led to the triumphal discovery of the omega minus resonance. The weak interaction currents were identified as octet tensors in SU(3) and the embarrassment of too few strangeness violating decays was parametrized by the Cabibbo angle within the SU(3) scheme. Since by this time analytic S-matrix techniques were well developed SU(3) was infused into them with indifferent triumphs.

The study of weak interaction had shown that like electric charge they too cut across hadrons and leptons. The equality of the vector weak coupling constant of nucleons and leptons prompted Feynman and Gell-Mann to replace the weak interaction vector current of the nucleons by the conserved isotopic spin current. To the extent one could neglect electromagnetism there was no renormalizaation needed for the vector coupling constant. However, the axial vector coupling constant was renormalized but only by a small amount. How come ?

To give a possible explanation for this Gell-Mann suggested that the currents were all normalized by their mutual commutation relations and that the divergence of the axial vector current was proportional to the pion field. Once this choice was made the proportionality factor could be deduced from the amplitude for pion decay. Thus the axial vector renormalized coupling constant could be related to pion-nucleon scattering. Using this method Adler, Tomozawa and Weisberger were able to calculate the axial vector renormalization.

6. Color

It was clear that while there were vector mesons amongst the hadronic resonances they had no right to be identified with the fundamental strong interaction agents. But the appearance of vector mesons on par with pseudoscalar mesons; and of spin $3/2^+$ resonance on par with spin $1/2^+$ baryons was suggestive of a structure for the hadrons, the more so since the octet of baryons were not in the fundamental 3-dimensional representation of SU(3). The mesons could be thought of as composites made of a triplet and its conjugate. But the baryons and the baryon resonances could be only identified with a third rank symmetric tensor in a triplet of spin 1/2 objects. So if Fermi statistics were to hold we had to have some other degree of freedom with respect to which the baryon complex was completely antisymmetric. This was identified with "color" (perhaps since Green and Greenberg were involved in this development). There should at least be 3 distinct colors and assuming that is all there is, it is natural to consider an SU(3) group acting on color. Since this is an exact symmetry one expects that it will play a role similar to the U(1) group of electromagnetism. One can consider still higher groups but in this process we may consider the left chiral and right chiral parts of a Dirac field as distinct fields.

7. Gauge Groups

If one were to identify the weak interactions and electromagnetic interactions the symmetry group connected with them must be SU(2) x U(1) with a four dimensional Lie algebra. But while the photon exchanged in electromagnetic interactions is massless the other vector mesons mediating weak interactions ought to be massive to give an effective low energy four-fermion interaction structure. so a neutral weak current - current interaction should be present. But if the masses of the vector mesons were primitive such a theory was not renormalizable. So Weinberg and Salam derived a theory in which the Lagrangian did not contain wither vector or spinor masses; and they were generated by virtue of a spontaneous symmetry violation using the "Higgs mechanism". It later turned out that `t Hooft was able to show that such theories were renormalizable if they were of the Yang-Mills type.

Since every Lie algebra seemed to seek gauge vector fields as its local expression one introduces vector fields with the color group SU(3) and allows these objects to generate the strong interaction. This theory is called quantum chromodynamics - according to it the hadrons are all composite and are primarily coupled to the "gluons" of the chromodynamic field. The forces between the hadron are the residual forces left after binding the colored "quark" triplets suitably. The nuclear force is then somewhat like the van der Waals force between neutral atoms. The internal symmetry group of particle physics is now SU(3) x SU(2) x U(1).

8. Grand Unification

It is natural to ask whether, like the SU(2) x U(1) of isospin-strangeness was subsumed in the SU(3) of the eightfold way, we could subsume SU(3) x SU(2) x U(1) in some "grand unification" group. Models involving SU(5) and SO(10) have been studied with some success. In such grand unified theories we have new interactions cutting across quarks and leptons leading to potential instability of the proton. Such proton decay experiments are now being pursued and evaluated.

To give quantitative expression to a broken grand unified theory we need to have some technique to evaluate the masses of vector mesons exchanged in lepton number changing interactions. For this one makes use of certain general field theoretic relations of renormalization group theory. to deduce these masses from the observed ratio of the fine structure constant to the chromodynamic coupling strength. Whether such a scheme is in accordance with nature or not is as yet unclear.

9. Supersymmetry and Gravity

Another area of development is supersymmetry. In place of a Lie group with commutators between the generators one can consider a graded Lie algebra with both commutators and anticommutators. The anticommuting generators are spinorial in character and thus the "superfield" contains fields with different spins. Such superfield theories have better convergence properties and are therefore being assiduously studied.

Gravitation has been with us for a long time but it has become part of particle physics only in the past decade. With its coming in the emphasis all along made regarding differential geometric methods has come into particle physics also. There is now the possibility that in the unification of gravity with other interactions both are modified - internal and space-time symmetries are part of a whole, the former arising from the latter by "spontaneous compactification".

10. Nature of Particle Theory

Particle physics in the past quarter century has developed a great deal and we see that what we considered before as accidents are essential parts of the design of the world. We see for example that the muon and its neutrino have their counterpart in the strange and charmed quarks just as the electron and its neutrino are the counterparts of the down and up quarks. A third pair of pairs is provided by the tau lepton and its partner and the top and bottom quarks. This close similarity of quarks and leptons is an update of the baryon-lepton symmetry; and it has prompted Pati and Salam to consider leptons as quarks of a fourth color.

Theoretical ideas and intuitive leaps are not lacking but there are many computational questions to be answered, in particular to know whether known systems have consistent and suitable solutions. For the lattice version of gauge theories Wilson and others are industriously pursuing computer solutions.

All in all, particle physics shows no signs of slowing down. Much of its renewed vigour has been due to accepting concepts and models from condensed matter physics. In other areas like the study of monopoles and other topological structures it is the insistence of fundamental quantum questions that give vigour. It is my expectation that as we come to the closing years of this century, particle physics has still more surprises for us.

STOCHASTIC QUANTIZATION*

B. Sakita
Department of Physics, City College of City University
of New York, New York, NY 10031, USA

I. INTRODUCTION
- Canonical Quantization
- Path-Integral Quantization
- Euclidean Path-Integral
- Parisi-Wu Stochastic Quantization

II. GENERAL THEORY

III. GENERALIZATIONS
- General Fokker Planck Hamiltonian and Corresponding Langevin Equation
- Stochastic Quantization of U(N) Lattice Gauge Theory
- Stochastic Quantization of Fermi Fields

IV. PATH-INTEGRAL FORMULATION OF STOCHASTIC PROCESSES AND HIDDEN SUPERSYMMETRY

V. STOCHASTIC QUANTIZATION OF GAUGE FIELDS

VI. FURTHER REMARKS

*Supported in part by a grant from the National Science Foundation under contract NSF-PHY-82-15364 and by CCNY Faculty Research Award Program #6-63264.

I. INTRODUCTION

The quantization methods used in field theories are i) canonical, ii) path-integral, and iii) stochastic quantization. The last method is relatively new. It was originated by Parisi and Wu[1)] in 1980. In this lecture I will try to review the stochastic quantization method without assuming reader's knowledge on this subject at all. Thus, I first recall the standard canonical quantization.

Canonical Quantization:

Let $L(q,\dot{q})$ be a Lagrangian of a system, q being a dynamical variable and $\dot{q}$ its time derivative. The canonical momentum p is defined by

$$p = \frac{\partial L}{\partial \dot{q}} \tag{1.1}$$

and Hamiltonian of the system is given by the following Lagrange transform:

$$H(p,q) = p\dot{q} - L(q,\dot{q}) \tag{1.2}$$

In canonical operator formalism of quantum mechanics the dynamical variable $\hat{q}$ and its canonical conjugate momentum $\hat{p}$ are operators in a Hilbert space (from here on operators are denoted with ^) and satisfy the following canonical commutation relation:

$$[\hat{q}, \hat{p}] = i \tag{1.3}$$

The state of the system is a time dependent vector in the Hilbert space (Schrödinger picture), and the mechanical equation of the state vector is the Schrödinger equation:

$$i \frac{\partial}{\partial t} \; |\psi(t)> \; = \; H(\hat{p},\hat{q})|\psi(t)> \tag{1.4}$$

where $\hat{H}$ is a Hamiltonian operator obtained from the classical Hamiltonian (1.2) by promoting the classical variable to the quantum operator variables. In this procedure there exists an ambiguity if p and q appear in a product form since $\hat{p}$ and $\hat{q}$ do not commute in quantum mechanics. This is called the ordering ambiguity. In this case one must define the quantum mechanics by specifying the order of the operators. Accordingly, to a classical system many quantum mechanical systems correspond.

A formal solution of Schrödinger equation (1.4) is given by:

$$|\psi(t)\rangle = \underbrace{e^{i\hat{H}t}}_{\text{evolution operator}} |\psi(0)\rangle \tag{1.5}$$

Using the coordinate representation

$$\hat{q}|q\rangle = |q\rangle q, \tag{1.6}$$

one defines the Feynman kernel

$$\langle q|\, e^{-i\hat{H}t}\, |q'\rangle \tag{1.7}$$

which is a transition amplitude of a particle in q' at t=0 to q at later time t.

The extension of this one variable quantum mechanics to many variables is formally done by adding an appropriate index to the canonical variables:

$$q_a\ ,\ p_a\ ;\ [\hat{q}_a, \hat{p}_b] = i\delta_{ab} \tag{1.8}$$

Field theories are the systems of many variables. The dynamical variables are on each space point. For example, for a scalar field theory one denotes the canonical coordinates and its conjugate momentum by $\phi(\vec{x})$, and $\pi(\vec{x})$ respectively. The canonical commutation is given by

$$[\hat{\phi}(\vec{x}), \hat{\pi}(\vec{x}')] = i\,\delta\,(\vec{x}-\vec{x}') \tag{1.9}$$

The coordinate representation in field theory is the $\phi(\vec{x})$-diagonal representation:

$$\hat{\phi}(\vec{x})\ |\phi\rangle = |\phi\rangle\ \phi(\vec{x}) \tag{1.10}$$

The wave function in this case is a functional of $\phi(\vec{x})$:

$$\langle \phi|\psi(t)\rangle = \psi\ [\phi\ (\cdot),\ t] \tag{1.11}$$

The Feynman kernel is similarly defined:

$$\langle \phi|\ e^{-i\hat{H}t}\ |\phi'\rangle \tag{1.12}$$

Path-Integral Quantization:

The path-integral quantization is defined by specifying the Feynman kernel as a sum over the paths i.e. the path-integral. Formally it is expressed as:

$$\begin{aligned} &\langle q|\ e^{-i\hat{H}t}\ |q'\rangle \\ &= \int..\int Dq\ \ e^{\,i\int_0^t L(q(t'),\ \dot{q}(t'))dt'} \\ &q(t) = q \\ &q(0) = q' \end{aligned} \tag{1.13}$$

The precise definition of the path-integral is as following. We divide the time interval into N small segments:

$$t > t_{N-1} > t_{N-2} > \dots > t_3 > t_2 > t_1 > 0 \qquad (1.14)$$
$$\underset{t_N}{\shortparallel} \qquad\qquad\qquad\qquad\qquad \underset{t_0}{\shortparallel}$$

Then consider a set of N-1 variables, which we write $q(t_i)$ $(i=1,2,\dots, N-1)$. We set $q(t_0) = q'$ and $q(t_N) = q$. Then, to a set of values of these variable a zigzag path from q' to q corresponds as shown in the following figure:

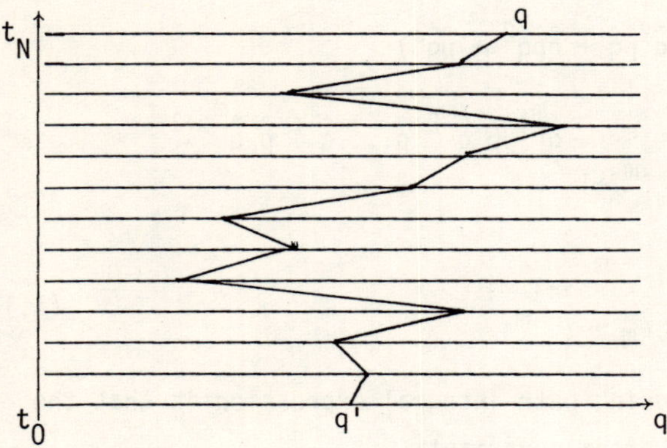

The action in (1.13) is replaced by

$$\int_0^t dt'\, L(q(t'), \dot{q}(t')) \Rightarrow \sum_{i=1}^{N-1} (t_{i+1}-t_i)$$

$$L\left(\frac{q(t_{i+1}) + q(t_i)}{2} \quad , \quad \frac{q(t_{i+1}) - q(t_i)}{t_{i+1} - t_i}\right) \qquad (1.15)$$

and the integration measure by

$$Dq \Rightarrow \prod_{i=1}^{N-1} dq(t_i) \left(\frac{2\pi i}{t_{i+1}-t_i} \right)^{\frac{1}{2}} \qquad (1.16)$$

The path-integral (1.13) is the $N=\infty$ limit of this multiple integral.

This prescription is called as the mid-point prescription since in (1.15) q(t') in the Lagrangian is replaced by the mid-point of $q(t_{i+1})$ and $q(t_i)$.

It is possible to obtain this path-integral expression from the operator formalism. The corresponding operator ordering to the mid-point prescription is the Weyl ordering defined by

$$(\hat{p}\hat{q})_W = \frac{1}{2}(\hat{p}\hat{q} + \hat{q}\hat{p})$$

$$(\hat{p}\hat{q}^3)_W = \frac{1}{4}(\hat{q}^3\hat{p} + \hat{q}^2\hat{p}\hat{q} + \hat{q}\hat{p}\hat{q}^2 + \hat{p}\hat{q}^3)$$

$$(\hat{p}^\ell\hat{q}^m)_W = \frac{1}{A}\sum_{\substack{m_1 m_2 \ldots m_{\ell+1} \\ \sum_i^{\ell+1} m_i = m}} \hat{q}^{m_1}\,\hat{p}\,\hat{q}^{m_2}\,\hat{p}\ldots\hat{q}^{m_\ell}\,\hat{p}\,\hat{q}^{m_{\ell+1}}$$

$$A = \sum_{m_1 m_2 \ldots m_\ell+1} \delta_{m\,,\,\sum_i^{\ell+1} m_i} \tag{1.17}$$

The advantage of using path-integral expression is that the expression is formally Lorenz covariant:

$$\begin{aligned} &< \phi|\, e^{-i\hat{H}t}\, |\phi' > \\ &= \int\ldots\int D\phi\, e^{i\int_0^t d\vec{x}\,\mathscr{L}(\phi(x),\, \partial_\mu\phi(x))} \end{aligned} \tag{1.18}$$

$$\phi(\vec{x},t) = \phi(\vec{x})$$
$$\phi(\vec{x},0) = \phi'(x)$$

Euclidean Path-Integral:

In quantum statistical mechanics one considers the partition function defined by

$$Z = \mathrm{tr}\, e^{-\beta\hat{H}} = \int dq < q|\, e^{-\beta\hat{H}}\, |q > . \tag{1.19}$$

So, the relevant matrix element to consider is

$$< q|\ e^{-\beta\hat{H}}\ |q' > \tag{1.20}$$

which can be obtained from the Feynman kernel (1.7) by replacing t by $-i\beta$. This is accomplished by considering a complex t-plane and by rotating the time axis by 90^0 clockwise to the negative imaginary axis (Wick rotation).

The action integral is then

$$i \int L\ dt \ \rightarrow\ -S \tag{1.21}$$

where

$$S = - \int_0^\beta d\tau\ L(q,\ i\dot{q}) \tag{1.22}$$

which we call Euclidean action.

For example, for the following Minkowsky Lagrangian

$$L = \frac{1}{2}\dot{q}^2 - V(q) \tag{1.23}$$

one obtains

$$S = \int\ d\tau\ [\ \frac{1}{2}\dot{q}^2 + V(q)\ ] \tag{1.24}$$

For a field theory in d-dimensional space

$$L = \int d\vec{x}\ \{\frac{1}{2}\dot{\phi}^2 - \underbrace{(\ \frac{1}{2}(\vec{\nabla}\phi)^2 + \frac{1}{2}m^2\phi^2 + \ldots.)}_{\substack{\parallel \\ V}}\} \tag{1.25}$$

one obtains

$$S[\phi] \equiv S = \int d\vec{x} \int_0^\beta dx_0\ \{\frac{1}{2}(\partial_0\phi)^2 + \frac{1}{2}(\vec{\nabla}\phi)^2 + \frac{1}{2}m^2\phi^2 + \ldots\} \tag{1.26}$$

The partition function (1.19) is then given by

$$Z = \int \cdots \int D\phi \quad e^{-S[\phi]}$$

$$\phi(\vec{x},0) = \phi(\vec{x},\beta) \tag{1.27}$$

which has a form of partition function of classical statistical mechanics in d + 1 dimensions. Thus, "d-dimensional quantum statistical mechanics = d + 1 dimensional classical statistical mechanics".

Parisi-Wu Stochastic Quantization:

The stochastic quantization of Parisi and Wu is a method of quantization for Bose fields in Euclidean space time. The correlation functions of scalar field defined by the standard Euclidean path integral are computed by the following prescription:

$$< \phi(x_1)\phi(x_2)\ldots\ldots\phi(x_n) > \equiv$$

$$\equiv \int \cdots \int D\phi \quad \phi(x_1)\phi_\eta(x_2)\ldots\ldots\phi(x_n)\ e^{-S[\phi]} / \int \cdots \int D\phi\ e^{-S[\phi]}$$

$$= \lim_{\tau\to\infty} < \phi_\eta(x_1)\phi_\eta(x_2)\ldots\ldots\phi_\eta(x_n,\tau) >_\eta \tag{1.28}$$

where $\phi_\eta(x,\tau)$ is a solution of Langevin equation

$$\frac{\partial}{\partial\tau}\phi_\eta(x,\tau) = -\left.\frac{\delta S[\phi]}{\partial\phi(x)}\right|_{\phi(x) = \phi_\eta(x,\tau)} + \eta(x,\tau) \tag{1.29}$$

with a given initial condition

$$\phi_\eta(x,0) = \phi_0(x) \tag{1.30}$$

and $< \ldots\ldots\ldots >_\eta$ is a white noise average:

$$< \eta(x,\tau) >_\eta = 0$$

$$< \eta(x,\tau)\ \eta(x',\ \tau') >_\eta = 2\delta(x-x')\delta(\tau-\tau')$$

(1.31)

In order to avoid some mathematical inconsistency we implicitly assume that the η average (4) is defined by the following limit:

$$< \ldots\ldots >_\eta =$$

$$= \lim_{\Lambda\to\infty} \frac{\int\cdots\int D\eta \ \ldots\ \ e^{\frac{-1}{4}\int dx d\tau d\tau'\ a_\Lambda(\tau-\tau')\ \eta(x,\tau)\ \eta(x,\tau')}}{\int\cdots\int D\eta\ e^{\frac{-1}{4}\int dx d\tau d\tau'\ a_\Lambda(\tau-\tau')\ \eta(x,\tau)\ \eta(x',\ ')}}$$

(1.32)

where $a_\Lambda(\tau-\tau')$ is a symmetric regulator function such that

$$a_\Lambda(\tau) = a_\Lambda(-\tau)$$

$$\int d\tau'\ a_\Lambda(\tau-\tau') = 1$$

$$\lim_{\Lambda\to\infty}\ a_\Lambda(\tau-\tau') = \delta(\tau-\tau') \qquad (1.33)$$

In order to show how this quantization works, let us consider a simple example of a free scalar field. The Langevin equation is given by

$$\frac{\partial}{\partial\tau}\,\phi(x,\tau) = (\Box - m^2)\phi(x,\tau) + \eta(x,\tau) \qquad (1.34)$$

The solution $\phi_\eta(x,\tau)$ with the initial condition $\phi_\eta(x,0) = 0$ is given by

$$\tilde{\phi}_\eta (k,\tau) = \int_0^\tau d\tau' \, e^{(\tau'-\tau)(k^2+m^2)} \tilde{\eta}(k,\tau') \quad (1.35)$$

where $\tilde{\phi}_\eta(k,\tau)$ and $\tilde{\eta}(k,\tau)$ are Fourier transform, accordingly from (1.31)

$$< \tilde{\eta}(k,\tau) \, \tilde{\eta}(k',\tau') >_\eta = 2\delta(k+k') \, \delta(\tau-\tau') \quad (1.36)$$

Thus,

$$< \tilde{\phi}(k,\tau) \, \tilde{\phi}(k',\tau') >_\eta =$$

$$= \delta(k+k') \frac{1}{k^2+m^2} \left(e^{|\tau-\tau'| (k^2+m^2)} - e^{-|\tau+\tau'| (k^2+m^2)}\right) \quad (1.37)$$

which becomes the standard Feynman propagator in the limit $\tau=\tau' \to \infty$.

For the interacting field such as ϕ^4 theory the Langevin equation has an additional term, $-g\phi^3$. The perturbative solution is expressed by tree graphs:

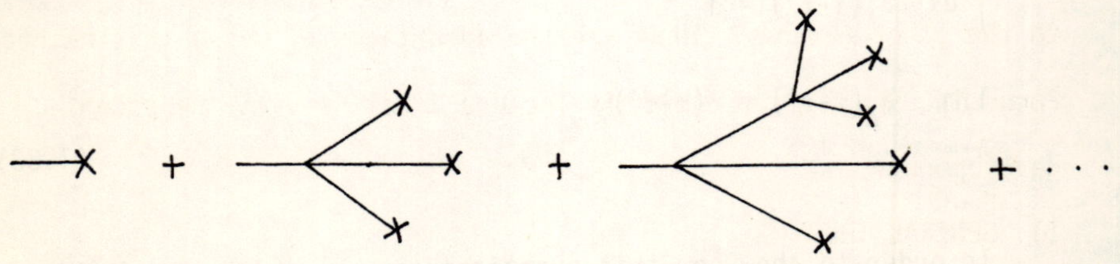

where x represents the random source η. The η average connects a pair of crosses as

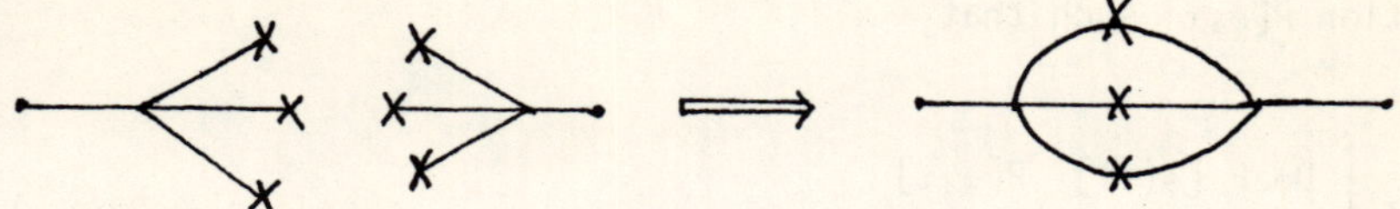

η - average

Thus, one obtains a similar expression to the standard Feynman diagram.

The stochastic quantization is based on the Langevin equation (1.29), which is formally obtained by considering the field as a function of a fictitious time τ as well as the ordinary space time x and by adding $\dot{\phi}$ and a random source term η to the classical equation $\frac{\delta S[\phi]}{\delta\phi(x)} = 0$. In this sense the quantization is carried out once the classical equation of motion is known. This feature contrasts with other quantization methods, the canonical quantization based on the Hamiltonian formalism and the path-integral quantization by means of the Lagrangian.

To what extent is the stochastic quantization method equivalent to the other methods? What are the advantages of using this method? Focussing these points in this lecture I will review the recent developments.

II. GENERAL THEORY[2]

In this section we formulate a general theory of stochastic processes in such a way to serve for the extension of the stochastic quantization described in the previous section to a wider class of field theories.

We assume first the existence of probability distribution

function $P[\phi,\tau]$ such that

$$\int D\phi \; F\,[\phi(\cdot)] \;\; P[\phi,\tau]$$

$$= \langle F\,[\phi_\eta(\cdot,\tau)]\,\rangle_\eta \tag{2.1}$$

where $F[\phi(\cdot)]$ is an arbitrary functional of $\phi(x)$. $\phi_\eta(x,\tau)$ is a solution of Langevin equation which is yet to be determined. In order to be the stochastic quantization (1.1) we demand

$$\lim_{\tau\to\infty} P[\phi,\tau] = e^{-S[\phi]} \Big/ \int\!..\!\int D\phi \; e^{-S[\phi]} \tag{2.2}$$

and

$$\int D\phi \; P[\phi,\tau] = 1 \tag{2.3}$$

(2.3) is merely the statement $P[\phi,\tau]$ being a probability distribution.

Next we assume that the time development of $P[\phi,\tau]$ is governed by a Schrödinger-type equation, the Fokker-Planck equation:

$$-\frac{\partial}{\partial\tau} P[\phi,\tau] = H_{FP} \; P[\phi,\tau] \tag{2.4}$$

We then ask what are the possible forms of H_{FP} which leads to (2.2) and (2.3), and what are the corresponding Langevin equations which assure the relation (2.1).

It is advantageous to use Dirac's canonical operator formalism[3]:

$$[\hat{\pi}(x),\, \hat{\phi}(x')] = -i\delta(x-x') \tag{2.5}$$

$$\hat{\phi}(x) \; |\phi> \; = \; |\phi> \, \phi(x) \tag{2.6}$$

$$P[\phi,\tau] \equiv \; < \phi | \; P(\tau) > \tag{2.7}$$

$$< \phi | \; \hat{\pi}(x) \; | \; P(\tau)> = - \, i \frac{\delta}{\delta\phi(x)} \; < \phi | \; P(\tau) > \tag{2.8}$$

The Fokker-Planck equation is then

$$- \frac{\partial}{\partial\tau} \; | \; P(\tau) > \; = \hat{H}_{FP} \; |P(\tau) > \tag{2.9}$$

A formal solution of $|P(\tau) >$ is given by

$$|P(\tau) > \; = e^{-\hat{H}_{FP}\tau} \; |P(0) > \tag{2.10}$$

Integrating (2.8) functionally over ϕ we obtain

$$\int D\phi \; < \phi | \; \hat{\pi}(x) \; |P(\tau) > \; = 0$$

for any $|P(\tau) >$. Namely,

$$\int D\phi \; |\phi > \; \equiv \; |0 > \quad , \quad \hat{\pi}(x) \; |0 > \; = 0 \tag{2.11}$$

With this notation (2.1) and (2.3) are expressed as

$$< 0 | \; F[\hat{\phi}(\cdot)] \; |P(\tau)> \; = \; < F[\phi_\eta(\cdot,\tau)]>_\eta \tag{2.12}$$

$$< 0 | \; P(\tau) > \; = 1 \tag{2.13}$$

Taking time derivative of (2.13) and using (2.9) we obtain

$$< 0|\ \hat{H}_{FP}\ |P(\tau)> = 0 \tag{2.14}$$

This is satisfied due to (2.11) if $\hat{H}_{FP}$ contains $\hat{\pi}(x)$ on its left.

Let E_n be an eigenvalue of $\hat{H}_{FP}$, and $|\chi_n>$ and $<\xi_n|$ be the corresponding right and left eigenvector respectively.

$$\hat{H}_{FP}\ |\chi_n> \ = \ E_n\ |\chi_n> \quad , \quad <\xi_n|\ \hat{H}_{FP} = E_n\ <\xi_n|$$

$$<\xi_n|\chi_m> \ = \delta_{nm}$$

$$\sum_n |\chi_n><\xi_n| = 1 \tag{2.15}$$

Using (2.10) we obtain

$$P[\Phi,\tau] = \sum_n e^{-E_n\tau}\ \chi_n[\phi]\ \ <\xi_n|\ P(0)> \tag{2.16}$$

If

$$R_e\ E_n > 0 \qquad (n \neq 0) \tag{2.17}$$

and

$$E_0 = 0$$

namely, there is an energy gap above the ground state, and further the ground state is non-degenerate, we obtain

$$\lim_{\tau\to\infty} P[\phi,\tau] = \chi_0[\phi] < \xi_0|P(0) > \tag{2.18}$$

Comparing (2.2) and (2.18) we require

$$\chi_0[\phi] \quad < \xi_0| \, P(0) > = \, e^{-S[\phi]} \Big/ \int D\phi \; e^{-S[\phi]} \tag{2.19}$$

We note

$$e^{-S[\phi]} = < \phi| \, e^{-S[\hat{\phi}]} \, |0 > \tag{2.20}$$

Thus

$$|\chi_0 > \propto e^{-S[\hat{\phi}]} |0 > \tag{2.21}$$

Therefore, using (2.11) we obtain

$$(\hat{\pi}(x) + [\hat{\pi}(x), \, S[\hat{\phi}]]) \quad |\chi_0 > \; =$$

$$= (\hat{\pi}(x) - i \, \frac{\delta S[\hat{\phi}]}{\delta \, \hat{\phi}(x)}) \quad |\chi_0 > \; = 0 \tag{2.22}$$

This expression and $\hat{H}_{FP}|\chi_0> = 0$ is compatible if $\hat{H}_{FP}$ has $\hat{\pi}(x) + [\pi(x), S[\hat{\phi}]]$ on its right. Thus, we obtain the following general form for $\hat{H}_{FP}$

$$\hat{H}_{FP} = \iint dxdx' \; \hat{\pi}(x) \; \hat{K}(x,x') \; (\hat{\pi}(x') + [\hat{\pi}(x'), \, S[\hat{\phi}]]) \tag{2.23}$$

where $\hat{K}(x,x')$ is in general a functional of $\hat{\phi}(x)$ and $\hat{\pi}(x)$. If we can choose $\hat{K}(x,x')$ such that the real part of eigenvalues of $\hat{H}_{FP}$ is positive definite, then (2.23) is a possible form.

For the case of Hermitean, $S^\dagger = S$, we choose $\hat{K}(x,x') = \delta(x-x')$

and make a similarity transformation to obtain the following positive definite form:

$$e^{\frac{1}{2}\hat{S}}\ \hat{H}_{FP}\ e^{-\frac{1}{2}\hat{S}} = \int dx\ \hat{Q}(x)\ \hat{Q}(x) \tag{2.24}$$

$$\hat{Q}(x) = \hat{\pi}(x) + \frac{1}{2}[\hat{\pi}(x),\ \hat{S}] \tag{2.25}$$

Even for non-Hermitian S, if we can choose $\hat{K}(x,x')$ such that after making an appropriate similarity transformation $\hat{H}_{FP}$ becomes a positive form, then we can use (2.23) to derive the corresponding Langevin equation. However, whether this can done for a general S or not is not known yet.

Using the identity (2.12) we like to obtain the equation that $\phi_\eta(x,\tau)$ should satisfy. We restrict here to the case of $K(x,x') = \delta(x-x')$, but the extension to the general case is straightforward.

First define the following Heisenberg operator

$$\hat{\phi}(x,\tau) = e^{\hat{H}_{FP}\tau}\ \hat{\phi}(x)\ e^{-\hat{H}_{FP}\tau} \tag{2.26}$$

then

$$<0|\ F[\hat{\phi}(\cdot)]\ |P(\tau)> \ = \ <0|\ F[\hat{\phi}(\cdot,\tau)]\ \ |P(0)> \tag{2.27}$$

Taking time derivative we obtain

$$\frac{\partial}{\partial\tau} <0|\ F[\hat{\phi}(\cdot,\tau)]\ |P(0)>$$

$$= <0| \int \left(\frac{\delta\hat{F}}{\delta\phi(x,\tau)}\ \hat{\phi}(x,\tau)\right)_W dx\ |P(0)> \tag{2.28}$$

where $(\quad)_W$ is the Weyl ordering defined in (1.17).

Since $\hat{\phi}(x,\tau)$ is defined by (2.26) and since $\hat{H}_{FP}$ is given by (2.23), we obtain

$$\dot{\hat{\phi}}(x,\tau) = - i[\hat{\pi}(x,\tau), \hat{S}] - 2i\hat{\pi}(x,\tau) \tag{2.29}$$

It is not difficult to prove

$$-2i < 0| \int (\frac{\delta\hat{F}}{\delta\phi(x,\tau)} \hat{\pi}(\vec{x},\tau))_W dx |P(0) >$$

$$= < 0| \int \frac{\delta^2\hat{F}}{\delta\phi^2(x,\tau)} dx |P(0) > \tag{2.30}$$

Inserting (2.29) into (2.28) and using (2.30) we obtain

$$\frac{\partial}{\partial\tau} < 0| F[\phi(\cdot,\tau)] |P(0) > =$$

$$= -i < 0| \int \frac{\delta\hat{F}}{\delta\phi(x,\tau)} [\hat{\pi}(x,\tau), S] dx |P(0) > +$$

$$+ < 0| \int dx \frac{\delta^2 F}{\delta\phi^2(x,\tau)} |P(0) > \tag{2.31}$$

We note that in the right hand side of (2.31) only a product of $\hat{\phi}$ appears. Thus we can use the identity (2.12). We obtain

$$\frac{\partial}{\partial\tau} < F[\phi_\eta(\cdot,\tau)] >_\eta =$$

$$= - <\int \frac{\delta F}{\delta\phi_\eta(x,\tau)} \frac{\delta S}{\delta\phi_\eta(x,\tau)} >_\eta + < \frac{\delta^2 F}{\delta\phi_\eta^2(x,\tau)} dx >_\eta \tag{2.32}$$

Since

$$\frac{\partial}{\partial\tau} < F[\phi_\eta(\cdot,\tau)] >_\eta = \int dx < \frac{\delta F}{\delta\phi_\eta(x,\tau)} \frac{\partial}{\partial\tau} \phi_\eta(x,\tau) >_\eta \tag{2.33}$$

and since we shall prove

$$< \int \frac{\delta^2 F}{\delta\phi^2(x,\tau)} dx >_\eta = < \int dx \frac{\delta F}{\delta\phi(x,\tau)} \eta(x,\tau) >_\eta \tag{2.34}$$

we obtain the Langevin equation:

$$\frac{\partial\phi_\eta(x,\tau)}{\partial\tau} = - \frac{\delta S[\phi_\eta(\cdot,\tau)]}{\delta\phi_\eta(x,\tau)} + \eta(x,\tau) \tag{2.35}$$

Proof of (2.34):

$$< \int dx \frac{\delta F}{\delta\phi(x,\tau)} \eta(x,\tau) >_\eta = \lim_{\Lambda\to\infty} \int d\tau' \, a_\Lambda^{-1}(\tau-\tau') \times$$

$$\times < \int dx \frac{\delta}{\delta\eta(x,\tau')} \frac{\delta F}{\delta\phi_\eta(x,\tau)} >_\eta$$

$$= \lim_{\Lambda\to\infty} \int d\tau' \, 2 \, a_\Lambda^{-1}(\tau-\tau') < \int dxdy \frac{\delta\phi_\eta(y,\tau)}{\delta\,\eta(x,\tau')} \frac{\delta^2 F}{\delta\phi_\eta(x,\tau)\delta\phi_\eta(y,\tau)} >_\eta \tag{2.36}$$

where a_Λ^{-1} is an inverse function of a_Λ:

$$\int a_\Lambda^{-1} (\tau-\tau') \, a_\Lambda(\tau'-\tau'') \, d\tau' = \delta(\tau-\tau'') \tag{2.37}$$

Next we prove

$$\lim_{\Lambda\to\infty} \int a_\Lambda^{-1}(\tau-\tau') \, d\tau' \frac{\delta\phi_\eta(y,\tau)}{\delta\eta(x,\tau')} = \frac{1}{2} \delta(x-y) \tag{2.38}$$

then (2.34) follows. Integrating langevin equation we obtain

$$\phi_\eta(x,\tau) = \phi_\eta(x,0) - \int_0^\tau d\tau'' \frac{\delta S[\phi_\eta(\cdot,\tau'')]}{\delta\phi_\eta(x,\tau'')} + \int_0^\tau d\tau' \; \eta(x,\tau') \tag{2.39}$$

Thus,

$$\frac{\delta\phi_\eta(x,\tau)}{\delta\eta(y,\tau)} = \theta(\tau-\tau') - \int_0^\tau d\tau \int \frac{\delta\phi_\eta(y,\tau'')}{\delta\eta(x,\tau)} \frac{\delta^2 S}{\delta\phi_\eta(x,\tau'')\delta\phi_\eta(y,\tau')} \tag{2.40}$$

Inserting it into (2.38) and using

$$\lim_{\Lambda\to\infty} \int_0^\tau a_\Lambda^{-1}(\tau-\tau') \, \theta(\tau-\tau') \, d\tau' = \frac{1}{2}$$

$$\lim_{\Lambda\to\infty} \int^\tau d\tau' \int_0^\tau d\tau'' \, a_\Lambda^{-1}(\tau-\tau') \, \theta(\tau''-\tau') = 0 \tag{2.41}$$

we obtain (2.38).

III. GENERALIZATIONS

In this section we generalize the stochastic quantization method to include U(N) lattice gauge theory and Fermi fields.

General Fokker-Planck Hamiltonian and Corresponding Langevin Equation[2,4]:

We consider a special case of the Hamiltonian (2.23) in which $K(x,x')$ depends only on $\phi(x)$ and factorizable:

$$K(x,x';\phi) = \int dz \; G(x,z;\phi) \; G(y,z;\phi) \tag{3.1}$$

In this case if S is real one can prove that eigenvalues of $\hat{H}_{FP}$ are semi-positive definite. Using the method of previous section we obtain the following Langevin equation:

$$\frac{\partial}{\partial\tau}\phi_\eta(x,\tau) = -\int dy\, K(x,y;\phi_\eta)\frac{\delta S}{\partial\phi_\eta(y,\tau)} + \int dydz\, G(x,z;\phi_\eta)\frac{\delta G}{\delta\phi_\eta(y,\tau)} + \int dy\, G(x,y;\phi_\eta)\,\eta(y,\tau) \tag{3.2}$$

This is a generalized Langevin equation by which one obtains the same correlation functions as previous case (a special case of $G(x,y;\phi) = \delta(x-y)$) in the $\tau\to\infty$ limit. Since G is arbitrary one can choose it as one's convenience.

Stochastic Quantization of U(N) Lattice Gauge Theory[5]:

In Wilson's lattice gauge theory one considers an SU(N) matrix on each link of square lattice. A link is specified by x and μ, where x is the position of a lattice point and μ is the direction along the link. Let $U_\mu(x)$ be an NxN unitary matrix on the link $x\mu$ and $dU_\mu(x)$ be the Haar measure. The partition function of the model is defined by the following path-integral:

$$Z = \int \prod_{x,\mu} dU_\mu(x)\, e^{-S[U]} \tag{3.3}$$

where S[U] is given by

$$S[U] = \frac{1}{4g^2}\sum_{x,\mu,\nu}\left[(1 - \mathrm{tr}(U_\mu(x)\, U_\nu(x+a\hat{\mu})\, U^\dagger_\mu(x+a\hat{\nu})\, U^\dagger_\nu(x))) + \text{h.c.}\right] \tag{3.4}$$

In this expression $x + a\hat{\mu}$ refer to the lattice point next to x in the μ direction.

In (3.3), the integration is in a product of U(N) group spaces, on the other hand in our previous Eculidean path-integral (1.27) the integration was in a product of real open spaces. Thus, appropriately modifying the derivative by Lie derivative we still use the general theory of the previous section.

$$D\phi \rightarrow \Pi dU_\mu(x)$$

$$\pi(\vec{x}) = - i \frac{\delta}{\delta\phi(x)} \rightarrow E^\alpha_\mu \tag{3.5}$$

$E^\alpha_\mu(x)$ is a Lie derivative of matrix $U_\mu(x)$, namely

$$E^\alpha_\mu(x) (U_{\mu'}(x')_{ij} = \delta_{xx'}\, \delta_{\mu\mu'}\, (t^\alpha\, U_\mu(x))_{ij} \tag{3.6}$$

where t^α is a fundamental representation of U(N) Lie algebra. We normalize t^α such that

$$\mathrm{tr}(t^\alpha t^\beta) = \delta_{\alpha\beta}\,, \quad \sum_\alpha (t^\alpha)_{ij}\, (t^\alpha)_{k\ell} = \delta_{i\ell}\, \delta_{jk}$$

$$[t^\alpha, t^\beta] = \sum_\gamma C^{\alpha\beta}_\gamma\, t^\gamma \tag{3.7}$$

(Caution: Our normalization is different from the standard one by factor $\frac{1}{\sqrt{2}}$, accordingly the structure constant $C^{\alpha\beta}_\gamma$ also.)

The Fokker-Planck Hamiltonian we use is

$$H_{FP} = \sum_{x\mu} E^\alpha_\mu(x)\, (E^\alpha_\mu(x) + (E^\alpha_\mu(x)\, S[U])) \tag{3.8}$$

which corresponds to (2.23) with K = 1. It is straightforward by using the method of previous section to obtain the following Langevin equation:

$$- i(\frac{\partial}{\partial\tau} U_\mu(x,\tau)) U_\mu^\dagger(x,\tau)$$

$$= - i(E_\mu(x) S[U(\cdot,\tau)]) + \eta_\mu(x,\tau) \tag{3.9}$$

where

$$E_\mu = \sum_\alpha E_\mu^\alpha t^\alpha , \eta_\mu = \sum_\alpha \eta_\mu^\alpha t^\alpha \tag{3.10}$$

and

$$< \eta_\mu^\alpha(x,\tau) \eta_\nu^\beta (x',\tau') >_\eta = 2 \delta_{\mu\mu'} \delta_{xx'} \delta_{\alpha\alpha'} \delta(\tau-\tau') \tag{3.11}$$

The Langevin equation (3.9) is used in the talk given by Dr. A. Guha.

Stochastic Quantization of Fermi Fields[6]:

The path-integral expression of Fermi fields to be considered is given by

$$\int D\psi D\bar{\psi}(.....) e^{-S[\psi,\bar{\psi}]} / \int D\psi D\bar{\psi} e^{-S[\psi,\bar{\psi}]} \tag{3.12}$$

where ψ and $\bar{\psi}$ are independent Grassmann variables. We restrict S to be bi-linear form, since in the most of interesting theories the Fermions appear bi-linearly in the Lagrangian and since even if Fermions appear in quartic form one can bi-linearize it by introducing a set of auxiliary fields:

$$S = \int \bar{\psi}(x) G \psi(x) dx \tag{3.13}$$

G may contain not only the derivative operators but also other fields, which we treat here as external fields. In general,

$$G^\dagger \neq G \tag{3.14}$$

The Fokker Planck Hamiltonian to be used is

$$H_{FP} = \int dx \, [\frac{\delta}{\delta\psi(x)} \; G^{\dagger} \; (\frac{\delta}{\delta\bar{\psi}(x)} + \frac{\delta S}{\delta\bar{\psi}(x)}) - \frac{\delta}{\delta\bar{\psi}(x)} \; G^{T\dagger} \; (\frac{\delta}{\delta\psi(x)} + \frac{\delta S}{\delta\psi(x)})]$$

$$= \int dx \, [\frac{\delta}{\delta\psi(x)} \; G^{\dagger} \; (\frac{\delta}{\delta\bar{\psi}(x)} + G \, \psi(x)) - \frac{\delta}{\delta\bar{\psi}} \; G^{T\dagger} \; (\frac{\delta}{\delta\psi(x)} - G^{T}\bar{\psi}(x))] \tag{3.15}$$

This was constructed based on the general discussion of section II. The arbitrary function and sign are fixed in such a way H_{FP} has semi-positive definite eigenvalues, which we shall prove later.

Following the general discussion of II, we derive the corresponding Langevin equation:

$$\frac{\partial}{\partial\tau}\psi(x,\tau) = - G^{\dagger} \, G \, \psi(x,\tau) + \frac{1}{2} G^{\dagger}\eta_1 + \eta_2$$

$$\frac{\partial}{\partial\tau}\bar{\psi}(x,\tau) = - (GG^{\dagger})^{T} \, \bar{\psi}(x,\tau) + \bar{\eta}_1 + \frac{1}{2} G^{\dagger} \, \eta_2 \tag{3.16}$$

$$< \eta_{\alpha}(x,\tau) \, \bar{\eta}_{\beta}(x',\tau') >_{\eta} = - < \bar{\eta}_{\beta}(x',\tau') \, \eta_{\alpha}(x,\tau) >_{\eta}$$

$$= 2 \, \delta_{\alpha\beta} \, \delta(x-x') \, a_{\Lambda}(\tau-\tau'). \tag{3.17}$$

Next we prove the positivity of H_{FP}. The method we use is based on the work of Fukai et al[7]. Let us consider two sets of Fermi operators:

$$\{\hat{\psi}, \hat{\psi}^\dagger\} = 1 \quad , \quad \{\hat{\bar{\psi}}, \hat{\bar{\psi}}^\dagger\} = 1$$

$$\{\hat{\psi}, \hat{\bar{\psi}}^\dagger\} = 0, \text{ etc.} \tag{3.18}$$

The coherent state representation defined by

$$|\psi,\bar{\psi}> \; = \; |0,0> + \psi|1,0> + \bar{\psi}|0,1> + \psi\bar{\psi}|1,1> \tag{3.19}$$

represents $\hat{\psi}$ and $\hat{\psi}^\dagger$ as follows:

$$\hat{\psi}^\dagger|\psi,\bar{\psi}> \; = \frac{\delta}{\delta\psi}|\psi,\bar{\psi}> , \quad \hat{\bar{\psi}}^\dagger|\psi,\bar{\psi}> \; = \frac{\delta}{\delta\bar{\psi}}|\psi,\bar{\psi}>$$

$$\hat{\psi}\,|\psi,\psi> \; = \psi\,|\psi,\bar{\psi}> \quad , \quad \hat{\bar{\psi}}\,|\psi,\bar{\psi}> \; = \bar{\psi}\,|\psi,\bar{\psi}> \tag{3.20}$$

$|0,0>$, $|0,1>$ etc. in (3.19) are the Fermion number representation of $\hat{\psi}^\dagger\hat{\psi}$ and $\hat{\bar{\psi}}^\dagger\hat{\bar{\psi}}$.

The Fokker Planck Hamiltonian (3.15) can be considered as a coherent state representation of the following operator Hamiltonian.

$$H_{FP} = \int dx\, [\hat{\psi}^\dagger\, G^\dagger\, (\hat{\bar{\psi}}^\dagger + G\,\hat{\psi}) - \hat{\bar{\psi}}\;\; G^{T\dagger}\, (\hat{\psi}^\dagger - G^T\hat{\bar{\psi}})] \tag{3.21}$$

By the following similarity transformation

$$\hat{H}_{FP} = e^{-\int dx\; \hat{\psi}^\dagger\; G^{-1}\hat{\bar{\psi}}^\dagger} \quad H_{FP}\; e^{+\int dx\; \hat{\psi}^\dagger\; G^{-1}\; \hat{\bar{\psi}}^\dagger} \tag{3.22}$$

we obtain the positive definite form:

$$\hat{H}_{FP} = \int dx\, (\hat{\psi}^\dagger\; G^\dagger G\; \hat{\psi} + \hat{\bar{\psi}}^\dagger (GG^\dagger)^T \hat{\bar{\psi}}) \tag{3.23}$$

IV. PATH-INTEGRAL FORMULATION OF STOCHASTIC PROCESSES AND HIDDEN SUPERSYMMETRY[8]

Let us define a generating functional

$$Z[j] = < \exp \int dx \; j(x) \; \phi_\eta(x,\tau) >_\eta \tag{4.1}$$

and then convert it to path-integral form by using Faddeev-Popov technique. (Since x is a spectator variable we omit it in the following expression.)

$$Z[j] = \int \ldots \int D\phi \; \underbrace{\underbrace{\delta(\phi - \phi_\eta)}_{\delta(\dot{\phi} + \frac{\partial S}{\partial \phi} - \eta) \left|\left|\frac{\delta \eta}{\delta \phi}\right|\right|} \; e^{ij\phi(\tau)} \; D\eta}\; e^{-\frac{1}{4}\int_0^\tau d\tau \; \eta^2(\tau)} \tag{4.2}$$

The Jacobian with the retarded boundary condition (1.30) can be computed as

$$\left|\left|\frac{\delta \eta}{\delta \phi}\right|\right| = \exp\left(\frac{1}{2}\int_0^\tau d\tau \; \frac{\partial^2 S}{\partial \phi^2(\tau)}\right) \tag{4.3}$$

We shall give details of calculation later. Inserting this into (4.2) we obtain

$$\begin{aligned} Z[j] &= \int_{\phi(0) = \phi_0} \ldots\ldots \int D\phi \; e^{j\phi(\tau)} \; e^{-\int_0^\tau d\tau \left[\left(\dot{\phi} + \frac{\partial S}{\partial \phi}\right)^2 - \frac{1}{2}\frac{\partial^2 S}{\partial \phi^2}\right]} \\ &= \int d\phi \; e^{j\phi - \frac{1}{2}(S[\phi] - S[\phi_0])} \int_{\substack{\phi(\tau) = \phi \\ \phi(0) = \phi_0}} \ldots\ldots \int D\phi \; e^{-\int_0^\tau d\tau \; L} \end{aligned} \tag{4.4}$$

where L is Fokker Planck Lagrangian given by

$$L = \frac{1}{4}\dot{\phi}^2 + \frac{1}{4}\left(\frac{\partial S}{\partial \phi}\right)^2 - \frac{1}{2}\frac{\partial^2 S}{\partial \phi^2} \equiv \frac{1}{4}\dot{\phi}^2 + V[\phi] \tag{4.5}$$

Using the relation between the path-integral and the operator matrix element, the last factor of (4.4) can be written as

$$\int \ldots \int D\phi \; e^{-\int_0^\tau d\tau L} = \langle \phi | e^{-\tau \hat{H}} | \phi_0 \rangle \tag{4.6}$$

$$\phi(\tau) = \phi$$
$$\phi(0) = \phi_0$$

where $\hat{H}$ is the Fokker Planck Hamiltonian

$$\hat{H} = \hat{\pi}^2 + V[\phi] = \left(\hat{\pi} + \frac{i}{2}\frac{\partial S}{\partial \phi}\right)\left(\hat{\pi} - \frac{i}{2}\frac{\partial S}{\partial \phi}\right) \equiv \hat{Q}^\dagger \hat{Q} \tag{4.7}$$

which has semi-positive definite form. The zero energy state of $\hat{H}$ is given by

$$\hat{H}\, |0\rangle = 0, \quad \langle \phi | 0 \rangle = e^{-\frac{1}{2} S[\phi]} / \left(\int d\phi \; e^{-S}\right)^{\frac{1}{2}} \tag{4.8}$$

which we assume to be normalizable. (i.e. $\int d\phi \; e^{-S} < \infty$). Inserting (4.5) into (4.4) and expanding by a complete set of eigenfunctions of $\hat{H}$ we obtain

$$Z[j] = \int d\phi \; e^{j\phi - \frac{1}{2} S[\phi]} \sum_n e^{-E_n} \langle \phi | n \rangle \langle n | \phi_0 \rangle \, e^{\frac{1}{2} S[\phi_0]} \tag{4.9}$$

Since $E_n \geq 0$ for $n \neq 0$, we obtain

$$Z[j] \xrightarrow[\tau \to \infty]{} \int d\phi \; e^{j\phi} \, e^{-S[\phi]} / \left(\int d\phi \; e^{-S}\right)^{1/2} \tag{4.10}$$

Here we assumed a <u>non-degenerate</u> zero energy ground state. (4.1) and (4.10) consist of another non-perturbative proof of the stochastic quantization.

Calculation of Jacobian (4.3)

$$\left|\left|\frac{\delta\eta}{\delta\phi}\right|\right| = \det \left[\frac{\partial}{\partial\phi(\tau')} \left(\dot{\phi}(\tau) + \frac{\partial S}{\partial\phi(\tau)}\right)\right]$$

$$= \det\left[\left(\partial_\tau + \frac{\partial^2 S}{\partial\phi^2(\tau)}\right) \delta(\tau-\tau')\right] \equiv \det (<\tau|M|\tau'>)$$

$$\equiv \det M$$

(4.11)

where

$$M = \hat{\partial}_\tau + \frac{\partial^2 S}{\partial\phi^2(\hat{\tau})} = \hat{\partial}_\tau \left(1 + \hat{\partial}_\tau^{-1} \frac{\partial^2 S}{\partial\phi^2(\hat{\tau})}\right) \tag{4.12}$$

$$\ell n \det M = \mathrm{tr}\ \ell n\ M$$

$$= \mathrm{tr}\left[\ell n\ \hat{\partial}_\tau^{-1} + \ell n\left(1 + \hat{\partial}_\tau^{-1} \frac{\partial^2 S}{\partial\phi^2(\hat{\tau})}\right)\right] \tag{4.13}$$

The first term gives merely a constant so we will omit it. The second term is

$$\int d\tau \ <\tau|\ \ell n\left(1 + \hat{\partial}_\tau^{-1} \frac{\partial^2 S}{\partial\phi^2(\tau)}\right) |\tau>$$

$$= \int d\tau \left[<\tau|\ \hat{\partial}_\tau^{-1} \frac{\partial^2 S}{\partial\phi^2(\hat{\tau})} |\tau> + <\tau|\ \hat{\partial}_\tau^{-1} \frac{\partial^2 S}{\partial\phi^2(\hat{\tau})} \hat{\partial}_\tau^{-1} \frac{\partial^2 S}{\mu\phi^2(\hat{\tau})} |\tau> \right.$$

$$\left. + \ .....\right]$$

(4.14)

With the retarded boundary condition we have

$$<\tau|\ \hat{\partial}_\tau^{-1} |\tau'> = \theta(\tau-\tau'). \tag{4.15}$$

It is easy to convince ourselves that from second term on are zero. The first term of (4.15) is then

$$\int d\tau d\tau' \; \theta(\tau-\tau') \;\; \delta(\tau-\tau') \frac{\delta^2 S}{\delta\phi^2(\tau')} = \underset{\displaystyle 1/2}{\underset{\|}{\theta(0)}} \int d\tau \;\; \frac{\partial^2 S}{\partial\phi^2(\tau)} \tag{4.16}$$

Thus, we obtain (4.3).

For the case of advanced boundary condition, the same calculation goes but we must use

$$< \tau | \; \hat{\partial}_\tau^{-1} \; |\tau' > = - \theta(\tau'-\tau) \tag{4.17}$$

accordingly

$$\left|\left|\frac{\delta\eta}{\delta\phi}\right|\right| = \exp\left(- \frac{1}{2} \int d\tau \frac{\partial^2 S}{\partial\phi^2(\tau)} \right) \tag{4.18}$$

Hidden Supersymmetry[9]:

Let us go back to the expression (4.2):

$$Z = \int \ldots \int D\phi \; \left|\left|\frac{\delta\eta}{\delta\phi}\right|\right| \; e^{- \int_0^\tau \frac{1}{4} (\dot{\phi} + \frac{\partial S}{\partial\phi})^2 \, d\tau} \tag{4.19}$$

where $\left|\left|\frac{\delta\eta}{\delta\phi}\right|\right|$ is a Jacobian, which depends on the boundary condition imposed on the Langevin equation. One formally replaces this Jacobian by Fermion integration:

$$\left|\left|\frac{\delta\eta}{\delta\phi}\right|\right| = \det \left[(\partial_\tau + \frac{\partial^2 S}{\partial\phi^2(\tau)}) \; \delta(\tau-\tau') \right]$$

$$= \int D\psi D\bar{\psi} \;\; e^{- \int_0^\tau \bar{\psi}(\tau) \, (\partial_\tau + \frac{\partial^2 S}{\partial\phi^2}) \, \psi(\tau) \, d\tau} \tag{4.20}$$

Then the partititon function (4.19) becomes

$$Z = \int\dots\int D\phi D\bar{\psi} \; e^{-\int_0^\tau d\tau \, L} \tag{4.21}$$

where

$$L = \frac{1}{4}\dot{\phi}^2 + \left(\frac{\partial S}{\partial\phi}\right)^2 + \bar{\psi}\left(\partial_\tau + \frac{\partial^2 S}{\partial\phi^2}\right)\psi \tag{4.22}$$

This is the same Lagrangian considered by E. Witten[10] as an example of non-relativistic supersymmetry. The Lagrangian admits a supersymmetry

$$\delta\phi = \bar{\varepsilon}\psi + \varepsilon\bar{\psi}$$

$$\delta\psi = \frac{\varepsilon}{2}\left(\dot{\phi} + \frac{\partial S}{\partial\phi}\right)$$

$$\delta\bar{\psi} = \frac{\bar{\varepsilon}}{2}\left(-\dot{\phi} + \frac{\partial S}{\partial\phi}\right)$$

(4.23)

provided the boundary condition of ϕ, ψ and $\bar{\psi}$ are the same. Thus, we impose the periodic boundary condition on the path-integral (4.21) and denote it by Z_{ss}. By using the relation between path-integral and operator expression we obtain

$$Z_{ss} = \mathrm{tr}\,[(-)^F e^{-\tau\hat{H}_{ss}}] \tag{4.24}$$

where F is the Fermion number operator and $\hat{H}_{ss}$ is given by

$$\hat{H}_{ss} = \hat{\pi}^2 + \frac{1}{4}\left(\frac{\partial S}{\partial\phi}\right)^2 + \frac{1}{2}[\hat{\psi}^\dagger, \hat{\psi}]\frac{\partial^2 S}{\partial\phi^2} \tag{4.25}$$

Since in the two component vector space of Fermion number 0 and 1, $(-)^F = [\hat{\psi}^\dagger, \hat{\psi}] = \sigma_3$ one obtains

$$Z_{ss} = Z_F - Z_B \tag{4.26}$$

where

$$Z_F = \mathrm{tr}\, e^{-\tau \hat{H}_F},\ Z_B = \mathrm{tr}\, e^{-\tau \hat{H}_B}$$

$$\hat{H}_F = \hat{Q}^\dagger \hat{Q} \quad ,\ \hat{H}_B = \hat{Q}\hat{Q}^\dagger \tag{4.27}$$

Note $\hat{H}_F$ is the same as (4.7).

The energy spectra of $\hat{H}_F$ and $\hat{H}_B$ are semi-positive definite. If zero energy state of $\hat{H}_F$ is normalizable and that of $\hat{H}_B$ is non-normalizable one obtains

$$\lim_{\tau\to\infty} Z_{ss} = \lim_{\tau\to\infty} Z_F \tag{4.28}$$

Thus, only in the case of unbroken supersymmetry the supersymmetric formalism approahces the form of stochastic quantization.

V. STOCHASTIC QUANTIZATION OF GAUGE FIELDS

One of the main reasons condisered by Parisi and Wu[1] to invent the stochastic quantization is the advantage of this method for gauge theories, since the gauge fixing is not necessary for the perturbation calculation. We shall examine this by considering a simple example of a pure Abelian gauge field.

The action is given by

$$S = \frac{1}{4}\int dx\ F^2_{\mu\nu} = \frac{1}{2}\int dx\ A_\mu\ [-\Box\, \delta_{\mu\nu} + \partial_\mu\partial_\nu]\ A_\nu \tag{5.1}$$

The Langevin equation is then obtained by using presctiption of I:

$$\frac{\partial}{\partial\tau}\ \tilde{A}_\mu\ (k,\tau) = -\ k^2(\delta_{\mu\nu} - \frac{k_\mu k_\nu}{k^2}\ \tilde{A}_\nu(k,\tau)) + \tilde{\eta}_\mu(k,\tau) \tag{5.2}$$

where $\tilde{A}_\mu$, $\tilde{\eta}_\mu$.... are Fourier transform of A_μ, η_μ....., respectively.

The solution of Langevin equation (5.2) is given by

$$\tilde{A}_\mu(k,\tau) = \int_0^\tau G_{\mu\nu}(k;\ \tau-\tau')\,\tilde{\eta}_\nu(k,\ '\tau)\,d\tau' + A^0_\mu(k) \tag{5.3}$$

where $G_{\mu\nu}$ is a Green's function:

$$[\delta_{\mu\nu}\frac{\partial}{\partial\tau} + (k^2\,\delta_{\mu\nu} - k_\mu k_\nu)]G_{\nu\sigma}(k;\tau-\tau') = \delta_{\mu\sigma}\,\delta(\tau-\tau') \tag{5.4}$$

Because of the additional time term it is possible to invert without fixing gauge and we obtain

$$G_{\mu\nu}(k;\tau-\tau') = \{\ (\delta_{\mu\nu} - \frac{k_\mu k_\nu}{k^2})e^{-\,k^2(\tau-\tau')} + \frac{k_\mu k_\nu}{k^2}\ \}\ \theta(\tau-\tau') \tag{5.5}$$

In order to obtain a two point correlation function we compute the η- average of $\tilde{A}_\mu(k,\tau)\ \tilde{A}_\nu(k',\tau')$ and obtain

$$\begin{aligned} &<\ \tilde{A}_\mu(k,\tau)\ \tilde{A}_\mu(k',\tau')\ >_\eta \\ &= \delta(k+k')\ \{\ \frac{1}{k^2}\,(\delta_{\mu\nu} - \frac{k_\mu k_\nu}{k^2})\ [e^{-k^2|\tau-\tau'|} - e^{-k^2(\tau+\tau')}] + \\ &+ 2\tau_<\ \frac{k_\mu k_\nu}{k^2}\ \} \\ &+ \frac{k_\mu k'_\nu}{k^2 k'^2}(k_\mu\ \tilde{A}^0_\mu(k)\ )\ (k_\nu\ \tilde{A}^0_\nu(k)) \end{aligned} \tag{5.6}$$

where $\tau_<$ stands for the smaller between τ and τ'.

From this expression immediately one can conclude that i) the longitudinal component of two point function diverges at $\tau\to\infty$, ii) the longitudinal component of the initial field configuration remains, iii) the gauge invariant correlation functions such as

$< F_{\mu\nu}(x) F_{\rho\sigma}(y) >$ are finite, and iv) with the choice of $A^0_\mu = 0$ one obtains the Landau gauge results. These are essentially the observation due to Parisi and Wu.[1]

Let us try to understand these features by the path-integral formalism described in IV. The corresponding expression to (4.4) and (4.5) are

$$P[A,t] = e^{-\frac{1}{2} S[A]} < A| \, e^{-\tau\hat{H}} \, |A^0 > e^{\frac{1}{2} S[A^0]} \tag{5.7}$$

and

$$< A| \, e^{-\tau\hat{H}} \, |A^0 > = \underset{\substack{A_\mu(x,\tau) = A_\mu(x) \\ A_\mu(x,0) = A^0_\mu(x)}}{\int..\int} DA_\mu \exp[- \int_0^\tau dxd\tau \, \{\frac{1}{4} \dot{A}^2_\mu + \frac{1}{4} (\frac{\delta^2 S}{\delta A_\mu}) - \frac{1}{2} \frac{\delta^2 S}{\delta A^2_\mu} \}] \tag{5.8}$$

We note that the last two terms of Fokker Planck Lagrangian in (5.8) are gauge invariant. For Abelian case S is given by (5.1) which contains only the transverse components of field:

$$S = \frac{1}{2} \int d^4k \, k^2 |\tilde{A}^T_\mu(k)|^2 \tag{5.9}$$

where

$$\tilde{A}^T_\mu (k) = (\delta_{\mu\nu} - \frac{k_\mu k_\nu}{k^2}) \tilde{A}_\nu(k)$$

$$A^L_\mu (k) = \frac{k_\mu k_\nu}{k^2} \tilde{A}_\nu(k) \tag{5.10}$$

Therefore the corresponding Fokker Planck Hamiltonian is separated into a sum of longitudinal and transverse part. The longitudinal part does not contain the potential term, namely the longitudinal component is cyclic in the Fokker Planck dynamics:

$$< A| e^{-\tau\hat{H}} |A^0 > = < A^T| e^{-\tau\hat{H}_T} |A^{0T} > \; < A^L| e^{-\tau\hat{H}_L}|A^{0L} > \tag{5.11}$$

$$\hat{H}_L = \int d^4k \, |\tilde{\pi}_L(k)|^2 \tag{5.12}$$

In the Langevin language this means there is no drift force.

Since $\hat{H}_L$ is equivalent to a sum of free particle Hamiltonians Feynman kernel is obtained as a product of free particle kernels:

$$< A^L| e^{-\tau\hat{H}_L}|A^{0L} > = \text{const } e^{-\frac{1}{4\tau} \int d^4x \, (A^L_\mu - A^{0L}_\mu(x))^2} \tag{5.13}$$

Thus, after a simple calculation one obtains

$$\int DA^L \, \tilde{A}^L(k) \, \tilde{A}^L(-k') \quad < A^L| \, e^{-\tau\hat{H}_L} \, |A^L_0 >$$

$$= \; 2\delta_{kk'}\tau + \tilde{A}^{0L}(k) \, \tilde{A}^{0L}(-k') \tag{5.14}$$

The machinery used here is not particularly useful for Abelian case. One can obtain the result much more quickly by using the Langevin equation. For the non-Abelian gauge theories the gauge invariant separation of longitudinal and transverse mode is not possible so that the problem becomes much more complicated. The Fokker Planck formalism used here may be useful in this case because the Fokker Planck Lagrangian resembles the standard non-Abelian gauge theories in $A_0 = 0$ gauge[11] and the collective coordinate technique used for the separation of variables for that problem may be used.

We add here the important obserbations noted by Namiki et al.[12]
i) The result of Landau gauge obtained by Parisi-Wu is due to a specific initial condition ($A^0_\mu = 0$). Keeping the initial configura-

tion finite and appropriately averaging over it they showed the gauge can be changed. Then the Landau gauge of Parisi and Wu is due to the initial condition. ii) The stochastic quantization of non-Abelian gauge fields yields correctly the effects of Faddeev-Popov ghost fields without introducing them.

Stochastic Gauge Fixing:

It is not necessary to fix gauge in stochastic quantization provided one computes gauge invariant quantities. However, for the purpose of computer simulation it is wise to fix gauge so that damping force acts on the non-gauge invariant modes also. This problem was first considered by Zwanziger[13].

Let F[A] be a gauge invariant quantity of non-Abelian gauge field (such as tr $(F_{\mu\nu}(x)\, F_{\rho\sigma}(x))$, trP $e^{i \oint A_\mu(x)\, dx_\mu}$ (Wilson loop) etc.:

$$F[A^\Omega] = F[A] \tag{5.15}$$

where A^Ω is defined by

$$A^\Omega_\mu(x) = u(\Omega)\, A_\mu(x)\, u^\dagger(\Omega(x)) + i\, u(\Omega(x))\, \partial_\mu u^\dagger(\Omega(x)) \tag{5.16}$$

The average of F[A] at τ is given by

$$\begin{aligned} < F[A] >_\tau &\equiv \int\!\!\ldots\!\!\int DA\, F[A]\, P[A,\tau] \\ &= \int\!\!\ldots\!\!\int DA\, F[A] < A|\, e^{-\tau \tilde{H}_{FP}} |A^0 > \end{aligned} \tag{5.17}$$

where

$$\tilde{H}_{FP} = - \int dx \frac{\delta}{\delta A^a_\mu(x)} \left(\frac{\delta}{\delta A^a_\mu(x)} + \frac{\delta S[A]}{\delta\ _\mu(x)} \right) \tag{5.18}$$

The measure DA in (5.17) is gauge invariant. Thus, one obtains

$$< F[A] >_\tau = \int..\int DA\ F[A] < A^\Omega | e^{-\tau\tilde{H}_{FP}} |A^0 >$$

$$= \int..\int DA\ F[A] <A| e^{-i\int G_a(x)\Omega^a(x,\tau)dx} e^{-\tau\tilde{H}_{FP}} |A^0 > \tag{5.19}$$

where $G_a(x)$ is a generator of the gauge transformation. Since the operators inside the bracket can be obtained by an appropriate τ dependent gauge transformation from $e^{-\tau\tilde{H}_{FP}}$, the corresponding Langevin equation is also obtained by this τ dependent gauge transformation[14].

$$\frac{\partial}{\partial\tau} A_\mu(x,\tau) = - \frac{\delta S}{\delta A_\mu} + D_\mu v + \eta_\mu \tag{5.20}$$

where

$$v = - i\ (\frac{\partial}{\partial\tau} u(\Omega(x,\tau))\, u^{-1} \tag{5.21}$$

Since Ω is arbitrary v is arbitrary. The second term (5.20) is the desired damping force along the gauge orbits.

VI. FURTHER REMARKS

The stochastic quantization reviewed in the previous sections is another quantization method which is designed to be equivalent to the Euclidean path-integral quantization. It is useful for the numerical simulations, although it does not give any great advantage compared to other methods[15]. Theoretically, however, the stochastic quantization seems to provide some advantages to some problems such as gauge theories discussed in the previous section. Further, it was shown that the quenched Eguchi-Kawai model at large N can be derived more elegantly by this method[16]. Nonetheless, I must say that no essentially new results came out of this method yet.

The other topics studied in stochastic quantization are the stochastic regularization and renormalization. Keeping τ finite makes the theory regularize to some extent but this is not sufficient. Further regularization is necessary. The stochastic regularization was proposed by Breit et al[4]. In the η average one does not take $\Lambda \to \infty$ and keep finite. This regularizes the expression finite. Since this regulates only the fictitious time direction it does not affect invariance of the theory such as gauge and Lorentz invariance. A systematic study of this regularization with finite τ and the renormalization was done by Alfaro[17]. He concludes that the stochastic regularization works for renormalizable theories but for unrenormalizable theories not only the stochastic regularization does not work but also the stochastic quantization itself is inconsistent.

An interesting question still unsolved is the question raised by Parisi[18]: "Can stochastic quantization be generalized to complex S?" He suggested the use of Langevin equation for complex S by considering a complex valued probability density. This problem had already been considered by Klauder[19] to some extent in the study of coherent-state Langevin equation. Let us elaborate the point.

Let us consider the Euclidean path-integral:

$$Z = \int .. \int Dq \;\; e^{-S[q]} \tag{6.1}$$

where

$$S = \int d\tau \; [\tfrac{1}{2} \dot{q}^2 + V(q)] \tag{6.2}$$

One can develop the stochastic quantization. No problem. However, if one expresses (6.1) in terms of the Euclidean phase space path-integral

$$Z = \int .. \int DpDq \; e^{-S[p,q]} \tag{6.3}$$

$$S[p,q] = \int d\tau \; [-ip\dot{q} + \tfrac{1}{2} p^2 + V(q)] \tag{6.4}$$

One finds a complex S, accordingly complex probability problem. However, this is merely an artifact due to a Fourier transform of the Gaussian integral. Therefore, this kind of difficulty associated with the complex probabilities should be resolved. If one goes back to the discussion of Fermi fields in III, one realizes a similar complex probability problem appeared then already in a different form. However, there exists no general study of this problem yet to my knowledge.

REFERENCES

1. G. Parisi and Wu Yong-Shi, Scientia Sinica, 24, 483 (1981).
2. B. Sakita, 7th Johns Hopkins Workshop, ed. G. Domokos and S. Kovesi-Domokos (World Scientific, 1983).
3. M. Namiki and Y. Yamanaka, Prog. Theor. Phys. 69, 1764 (1983).
4. J.D. Breit, S. Gupta and A. Zaks, Nucl. Phys. B233, 61 (1984).
5. G. Parisi (unpublished)
   I.T. Drummond, S. Duane, R.R. Gogan, Nucl. Phys. B220, 119 (1983).
   A. Guha and S.-C. Lee, Phys. Lett. 134B, 216 (1984).
   J. Alfaro and B. Sakita, Proc. Topical Symp. High Energy Phys., ed. T. Eguchi and Y. Yamaguchi (World Scientific, 1983).
   M.B. Halpern, UCB-PTH - 83/1.
6. There were errors in reference 2. The editors disregarded the corrections sent a few days later.] I corrected them here. Similar Langevin euations were written and justified by perturbative calculations:
   J.D. Breit, S. Gupta and A. Zaks, Nucl. Phys. B233, 61 (1984).
   P.M. Damgaard, K. Tsokos, Nucl. Phys. B235, 75 (1984).
7. T. Fukai, H. Nakazato, I Ohba, K. Okano, Y. Yamahaka, Prog. Theor. Phys. 69, 1600 (1983).
8. E. Gozzi, Phys. Rev. D28, 1922 (1983).
   See also B. Sakita, Proc. Leipzig Conference (1984).
9. G. Parisi and N. Sourlas, Phys. Rev. Lett. 43, 744 (1979);
   Nucl. Phys. B206, 321 (1982).
   E. Gozzi, Phys. Rev. D28, 1922 (1983).
   E.S. Egorian, S. Kalitzin, Phys. Lett. 129B, 320 (1983).

R. Kirshner, Phys. Lett. B139, 180 (1984).

10. E. Witten, Nucl. Phys. 188B, 513 (1981).
11. J.L. Gervais and B. Sakita, Phys. Rev. D18, 453 (1978).
12. M. Namiki, I. Ohba, K. Okano, Y. Yamanaka, Prog. Theor. Phys. 69, 1580 (1983).
13. D. Zwanziger, Nucl. Phys. B192, 259 (1981).
14. M. Horibe, A. Hosoya, J. Sakamoto, Prog. Theor. Phys. 70, 1636 (1983).
15. A. Guha, Talk given in this Workshop.
16. J. Alfaro and B. Sakita, Phys. Lett. 121B, 339 (1983).
    J. Alfaro, Phys. Rev. D28, 1001 (1983).
    J. Greensite and M.B. Halpern, Nucl. Phys. B211, 339 (1983).
17. J. Alfaro, LPTENS 84/4 - Trieste preprint (July 1984).
18. G. Parisi, Phys. Lett. 131B, 393 (1983).
19. J.R. Klauder, Acta Physica Austriaca, Suppl. XXV251 (1983); Phys. Rev. A29, 2036 (1984).

Numerical Studies of Lattice Systems by Stochastic Quantisation

Arunabha Guha
Department of Physics and Astronomy
University of Rochester
Rochester, NY 14627

1. INTRODUCTION

In the last few lectures, Prof. Sakita has introduced you to the method of Stochastic Quantisation of field theories. Now that you have seen the formalism, let me talk about a practical application of the method to investigate field theories where non-perturbative effects are important.

In recent years numerical simulation of statistical systems (euclidian field theories) has become popular among high energy physicists for various studies such as phase transition behavior in model systems, properties of hadrons in lattice gauge theories, etc. The simplest example is computing the correlation length ξ (= inverse mass-gap) of a system where one starts with a suitable local operator $\phi(x)$ and calculates the connected function

$$\langle\phi(x)\phi(o)\rangle_c \underset{|x|\to\infty}{\sim} |x|^{-\frac{D-1}{2}} \exp(-m|x|) \tag{1.1}$$

where D is the space-time dimension and m is the minimum mass of the state created from vacuum by the operator $\phi(x)$. The power-law prefactor assumes that the minimum mass state is a one-particle state.

The main problem in extracting the mass m from equation (1.1) is that one has to compute the connected function for large $|x|$ where it is small. So in typical Monte Carlo simulations it is difficult to give a reliable estimate of mass even with very high statistics.

In the next section, I will discuss a new algorithm based on Langevin equation which allows us to compute the connected correlation function directly. I will also discuss numerical results obtained in various models such as ϕ^4 theory,[1)] non-linear sigma model,[2)] SU(N) chiral spin models[4)] and gauge theories[5,6)].

2. LANGEVIN EQUATION

Consider a statistical system with Hamiltonian $H(\phi)$ where the dynamical variables are ϕ_i (i includes both space-time and internal indices). The expectation value of a function $F(\phi)$ is given by

$$\langle F(\phi)\rangle = \int d[\phi]F(\phi)\exp[-H(\phi)]/\int d[\phi]\exp[-H(\phi)] \tag{2.1}$$

In stochastic quantisation, one starts with the Langevin equation

$$\dot{\phi}_i(t) = -\frac{\partial H}{\partial \phi_i} + \eta_i(t) \quad , \quad \langle \eta_i(t)\eta_j(t')\rangle = 2\delta_{ij}\delta(t-t') \tag{2.2}$$

where $\phi_i(t)$ is now a function of fictitious time "t" and $\eta_i(t)$ is a gaussian random variable. Then $\langle F(\phi)\rangle$ in equation (2.1) can be computed as

$$\langle F(\phi)\rangle = \lim_{T\to\infty} \frac{1}{T}\int_t^{T+t} dt' F(\phi(t')) \tag{2.3}$$

where t is at least the "thermalisation" time. To actually solve the langevin equation (2.2) numerically, we first discretise time "t" as

$$t = n\varepsilon \quad , \quad n = \text{integer}$$

$$\phi_i(n+1) = -\varepsilon \frac{\partial H}{\partial \phi_i} + \phi_i(n) + \sqrt{2\varepsilon}\,\xi_i(n) \tag{2.4}$$

$$\langle \xi_i(m)\xi_j(n)\rangle = \delta_{ij}\delta_{mn}$$

The computer time is proportional to ε^{-1} and one can show that the error in the final result is proportional to ε (for small ε), i.e.,

$$\langle F(\phi)\rangle\big|_{\varepsilon} = \langle F(\phi)\rangle + O(\varepsilon) \tag{2.5}$$

This is a definite advantage over typical Monte Carlo procedures because now we can practically remove the systematic error from the data by varying ϵ and extrapolating to $\epsilon=0$. We shall come back to this point later.

Let us now discuss some specific examples.

Example (i) ϕ^4 theory in 1 dimension[1)]

Consider

$$H(\phi) = \sum_{i=1}^{N} [\frac{1}{4} \phi_i^4 + 4(\phi_i - \phi_{i+1})^2] \tag{2.6}$$

The system was simulated [1] by Langevin equation (2.4) with $\epsilon=.0025$ and N=98. The correlation function $\langle\phi_i \phi_k\rangle$ is calculated from equation (2.3) with T=125. Results are shown in figure (1). The mass is estimated to be 0.289.

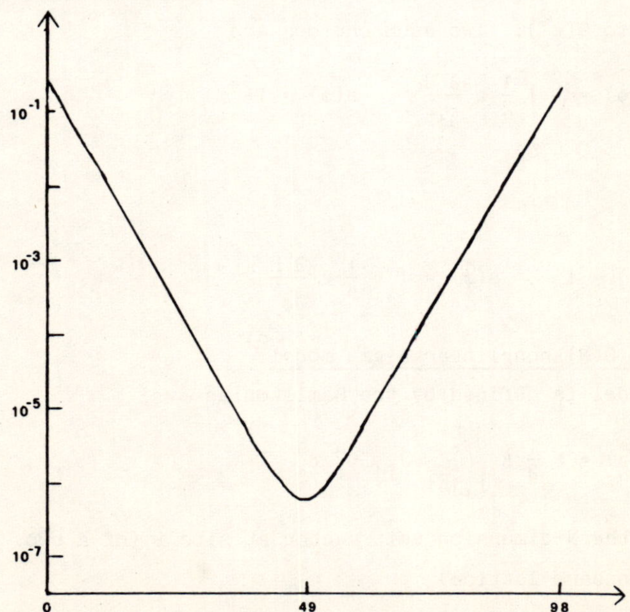

Figure 1. The correlation function $\langle\phi(i)\phi(k)\rangle$ as function of the distance i-k. The periodic boundary condition implies the invariance under the transformation $i-k \to N-(i-k)$; (N=98).

A note:

As mentioned before, using a non-zero ε results in an error which is of order ε. However, by a clever choice, one can reduce the error to $O(\varepsilon^2)$ as follows.

In order to reach thermodynamic equilibrium, one has to enforce the detailed balance principle for the transition probabilities $P(x,x')$:

$$\exp(-H(x))P(x,x') = P(x',x)\exp(-H(x')) \tag{2.7}$$

For the one dimensional case (2.6), we can modify the Langevin equation (2.4) as [1]

$$\phi(n+1) \simeq \phi(n) - \varepsilon \frac{\partial H}{\partial \phi} c(\phi) + s(\phi) \sqrt{2\varepsilon}\, \xi(n) \tag{2.8}$$

Now we can choose $c(\phi)$ and $s(\phi)$ in such a way that detailed balance is enforced up to $O(\varepsilon^2)$. Two such choices are

$$c(\phi) = 1 + \frac{1}{2} \varepsilon \frac{\partial^2 H}{\partial \phi^2} \quad , \quad s(\phi) = 1$$

or

$$c(\phi) = 1 \quad , \quad s(\phi) = 1 - \frac{1}{4} \varepsilon \frac{\partial^2 H}{\partial \phi^2} \tag{2.9}$$

Example (ii) O(N) non-linear sigma model[2)]

This model is defined by the Hamiltonian

$$H(\vec{\sigma}) = -\frac{1}{2} \beta \sum_{i,j=1}^{L} J_{ij}\, \sigma_i^a\, \sigma_j^a \tag{2.10}$$

where $\vec{\sigma}_i$ is the N-dimension unit vector at site i (of a two dimensional square lattice)

$$\vec{\sigma}_i \cdot \vec{\sigma}_i = 1 \tag{2.11}$$

and J_{ij} is a nearest neighbor interaction. The discretised Langevin equation is [2]

$$\tilde{\sigma}_i^a(n+1) = \sigma_i^a(n) + \varepsilon\left(\delta^{ab} - \sigma_i^a(n)\,\sigma_i^b(n)\right)\left(\beta\sum_j J_{ij}\sigma_j^b + \xi_i^b\right)$$

$$\sigma_i^a(n+1) = \frac{\tilde{\sigma}_i^a(n+1)}{\left[\sum_a \tilde{\sigma}_i^a(n+1)^2\right]^{1/2}} \tag{2.12}$$

$$\langle \xi_i^a(m)\xi_j^b(n)\rangle = \frac{2}{\varepsilon}\,\delta_{ij}\,\delta^{ab}\,\delta_{mn}$$

The reader should note that the constraint (2.11) has automatically been taken into account in Langevin equation (2.12).

To extract the mass-gap, it is convenient to sum the spins along any axis (we can call it the spatial direction) and define the zero-momentum Green's function

$$\langle \vec{s}(i_1)\cdot\vec{s}(j_1)\rangle \underset{|i_1-j_1|\to\infty}{\simeq} \exp(-m|i_j-j_1|)$$

$$s^a(i_1) = \sqrt{\frac{N}{L}}\sum_{i_2=1}^{L}\sigma^a_{i_1,i_2} \tag{2.13}$$

Figure (2) shows the result for O(4) at β=0.4[2].

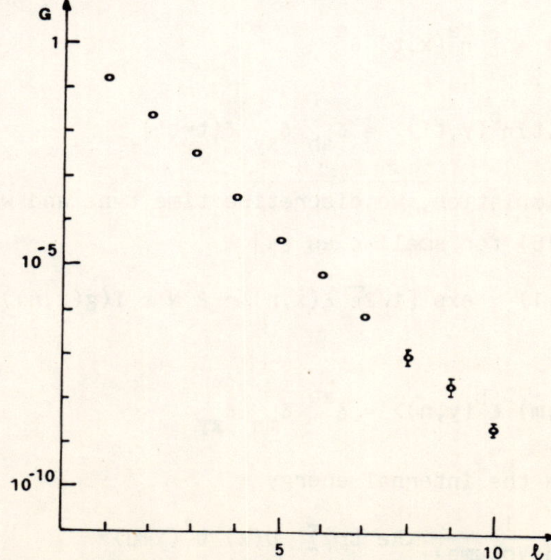

Figure 2. The (S(1)S(1+ℓ)) correlation function in a logarithmic scale on a 30×30 lattice for n=4, β=0.4. The errors are only statistical.

Example (iii) SU(N) chiral spin model[4)]

In this case, the dynamical variables are SU(N) matrices U(x) (x is the lattice site in a d-dimensional euclidian space-time). The Hamiltonian is

$$H(U) = \beta N \operatorname{tr} \sum_{x} \sum_{\mu=1}^{d} [U(x)U^{+}(x+\mu) + h.c.] \tag{2.14}$$

Unlike the case of scalar fields, one now has to write down the Langevin equation on the group manifold (we could also express the matrix $U(x) = \exp i\theta^a\phi^a(x)$, θ^a = generators of SU(N), and write a Langevin equation for $\phi^a(x)$ but that is quite complicated to simulate). This case is quite similar to lattice gauge theories and the correct Langevin equation is [3]

$$\partial_t U(x,t) = i\eta(x,t)\, U(x,t) - \beta N\, T(g(x,t)\, U(x,t)$$

$$T(g) = g - \frac{1}{N}(\operatorname{Tr} g)\mathbb{1} \quad , \quad g = UA^{+} - AU^{+}$$

$$A(x) = \sum_{\mu=1}^{d} [U(x+\mu) + U(x-\mu)] \tag{2.15}$$

$$\eta(x,t) = \sum_{a} \eta^a(x,t)\,\theta^a$$

$$\langle \eta^a(x,t)\eta^b(y,t')\rangle = \delta_{ab}\,\delta_{xy}\,\delta(t-t')$$

For numerical simulation, we discretise time $t=n\varepsilon$ and write down a solution of (2.15) for small ε as [4]

$$U(x,n+1) \simeq \exp[i\sqrt{2\varepsilon}\,\xi(x,n) - \beta N \varepsilon\, T(g(x,n))]\, U(x,n) \tag{2.16}$$

$$\langle \xi^a(x,m)\,\xi^b(y,n)\rangle = \delta^{ab}\,\delta_{mn}\,\delta_{xy}$$

Figure (3) shows the internal energy

$$E = \frac{1}{Nd(\text{Volume})}\langle \operatorname{Re} \operatorname{tr} \sum_{x\mu} U(x)\, U^{+}(x+\mu)\rangle \tag{2.17}$$

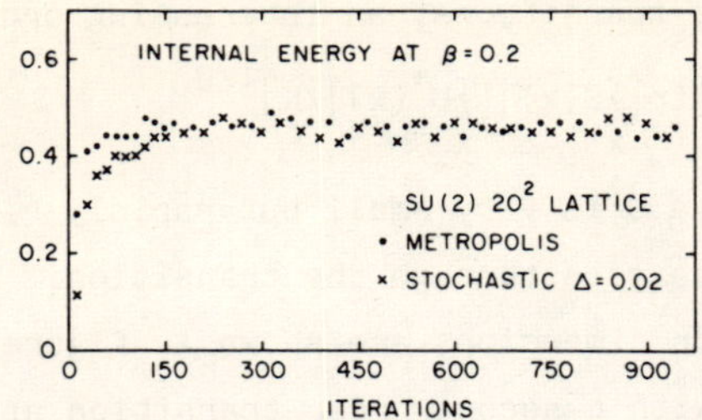

Figure 3. A comparison of stochastic quantization with metropolis run for SU(2).

at β=0.2 for SU(2) on a 20×20 lattice both by stochastic quantisation and metropolis algorithm [4]. As one can see the systematic error is within the statistical fluctuations which is comparable in both cases. The variation of E with β for SU(2), SU(3) and SU(4) is shown in figure (4).

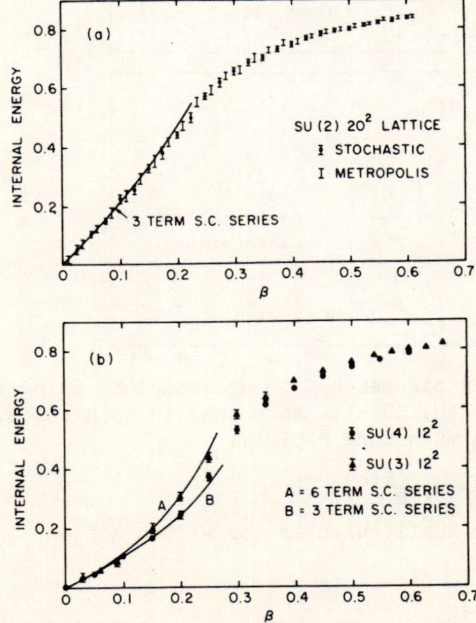

Figure 4. E in two dimensions for (a) SU(2) using MA and SQ. (b) SU(3) and SU(4) using SQ.

To look for phase transitions, an interesting order parameter is

$$C = \frac{1}{\text{Volume}} \left[tr [\sum_x U(x)][\sum_x U^+(x)]/N \right]^{1/2} \qquad (2.18)$$

In the disordered phase C is very small but rapidly rises to a large value near unity as one goes through the transition. Results for SU(2) and SU(3) in four dimensions are shown in figure (5) [4]. SU(2) result is consistent with a second order transition at $\beta_c \approx 0.074$ whereas SU(3) has a first order transition at $\beta_c \simeq .105 - .110$.

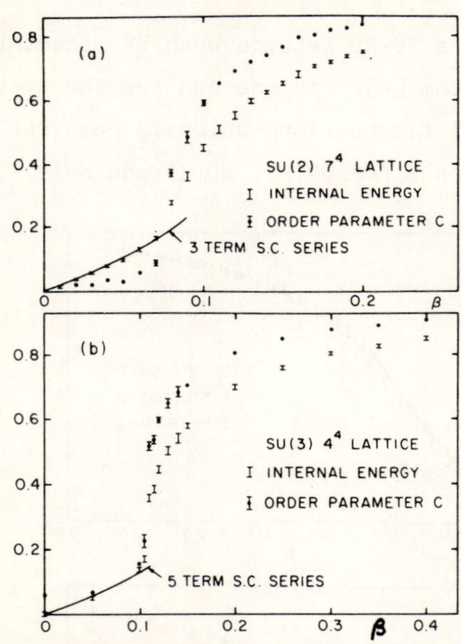

Figure 5. E and order parameter C in 4 dimensions using SQ for (a) SU(2), (b) SU(3). Note that in both cases, compared to E, C rises much more rapidly.

Example (iv) gauge theories[5,6]

In this case the Hamiltonian is the Wilson action

$$H(U) = \beta \sum_{plaq} (1 - \frac{1}{2} \text{Tr } U_p) \qquad (2.19)$$

where U_p is the product of link matrices around the plaquette p. Authors of reference [5] simulated SU(2) gauge theory with a langevin equation slightly different from (2.16). They extracted a glueball mass

$$m = (2.72 \pm 0.2)(\text{string tension}) \qquad (2.20)$$

Finally, authors of reference [6] have simulated a non-compact version of QCD using a Langevin equation for the gauge potentials A_μ^a directly instead of link variables $U_\mu(x)$. Readers are referred to their paper for details.

References

1. Parisi, G., Nucl. Phys. B180 [FS2] (1981) 378.
2. Parisi, G., Nucl. Phys. B205 [FS5] (1982) 337.
3. Guha, A. and Lee, S.-C., Phys. Rev. D27 (1983) 2412.
   Drummond, J. T. et al., Nucl. Phys. B220 (1983) 119.
   Alfaro, J. and Sakita, B., Phys. Lett. 121B (1983) 339.
   Sakita, B., Invited talk at Topical Syposium on High Energy Physics (University of Tokyo, Japan, 1982), unpublished.
   Halpern, M. B., Berkeley preprint UCB-PTH-83/1.
4. Guha, A. and Lee, S.-C., Physics Lett. 134B (1984) 216; Nucl. Phys. B240 [FS12] (1984) 141.
5. Falcioni, M. et al., Nucl. Phys. B215 [FS7] (1983) 265.
6. Stamatescu, I. O., Wolff, U. and Zwanziger, D., Nucl. Phys. B225 [FS9] (1983) 377.

THE SKYRME SOLITON

A.P. Balachandran
Physics Department, Syracuse University
Syracuse, N.Y. 13210, USA

1. INTRODUCTION

Quantum Chromodynamics (QCD) is currently believed to be the correct model to describe strong interactions at energy scales which are small compared to the grand unification scale. When the quarks are taken to be massless, QCD exhibits a global symmetry under the chiral group $G=SU(N_f)_L \times SU(N_f)_R$ where N_f is the number of flavours and $SU(N_f)_{L,R}$ act separately on the left and right quarks. There are good theoretical reasons to expect that this global group is spontaneously broken to the vector subgroup $H=SU(N_f)$. The symmetry breakdown $G \to H$ implies the existence of $N_f^2 - 1$ Goldstone bosons which are identified with the pseudoscalar mesons π , K, ... As is well known, at low energies, the interactions of these Goldstone modes can be described by a "nonlinear" Lagrangian where the fields are valued in the coset space G/H.

The description of the interactions of pseudoscalar mesons in terms of such nonlinear Lagrangians substantially predate the formulation of QCD. Already in the early 60's, such Lagrangians were being studied in depth by Gürsey and others[1)]. At that time, in a remarkable series of papers[2)], Skyrme observed that such Lagrangians admit soliton solutions. He proposed to interpret these solitons in terms of nucleon, delta, etc. Skyrme[2)] and Williams[3)] also gave general arguments to show that for two flavours, if a suitable quantization scheme is adopted, the spin of these particles will emerge correctly as half odd integer.

The ideas of Skyrme and Williams were ignored by particle physicists until they were revived by Pak and Tze[4] who studied the current algebraic and topological aspects of the chiral model in depth. In a subsequent paper, Gipson and Tze[5] also argued that the usual weak interaction model as well predicts "Skyrme" solitons with exotic properties.

The revival of interest in the proposals of Skyrme dates from these papers of Tze and coworkers and from the more recent work of the Syracuse-Austin group[6] and Witten[7]. In this talk I will review the Skyrme model with emphasis on its structural aspects. During the last two years, the subject has shown a big growth. I am not in a position to review it exhaustively both for reasons of time and due to my limitations.

2. CHIRAL SYMMETRIES OF QCD

In this section, we shall briefly review the global symmetries of QCD and indicate the arguments which lead one from QCD to the effective chiral Lagrangian for the description of strong interactions at low energies.

This set of lectures is meant to be a pedagogic review of solitons in the chiral model and is not meant to be a review of QCD. Thus this section is to be regarded as no more than a reminder of the material that the reader once knew well.

QCD is a gauge theory based on the "colour" group SU(3). This group is commonly denoted as $SU(3)_c$. The fields in QCD are the quark fields q_a^α ($\alpha = 1,2,3$; $a = 1,2 \ldots N_f$) and the Yang-Mills potentials A_μ^ρ ($\rho = 1,2, \ldots , 8$). The quark field is a colour triplet and its index α is transformed by SU(3) . The index a is a flavour index and we assume that there are N_f flavours altogether.

In the standard $SU(3)_c \times SU(2) \times U(1)$ model of strong and weak interactions, the masses of the quarks are induced by the Higgs which are responsible for the spontaneous breakdown of the weak group $SU(2) \times U(1)$ to the $U(1)$ group of electromagnetism. The quarks are therefore massless when these Higgs fields are ignored. Since the "current" masses of $u=q_1$ and $d=q_2$ are small compared to say the pion mass and that of $s=q_3$ is of the order of 200 MeV, there is reason to hope that the approximation which ignores the Higgs fields and quark masses is a reasonable first approximation to the dynamics involving only the first three quarks. This hope receives support from the success of the "eight fold way" which is based on the SU(3) group acting on u,d and s. The eight fold way assumes that in the leading approximation, this SU(3) is a good symmetry. Such an assumption requires that in the leading approximation, u,d and s have equal masses. The success of the eight fold way encourages the hope that in the first instance, it is permissible to treat u,d and s as massless and to include the corrections due to the strange quark mass at a later stage. It is doubtful however that we can similarly ignore the masses of the remaining quarks (c,b,t ...) in view of their very large masses in comparison with the pion mass. (The mass of c for instance is $\simeq$ 1.5 GeV.)

This article treats the chiral symmetry properties of QCD. These symmetries exist in QCD only in the zero quark mass limit. For reasons of generality, we shall often phrase our discussion for any N_f as though all quarks have zero mass and chiral symmetry is a good symmetry. However, in view of the preceding remarks, the resultant phenomenological inferences may not be realistic if they involve the heavy quarks.

We may note here that it is not quite accurate to state that quarks have no mass in the absence of Higgs fields if their bare masses are zero. Spontaneous breakdown of chiral symmetry will in general contribute to the masses of the quarks. However, in the absence of the Higgs fields, the QCD Lagrangian will have chiral symmetry and for us this is the significant point.

In the absence of the Higgs fields, the QCD Lagrangian density reads

$$L_{QCD} = - \bar{q}_a \ \gamma_\lambda \ D_\lambda \ q_a \quad - \quad 1/4 \ F^\rho_{\mu\nu} \ F^\rho_{\mu\nu}$$

where D_λ is the covariant derivative and $F^\rho_{\mu\nu}$ are the components of the field tensors for the colour group.

In terms of the left and right handed quark fields

$$q_L = \frac{1+\gamma_5}{2} \ q, \quad q_R = \frac{1-\gamma_5}{2} \ q,$$

the quark part of the Langrangian density L_{QCD} becomes

$$-\bar{q}_{La} \ \gamma_\lambda \ D_\lambda \ q_{La} \quad - \quad \bar{q}_{Ra} \ \gamma_\lambda \ D_\lambda \ q_{Ra}$$

This Lagrangian density and hence L_{QCD} are invariant under separate unitary transformations on the flavour indices of the left handed and right handed quarks: it has the symmetry of the group $U(N_f) \times U(N_f) = U(N_f)_L \times U(N_f)_R = \{(u_L,u_R)\}$ under which the quarks transform according to the rule

$$q_L \to u_L \ q_L, \ q_R \to u_R \ q_R$$

Let us briefly consider the various subgroups of this group.

(a) $U(1)_V$: This is the subgroup which transforms q_L and q_R by the same phase:

$$q_L \to e^{i\theta} \ q_L, \quad q_R \to e^{i\theta} \ q_R$$

The conserved charge associated with this symmetry is the baryon number. This symmetry is exact in QCD in the sense that it is not spoiled either by anomalies or by spontaneous symmetry breakdown in the full quantum theory.

b) $U(1)_A$: This is the so-called "chiral" U(1) group. Under this group, q_L and q_R transform by opposite phases:

$$q_L \rightarrow e^{i\theta} q_L \ , \ q_R \rightarrow e^{-i\theta} q_R$$

This symmetry however is not preserved when L_{QCD} is quantized, but is broken by an anomaly. More precisely, the current $J^5_\mu = i\bar{q}\,\gamma_\mu \gamma_5 q$ which is associated with this transformation, and which is conserved according to naive manipulations with L_{QCD}, is in fact not conserved in the full quantum theory. Rather, it fulfills the equation

$$\partial_\mu J^5_\mu = \text{constant} \times \varepsilon_{\mu\nu\lambda\rho} F^a_{\mu\nu} F^a_{\lambda\rho}$$

(c) $G = SU(N_f)_L \times SU(N_f)_R$: This is the subgroup $U(N_f)_L \times U(N_f)_R$ where L and R are restricted to be matrices of determinant unity. There are many indications that in the full quantum theory, this symmetry group G is spontaneously broken to the subgroup $H = \{(V,V)\}$. The group H is $SU(N_f)$. It is the vector subgroup of G which transforms q_L and q_R by the same amount. For two flavours, H is the SU(2) of isospin and for three flavours, it is the SU(3) of eight fold way.

Because G is spontaneously broken to H, by Goldstone's theorem, there are $N_f^2 - 1$ massless mesons in QCD. For two flavours, these Goldstone modes are identified with the three pions, while for three flavours, these modes are identified with the pseudoscalar octet.

In this picture, then, when the quarks have zero mass, the pseudoscalar mesons are exactly massless. They become massive when the interactions between the quark and Higgs fields are turned on, the quarks acquire mass and G gets explicitly broken in the Lagrangian.

The effective Lagrangian $\hat{L}$ emerges when we attempt to construct a model which describes the dynamics of these Goldstone modes. Let us list the properties we require of this Lagragian in the zero quark mass limit: (a) Firstly, $\hat{L}$ must be invariant under $G = SU(N_f)_L \times SU(N_f)_R$, this property being the anaologue of the G invariance of the

QCD Lagrangian. Thus $\hat{L}$ is to be constructed from a (multicomponent) field Φ which is transformed by G, $\hat{L}$ being invariant under these transformations. (b) Secondly, $\hat{L}$ should have exactly $N_f^2 - 1$ degrees of freedom per space time point. This is a requirement of minimality: we want $\hat{L}$ to describe the dynamics of the Goldstone modes and only of these modes. Thus the effective Lagrangian we discuss involves the approximation where the coupling of the Goldstone modes to other excitations are ignored. It is possible to improve this model by introducing for instance fields for vector and axial vector mesons, but we shall not discuss these modifications. (c) Finally, we require that the subgroup of G which leaves any value of the field invariant is (isomorphic to) exactly H and no more. If this can be arranged, then we would have nicely built in the spontaneous symmetry breakdown $G \to H$ in the geometry of the field space itself. For, in the ground state, the field will have some constant value, Φ_0 say, for all $\vec{x}$. The symmetry group of the ground state is thus H, signifying the breakdown $G \to H$.

Requirements (b) and (c) determine the space of values of the field to be the coset space G/H. Instead of explaining this theorem in generality, we exhibit G/H in the case of our interest, and verify that (b) and (c) are fulfilled.

For $G = SU(N_f)_L \times SU(N_f)_R = \{(L,R)\}$ and $H = \{(V,V)\}$, the manifold G/H is the same as the manifold of the group $SU(N_f)$. Thus the field Φ can be identified with the field U where U(x,t) is an element of $SU(N_f)$. The group G acts on U according to the rule

$$U \to L\,U\,R^\dagger$$

The dimension of $SU(N_f)$ is $N_f^2 - 1$. Therefore, requirement (b) is fulfilled.

As regards (c), we can observe that a typical value of U(x,t) is the unit matrix 1. The little group H leaving 1 invariant is $H = \{(V,V)\}$ since $V^\dagger V = 1$. There is nothing special about this value 1 for U, since the little group of U(x,t) is $\{(U(x,t)VU^\dagger(x,t),V)\}$

which is also isomorphic to H. Thus condition (c) is also verified.

We must now construct a chirally invariant action density for U in order to meet requirement (a). We can construct several such action densities. The general construction uses the "left invariant" Maurer-Cartan "form" (or covariant vector)

$$L_\mu = U^\dagger \partial_\mu U$$

It transforms under (L,R) according to

$$L_\mu \to R L_\mu R^\dagger$$

so that it is invariant under $SU(N_f)_L$.

Equally well, we can use the "right invariant" Maurer-Cartan covector $R_\mu = \partial_\mu U U^\dagger$ which under (L,R) is transformed to

$$R_\mu \to L R_\mu L^\dagger$$

Thus R_μ is invariant under $SU(N_f)_R$. As can be verified in the specific example below, any invariant function f(L) is equal to f(R) and conversely. This is because

$$R_\mu = U L_\mu U^\dagger$$

so that

$$f(R) = f(ULU^\dagger) = f(L)$$

where the last step follows because f is an invariant function.

Now we are free to write down any invariant Lagrangian density using L_μ. A typical example, and one of greatest relevance to us, is

$$\hat{L} = \frac{f_\pi^2}{2} \mathrm{Tr}\, L_\mu L_\mu + \frac{1}{32e^2} \mathrm{Tr}\, \{[L_\mu, L_\nu]^2\} \equiv \hat{L}_o + \hat{L}_{SK}$$

The constant f_π in this $\hat{L}$ can be determined using weak interaction theory from the $\pi^+ \to \mu^+ + \nu_\mu$ reaction rate. The result is

$$f_\pi \simeq 67 \text{ MeV}.$$

The second term $\hat{L}_{SK}$ in this action density is called the Skyrme term.

The energy integral for this $\hat{L}$ is, for time independent U,

$$E = \int d^3 x \left[-\frac{f_\pi^2}{2} Tr(L_i L_i) - \frac{1}{32e^2} Tr([L_i, L_j]^2) \right],$$

Thus the condition $U \to$ constant as $|\vec{x}| \to \infty$ ensures finiteness of energy. Without loss of generality, we may take this constant to be the unit matrix.

3. THE SOLITON

Since $U \to 1$ as $|\vec{x}| \to \infty$, we can take the physical space at constant time to be the three sphere S^3 instead of R^3. Also U has values in the group $SU(N_f)$. Since $\pi_3\{SU(N_f)\} = Z$ for $N_f \geq 2$, the model has an infinite number of solitonic sectors characterized by an integer valued winding number B. There is also an explicit formula for B:

$$B = \frac{1}{24\pi^2} \int d^3x \; \varepsilon_{ijk} \; Tr \; L_i \; L_j \; L_k$$

The "charge" B arises from the current

$$J_\mu = \frac{1}{24\pi^2} \; \varepsilon_{\mu\nu\lambda\rho} \; Tr \; L_\nu \; L_\lambda \; L_\rho$$

which is conserved purely for algebraic reasons (without the necessity for the use of equations of motion):

The static solution for the B=1 sector for $N_f = 3$ can be constructed using the ansatz

$$U_c = \begin{bmatrix} \cos\theta(r) + i\vec{\tau}\cdot\hat{x}\sin\theta(r) & & 0 \\ & & 0 \\ 0 & 0 & 1 \end{bmatrix}, \quad \hat{x}_i = x_i/r.$$

Here, τ_i are the Pauli matrices, they generate isospin rotations on the u and d quarks. U_c is "spherically symmetrical" under the combined spatial and isospin rotations:

$$-i(\vec{x} \times \vec{\nabla})_i U_c + \left[\frac{\lambda_i}{2}, U_c\right] = 0.$$

Here λ_i are the first three Gell-Mann matrices. The winding number B in terms of θ is

$$B = \frac{1}{\pi}\left[\theta(0) - \theta(\infty)\right]$$

Thus we can achieve B=1 with $\theta(o) = \pi$ and $\theta(\infty) = 0$.

In these considerations, $\hat{L}_{SK}$ stabilizes the soliton and prevents it from shrinking to zero size.

Skyrme and Williams also argued that the solitons for odd B can be quantized <u>either</u> as a boson <u>or</u> as a fermion when $N_f = 2$.

The suggestion of Skyrme was that the low lying states in the B=1 solitonic sector describe the nucleon and delta. He also proposed to identify J_μ with the baryonic current of the soliton and B with the baryon number.

These suggestions of Skyrme raised the following two outstanding qualitative questions which were resolved only recently: (a) Why is J_μ the baryonic current ? (b) Why should one quantize the odd B states as fermions ?

We shall first discuss the quantization ambiguities for $N_f = 2$. We shall see that besides the possibilities suggested by Skyrme and Williams, there is a third possibility in which both bosons and fermions occur in the spectrum. We shall then discuss the answers to the questions (a) and (b).

4. SEMICLASSICAL QUANTIZATION OF THE SKYRMION

The ansatz U_c is not invariant under spatial rotations. Under a spatial rotation $R \in SO(3)$, it becomes

$$\cos \theta(r) + i \ \tau_j R_{ji} \hat{x}_i \sin \theta(r) \qquad (\alpha)$$

(We now consider $N_f = 2$ so that $U_c = \cos \theta(r) + i \vec{\tau} \cdot \hat{x} \sin \theta(r)$). Since the Lagrangian is invariant under spatial rotations, the energy of the configuration for every such R is degenerate with the energy for $R = 1$.

The ansatz U_c is not invariant under the flavour SU(2) rotations either. If $A \in SU(2)$, the configuration

$$A\{\cos \theta(r) + i \vec{\tau} \cdot \hat{x} \sin \theta(r)\}A^{\dagger} \qquad (\beta)$$

is distinct from U_c. Further since the Lagrangian is invariant under flavour rotations, these configurations characterized by different A are degenerate in energy.

If A is an element of SU(2), there is a well known two to one map $A \to R(A)$ from SU(2) onto SO(3). It is given by the formula

$$A \tau_i A^{\dagger} = \tau_j R_{ji}(A) \qquad (\gamma)$$

Comparison of (α) and (β) now shows that these degenerate configurations are in fact the same.

These configurations fulfill the boundary condition $U \to 1$ at spatial infinity and are thus physically allowed configurations.

Let ψ_A be the quantum state peaked around the classical ground state $A U_c A^\dagger$. Then the quantum ground state is not any one of these ψ_A's, it is rather a superposition of all such states. We want to determine this quantum ground state by semiclassical reasoning.

Since the mode responsible for the existence of degenerate classical ground states and for the differentiation of classical and quantum ground states is the mode generated by flavour rotation, the simplest approximation to the quantum ground state is to retain just this mode and quantize it. This amounts to the substitution

$$U(x,t) = A(t)\, U_c(x)\, A(t)^\dagger$$

in the Lagrangian and quantization of A. That is, we retain just the "zero frequency mode", "zero mode" or "collective coordinate" A(t) and quantize it.

On substituting $U = AU_c A^\dagger$, we find that

$$\int d^3 x \ \hat{L}$$

reduces to

$$L = - I/2 \ \mathrm{Tr}(A^{-1} \dot{A})^2$$

where the "moment of inertia" I depends on U_c. Its specific form is not important for us.

The quantization of L is quite ambiguous. Thus it can be quantized in three different ways: (i) In method (i), the states of the system are all bosonic. (ii) In method (ii), they are all fermionic. (iii) In method (iii) both bosonic and fermionic states occur in the spectrum.

5. QUANTIZATION METHODS (i) AND (ii)

For methods (i) and (ii), we rewrite L in terms of R(A) using (γ). It is a small exercise in group theory to show that

$$R(A)^{-1}\, dR(A)/dt = \theta_i \, \mathrm{Tr}\, \tau_i \, A^{-1}\, dA/dt$$

where θ_i are the spin one angular momentum matrices:

$$[\theta_i]_{jk} = -i\, \varepsilon_{ijk}$$

Note also that

$$A^{-1} \dot{A} = \tau_i/2 \, \mathrm{Tr}\; \tau_i \, A^{-1} \dot{A},$$

$$\mathrm{Tr}(A^{-1} \dot{A})^2 = 1/2 \, \mathrm{Tr}\; \tau_i \, A^{-1} \dot{A}\, \mathrm{Tr}\, \tau_i \, A^{-1} \dot{A},$$

$$\mathrm{Tr}\{R(A)^{-1} \dot{R}(A)\}^2 = 2 \, \mathrm{Tr}\; \tau_i \, A^{-1}\dot{A}\, \mathrm{Tr}\; \tau_i \, A^{-1} \dot{A}.$$

Let us substitute from these identities in L to find a Lagrangian for R(A). As A runs over SU(2), R(A) runs over all of SO(3), so that we can drop the dependence on A of R(A) and call it R and regard this R (a generic element of SO(3) as the point of the configuration space. Thus the Lagrangian becomes

$$L = - I/8 \, \mathrm{Tr}(R^{-1} \dot{R})^2, \; R \in SO(3)$$

This Lagrangian describes a rigid rotator with its orientation specified by the SO(3) matrix R. The phase space for this system is well known. The role of the configuration space variables is played by the elements R_{ij} of R. Their Poisson brackets (PB´s) vanish:

$$\{R_{ij}, R_{kl}\} = 0$$

The role of the momentum variables is played by the "left" SU(2) generators L_α. Their PB`s are those of angular momenta:

$$\{L_\alpha , L_\beta \} = \varepsilon_{\alpha\beta\gamma} L_\gamma$$

Further they act from the left on R in the sense that their PB's with R correspond to left multiplication by $i\,\theta_\alpha$:

$$\{L_\alpha , R_{ij}\} = i\{ \theta_\alpha R\}_{ij}$$

Of course we can replace L_α by "right" SU(2) generators R_α which act from the right on R. Thus if

$$R_\alpha = L_\beta R_{\beta\alpha}$$

then

$$\{R_\alpha , R_\beta \} = - \varepsilon_{\alpha\beta\gamma} R_\gamma,$$
$$\{R_\alpha , R_{ij}\} = i \left[R\, \theta_\alpha\right]_{ij}$$

The last equation here follows easily on using the identity $R\,\theta_\alpha R^T = R\,\theta_\alpha R^{-1} = \theta_\beta R_{\beta\alpha}$. Using it, we can deduce that $\{ R_\alpha , R_\beta \}$ and $- \varepsilon_{\alpha\beta\gamma} R_\gamma$ have the same PB's with R_{ij} and thus deduce the expression for $\{ R_\alpha , R_\beta \}$. (The elements R_{ij} of the matrix R and the right rotation generators R are not to be confused with each other). In a similar way, we can deduce that

$$\{L_\alpha , R_\beta \} = 0$$

The physical meaning of L_α and R_β follow from their PB's with U. Since

$$U = \cos \theta(r) + i\tau_j R_{ji} \hat{x}_i \sin \theta(r)$$

we find, using $(\theta_\alpha)_{ij} = - i\, \varepsilon_{\alpha ij}$

$$\{L_\alpha , U(\vec{x})\} = i \left[\frac{\tau_\alpha}{2} , U(\vec{x}) \right] ,$$

$$\{R_\alpha , U(\vec{x})\} = (\vec{x} \times \vec{\nabla})_\alpha U(\vec{x})$$

In other words, L_α is the generator of SU(2) flavour (isospin) rotations and R_α is what is normally called the (negative of) angular momentum.

With this interpretation in hand, we can easily find their expressions in terms of R and $\dot{R}$. Thus L_α is the constant of motion associated with the variation

$$\delta R = \varepsilon i \theta_\alpha R$$

where ε is a small parameter. Making ε time dependent, we find

$$\delta L = - i \frac{I}{4} \mathrm{Tr}\ \theta_\alpha \dot{R} R^{-1} \dot{\varepsilon}$$

Thus by Noether`s theorem,

$$L_\alpha = - i \frac{I}{4} \mathrm{Tr}\ \theta_\alpha \dot{R} R^{-1}$$

Similarly, for R_α, the variation is

$$\delta R = - i R \theta_\alpha \varepsilon$$

Since

$$\delta L = i \frac{I}{4} \mathrm{Tr}\ \theta_\alpha R^{-1} \dot{R} \dot{\varepsilon}$$

for this variation,

$$R_\alpha = i \frac{I}{4} \mathrm{Tr}\ \theta_\alpha R^{-1} \dot{R}$$

We have thus got the "momentum" variables in terms of R_{ij} and $\dot{R}_{ij}$.

We have yet to write down the Hamiltonian. Since the Lagrangian consists only of kinetic energy terms, the Hamiltonian H is the Lagrangian written in terms of R and L_α or R and R_α. Thus

$$H = \frac{I}{4} L_\alpha L_\alpha = \frac{I}{4} R_\alpha R_\alpha$$

We now turn to the quantization of this system. On passage to quantum theory, we regard R, L_α and R_α as quantum operators $\hat{R}$, $\hat{L}_\alpha$ and $\hat{R}_\alpha$. Their commutators $[\cdot , \cdot]$ follow from the PB's of the classical variables by the usual prescription of Dirac:

$$[\hat{R}_{ij}, \hat{R}_{kl}] = 0$$

$$[\hat{L}_\alpha , \hat{R}_{ij}] = - \{ \theta_\alpha \hat{R}\}_{ij} ,$$

$$[\hat{L}_\alpha , \hat{L}_\beta] = i \ \varepsilon_{\alpha\beta\gamma}\hat{L}_\gamma ,$$

$$[\hat{R}_\alpha , \hat{R}_{ij}] = \{\hat{R} \ \theta_\alpha \}_{ij} ,$$

$$[\hat{R}_\alpha , \hat{R}_\beta] = -i \ \varepsilon_{\alpha\beta\gamma} \ \hat{R}_\gamma$$

$$[\hat{L}_\alpha , \hat{R}_\beta] = 0$$

Quantization consists in finding a representation of this algebra of observables on a Hilbert space H. It is generally assumed that this representation must be irreducible. There are two inequivalent IRR's of this algebra, so that its quantization is ambiguous. Let H_B and H_F be the Hilbert spaces on which this algebra is realised. The space H_B is spanned by states with J = I where both J and I are integral. (Here J(J+1) and I(I+1) are eigenvalues of $\hat{L}_\alpha \hat{L}_\alpha$ and $\hat{R}_\alpha \hat{R}_\alpha$ respectively). These states are bosonic. This quantization is the same as the usual quantization of the rigid rotor with J and I being interpreted as the body fixed and spacce fixed angular momenta. The space H_F on the other hand is spanned by states with J = I = 1/2, 3/2, ... These states are thus fermionic.

We now give the details of these quantizations.

5.1 Method 1 : Consider functions ψ from the group SO(3) to complex numbers. Thus for each R ε SO(3), ψ (R) is a complex number. The scalar product between two such functions is defined to be

$$(\psi, \psi') = \int dR \quad \psi(R)^* \ \psi'(R)$$

where d R is the invariant measure on SO(3). The Hilbert space H_B is the space of these functions with this scalar product.

Quantization consists in finding operators $\hat{R}_{ij}$, $\hat{L}_\alpha$ and $\hat{R}_\alpha$ such that the commutation relations (C.R.´s) are fulfilled. The first two operators are defined by the rules

$$\{\hat{R}_{ij}\psi\}(R) = R_{ij} \ \psi(R),$$

$$[e^{i\varepsilon_\alpha L_\alpha}](R) = \psi[e^{-i\varepsilon_\alpha \theta_\alpha} R]$$

where ε_α are arbitrary constants. The infinitesimal form of these equations gives the differential realisation of $\hat{L}_\alpha$. The rule for $\hat{L}_\alpha$ is suggested by the fact that $\hat{L}_\alpha$ generates left rotations on R. Similarly $\hat{R}_\alpha$ are defined by

$$[e^{i\varepsilon_\alpha R_\alpha} \psi] \ (R) = \psi[Re^{-i\varepsilon_\alpha \theta_\alpha}]$$

It is possible to check the C.R.´s from these definitions. An even more transparent way to understand these definitions, which makes the C.R.´s obvious, involves introducing a basis for H_B. We know from the Peter-Weyl theorem that any function $\psi \in H_B$ has the expansion

$$\psi(R) = \sum_{L=0}^{\infty} \ \sum_{M,N=-L}^{L} a_{MN}{}^L \, D_{MN}{}^L (R)$$

where the matrices $\{D^L(R)\}$ $(L = 0,1,2, \dots)$ furnish the $(2L + 1)$ dimensional irreducible representation of the rotation group. Thus the functions $D_{MN}{}^L$ form a basis for H_B. On these basis functions, the preceding rules reduce to

$$\hat{L}_\alpha \ D_{MN}{}^L = \{J_\alpha^{(L)}\}_{M'M} \ D_{M'N}{}^L$$

$$\hat{R}_\alpha \ D_{MN}{}^L = - \{J_\alpha^{(L)}\}_{N'N} \ D_{MN'}{}^L$$

where the matrices $J_\alpha^{(L)}$ (α = 1,2,3) represent the angular momenta in the (2L + 1) dimensional representation. We may in fact regard these equations as defining the operators on H_B.

It is easy to verify the C.R.'s from these equations. Further since $\hat{R}_{ij}$ transforms like a vector (in the index i for $\hat{L}_\alpha$, in the index j for $\hat{R}_\alpha$), the state $R_{ij}\ D^L{}_{MN}$ is a superposition of $D^{L'}$s with $L' = L$, L+1 and L−1. A state of integer L is thus mapped to states of integer L under the action of $\hat{R}$ and of $\hat{L}_\alpha$ and $\hat{R}_\alpha$ and the Hilbert space H_B is invariant under the action of all the observables.

From these equations, we see that $\hat{L}^2$ and $\hat{R}^2$ are equal on $D_{MN}{}^L$:

$$\hat{L}^2\ D_{MN}{}^L = \hat{R}^2\ D_{MN}{}^L = L(L+1)D_{MN}{}^L$$

Recalling that $\hat{L}_\alpha$ and $\hat{R}_\alpha$ are spin (or rather total angular momentum) and isospin operators, we thus see that L has the meaning of total angular momentum or isospin of the state $D_{MN}{}^L$ Since L is integral, all these states are bosonic.

5.2 Method 2 : This method is quite similar to method 1, except that we choose a different Hilbert space H_F. A basis for this space is $(D^J_{\rho\sigma})$ where the matrix D^J (g) (g ε SU(2), J = 1/2, 3/2, ...) represents g in the (2J+1) dimensional irreducible representation of g. Note that J runs only over half integral values. The scalar product is defined by

$$(D^J_{\rho\sigma}\ ,\ D^{J'}_{\rho'\sigma'}) = \int dg\ D^J_{\rho\sigma}\ (g)^*\ D^{J'}_{\rho'\sigma'}\ (g)$$

where dg is the invariant SU(2) measure.

The operator $\hat{R}_{ij}$ is defined using the correspondence $g \rightarrow R(g)$ from SU(2) to SO(3):

$$\{\hat{R}_{ij}\ D^J_{\rho\sigma}\}\ (g) = R(g)_{ij}\ D^J_{\rho\sigma}\ (g)$$

The remaining operators are defined in exact analogy to method 1:

$$\hat{L}_\alpha \; D^J_{\rho\sigma} = \{J^{(J)}_\alpha\}_{\rho'\rho} \; D^J_{\rho'\sigma} ,$$

$$\hat{R}_\alpha \; D^J_{\rho\sigma} = - \{J^{(J)}_\alpha\}_{\sigma'\sigma} \; D^J_{\rho\sigma'} .$$

The commutation rules are easily verified. Further since $R(g)_{ij}$ adds integer angular momentum to the states, $\hat{R}_{ij} H_F \subseteq H_F$. Thus, H_F is closed under the action of all the observables.

The spectra of spin and isospin are calculated as in method 1 with the result angular momentum J = isospin I = 1/2, 3/2 ... Thus H contains only fermionic states.

5.3 Method 3 : We now turn to the third method of quantization. In this method, we return to the original form of the Lagrangian where the configuration space Q is SU(2). The phase space T*Q is then spanned by A_{ij}, L_α where

$$\{A_{ij}, A_{kl}\} = 0 ,$$

$$\{L_\alpha , A_{ij}\} = i \left[\frac{\tau_\alpha}{2} A \right]_{ij} ,$$

$$\{L_\alpha , L_\beta \} = \varepsilon_{\alpha\beta\gamma} L_\gamma .$$

(The P.B.'s involving R_α can be deduced from these equations). The associated quantum operators fulfill the commutation relations

$$[\hat{A}_{ij}, \hat{A}_{kl}] = 0$$

$$[\hat{L}_\alpha , \hat{A}_{ij}] = - \{\frac{\tau_\alpha}{2} \hat{A}\}_{ij} ,$$

$$[\hat{L}_\alpha , \hat{L}_\beta] = i \, \varepsilon_{\alpha\beta\gamma} \hat{L}_\gamma .$$

In previous discussion, we have also listed the commutators involving the right angular momentum or isospin operators $\hat{R}_\alpha = \hat{L}_\beta \, R(\hat{A})_{\beta\alpha}$ from which the corresponding commutators relevant for the present case can be deduced.

The crucial difference of this approach from the first two approaches resides in the structure of the configuration space Q: it is now SU(2) and not SO(3). Thus if in imitation of our earlier activity, we try to realise the algebra of these observables on functions f(A) on SU(2) according to the equations

$$\{\hat{A}_{ij} f\}(A) = A_{ij} f(A),$$

$$\{\hat{L}_\alpha f\}(A) = \operatorname{Lm}_{\varepsilon_\alpha \to 0} \left[f\left[e^{-i\varepsilon_\alpha \tau_\alpha/2} A\right] - f(A)\right],$$

$$\{\hat{R}_\alpha f\}(A) = \operatorname{Lm}_{\varepsilon_\alpha \to 0} \left[f(A e^{-i\varepsilon_\alpha \tau_\alpha/2} A) - f(A)\right],$$

then we can <u>not</u> restrict these functions to have either integer values of J or half integer values of J. That is, they can not belong exclusively to either H_B or H_F. The reason is that $\hat{A}_{ij}$ has J = 1/2, being equal to $D^{1/2}_{ij}$ (A). Therefore $\hat{A}_{ij}$ maps H_B to H_F and H_F to H_B : $\hat{A}_{ij}$ $H_B \leq H_F$, $\hat{A}_{ij}$ $H_F \leq H_B$. We can see this result in another way. Functions in H_B can be thought of as even functions of A. This is because a function of a which is invariant under 2π rotation is an even function of A:

$$f(A) = f(Ae^{i2\pi\tau_3/2}) = f(-A)$$

Similarly, functions in H_F are odd functions of A. Since A_{ij} is odd in A, the stated results follow. As a consequence, we need the full space H of all square integrable functions on SU(2) to realize the algebra of observables in Method 3. This Hilbert space contains <u>both</u> bosons and fermions.

This concludes our discussion of the quantization ambiquities in the two flavour model. We now turn to the proofs of Skyrme's conjectures.

6. J_μ IS THE BARYONIC CURRENT

The demonstration of this result was carried out by the Syracuse-Austin group[6] using some results of Goldstone and Wilczek[8] (see also Gipson[5]). The idea of the proof is simple. Let us couple

U to a fermion field $\psi = (\psi_a)$ (where a is the flavour index) via the Lagrangian density

$$-\bar{\psi}\gamma_\lambda \partial_\lambda \psi - m\{\bar{\psi}_L U \psi_R + h.c\}$$

Then acccording to the results of Goldstone and Wilczek, the ground state of the Dirac fermion (the "Dirac sea") gets polarized in the presence of U and carries a nonzero baryon current:

$$< \bar{\psi} \gamma_\mu \psi > = J_\mu \quad .$$

If ψ is taken to be the quark or the nucleon, this formula shows that J_μ becomes exactly equal to the baryon current of the Dirac sea, thereby showing the result we were after.

7. THE B=1 SOLITON IS A FERMION WHEN $N_f \geqslant 3$

The proof of this remarkable result is due to Witten. He recalled that for $N_f \geq 3$, the action necessarily contains the so-called Wess-Zumino term[9]. The necessity of this term can be inferred for example from the flavour anomalies of the theory when the flavour group is gauged. He then showed that in the presence of this term, the odd B solitonic sectors have to be quantized as fermions and the even B sectors as bosons. More precisely, when the Skyrme soliton is quantized and $N_f \geq 3$, the Lagrangian implies a constraint on the allowed states which uniquely picks out one quantization scheme. This result is true whether or not there is a Wess-Zumino term. With the Wess-Zumino term, this constraint is such that for three colours the states are fermionic or bosonic according to whether B is odd or even.

Let me indicate a simplified explanation of these results based on the work of Witten[7] and others[10]. The "spherically symmetric" ansatz introduced earlier is not invariant under the unbroken $SU(N_f)$ flavour group. The action of this group on the ansatz then gives rise to "collective coordinates". We now argue that the results of

Witten can be understood in terms of the quantization of these coordinates.

The collective coordinates are introduced by setting

$$U = A(t)\, U_c\, A(t)^{\dagger}$$

where $A(t) \in SU(N_f)$ and U_c is the classical solution discussed earlier.

Now if h is an element of the SU(2) subgroup (with generators λ_i) of SU(3), we have the formula

$$h\ \lambda_i\ h^{-1} = R(h)_{ji}\ \lambda_j$$

where $R(h) \in SO(3)$. Thus under the transformation $g \to gh$, $U(x,t) \to U(R(h)x,t)$. In other words $g \to gh$, $h \in SU(2)$ is the action of the spatial rotation group on g. Further, since $U \to sUs^{\dagger}$ when $g \to sg$, the action of the flavour $SU(N_f)$ on A is just $A \to sA$.

Specialising to $N_f = 3$ it may be shown that after the substitution $U = A(t)\, U_c\, A(t)^{\dagger}$, the Wess-Zumino term reduces to

$$L_{WZ} = -(i\ BN_c/3\ \mathrm{Tr}\ Y^2)\ \mathrm{Tr}\ Y\ A^{-1}\ \dot{A}$$

where N_c is the number of colours and Y is the hypercharge in the triplet representation:

$$Y = 1/3 \begin{bmatrix} 1 & & 0 \\ & 1 & \\ 0 & & -2 \end{bmatrix}$$

An easy calculation shows that under the hypercharge transformation <u>on the right,</u>

$$A(t) \to A(t)e^{iY\,\theta(t)}\ ,$$

L_{WZ} changes as follows:

$$L_{WZ} \rightarrow L_{WZ} + B\,\dot{\theta}\,N_c/3$$

One can also show that none of the other terms in the Lagrangian are affected by this transformation. Therefore, for $N_c = 3$, if Y is the hypercharge operator on the quantum states, by Noether's theorem,

$$\hat{Y} = B$$

on all allowed states.

The wave functions are superpositions of matrix elements of SU(3) representation matrices which fulfill this constraint. For instance, for B = 1, the octet representation matrix gives wave functions of the form

$$D^{(8)}_{(I,I_3,Y)(I',I'_3,Y'=1)}(A)$$

where the matrix is in the standard (I,I_3,Y) diagonal basis. The constraint implies that the hypercharge in the column index is 1 since Y acts on the right on A. In the octet, the states with $Y'=1$ have "isospin" $I'=1/2$. Since spatial rotation acts on g by $g \rightarrow gh$, it follows that these wave functions have spin half. Since SU(3) acts on g on the left, these wave functions also describe an octet of baryons.

A similar calculation can be done for the decuplet where the wave functions are

$$D^{(10)}_{(I,I_3,Y)(I',I'_3,Y'=1)}(A)$$

In the decuplet, $Y'=1$ implies that $I'=3/2$ so that these wave functions describe a decuplet of spin 3/2 states.

In the discussion of the equivalence of the sine-Gordon and Thirring models, we are familiar with the explicit construction of the Fermi fields from the Bose fields. A similar construction exists for the Skyrme model as well and has been carried out by Rajeev[11].

The connection between the Skyrme model and strong coupling model has been studied by Bardakci and by Gervais and Sakita[11] who show that the two models coincide in the large colour ($N_c \to \infty$) limit. Professor Sakita will explain this connection in greater detail in one of his lectures.

8. PHENOMENOLOGY

This discussion will be very brief and will at best be an indication of the sort of research being done on the phenomenological implications of the Skyrme model.

The first important paper in this line of research is due to Adkins et al.[12] They quantized the zero mode A(t) and obtained a variety of predictions for the nucleon and the delta. The agreement with experiments was in the 20-30% range. Nappi and coworkers[13] have continued this line of study by introducing the ω meson and the breathing modes into the problem. The breathing modes have also been studied by other authors[13].

Another line of research[14] attempts to introduce the quarks into the Skyrme model by imagining the quark bag to be a defect in the Skyrmion field configuration. The U field does not live inside the bag while the quarks are confined to the bag. On the surface of the bag, the quark field is subjected to the Chodos-Thorn-Inoue-Maskawa boundary condition[15] which respects chiral symmetry. It has been shown that the defect does not spoil the interpretation of the topological charge as the baryonic charge, and that the baryonic charges from the inside and the outside of the bag always add up to the topological charge. In this description, as the radius of the bag goes to zero, we recover the pure Skyrme model while when it goes to infinity, we approach a pure bag description. The model appears to lead to very reasonable predictions. It may be noted that there are differences in the treatment of the boundary conditions between the Stony Brook group and Biedenharn et al.

An interesting variant of this idea is due to Nair and Rodgers[16]. There are good reasons to expect that a better description of the baryon might be as a quantum mechanical superposition of the states given by the Skyrme description and the bag description. The phenomenology of such a model has been studied by these authors with satisfactory results.

There have been calculations of the Skyrme constant e by Bhattacharya and Rajeev and by Rubakov[17] using different methods. Their calculations lead to a prediction for e and a reasonable soliton mass. Bhattacharya and Rajeev have also calculated other physical parameters like the F/D ratio with respectable agreement with experiment. The F/D ratio has also been calculated successfully by different methods by Sriram et al.[17]

The Syracuse-Austin group has also been involved in the study of the dibaryon states in the Skyrme model[18]. These come about as follows. Recall that the definition of "spherical symmetry" in the standard Skyrme ansatz involves the isospin SU(2) group. Now, for three flavours, there is also the SO(3) group acting on u,d and s with generators $\Lambda_i\,[\Lambda_1=\lambda_7,\ \Lambda_2=-\lambda_5, \Lambda_3=\lambda_2]$. Thus we can construct ansatze U with "spherical symmetry" based on this SO(3) subgroup of SU(3):

$$-\,i(\vec{x}\times\vec{\nabla})_i\,U + [\Lambda_i, U] = 0$$

The winding numbers for such U are constrained to be even numbers. In the B=2 sector, the least massive state is an SU(3) singlet with zero spin and mass of the order of 2 GeV. There are also octet excitations with spins 1 and 2. They seem to correspond to the six quark "H" states predicted by Jaffe using the MIT bag model.

This concludes my brief survey of the applications of the Skyrme model. A number of interesting papers have not been reviewed[20]. In particular, we have not discussed the criticisms which have been raised regarding the validity of the semiclassical method for quantizing the soliton[21].

9. PROGNOSIS

Although the ideas of Skyrme on the description of baryons as solitons are over twenty years old, their novelty and elegance are being appreciated only during the last few years. Their implications in particle physics and elsewhere are yet to be fully understood. A satisfactory field theoretic implementation of these ideas which goes beyond the semiclassical approximation is also lacking. We may therefore expect vigorous research in this field during the next few years.

ACKNOWLEDGEMENTS

I have benefited very much from discussions with the members of the Syracuse group and with A. Stern in the preparation of this review. I thank Harald Gomm and Fedele Lizzi for suggestions on the manuscript. This work was supported by the Department of Energy under contract number DE-AC02-76ER03533. This work was partially written up at the Centre for Theoretical Studies, Indian Institute of Science, Bangalore and Physics Department, Indian Institute of Technology, Kanpur. I am grateful for the hospitality I was offered at these institutions.

REFERENCES

The references are only meant to be indicative of the work being done and are not exhaustive.

1. Nishijima, K., Nuovo Cimento 11, 698 (1959); F. Gürsey, Nuovo Cimento 16, 230 (1960); Ann. Phys. 12, 91 (1961).
2. T.H.R. Skyrme, Proc. Roy. Soc. (London) A260, 127 (1961); Nucl. Phys. 31, 556 (1962); J. Math. Phys. 12, 1735 (1971).
3. Williams, J.G., J. Math. Phys. 11, 2611 (1970).

4. Pak, N.K. and Tze, H. Ch., Ann. Phys. 117, 164 (1979).
5. Gipson, J.M.. and Tze, H. Ch., Nucl. Phys. B183, 365 (1984).
6. Balachandran, A.P., Nair, V.P., Rajeev, S.G. and Stern, A., Phys. Rev. Lett. 49, 1124 (1982); Phys. Rev. D27, 1153 (1983).
7. Witten, E., Nucl. Phys. B223, 422, 433 (1983).
8. Goldstone, J. and Wilczek, F., Phys. Rev. Lett. 47, 986 (1981).
9. Wess, J. and Zumino, B., Phys. Lett. 37B, 95 (1971); Hari Dass, N.D., Phys. Rev. D5, 1542 (1972).
10. Guadagnini, E., Nucl. Phys. B236, 35 (1984); Ramdas, T.R., Comm. Math. Phys. 93, 355 (1983); Jain, S. and Wadia, S.R., Tata Institute of Fundamental Research, Bombay preprint (1984); Balachandran, A.P., Barducci, A., Lizzi, F., Rodgers, V.G.J. and Stern, A. (unpublished). (See the second paper of ref. 18 where the work of this group is discussed).
11. Rajeev, S.G., Phys. Rev. D29, 2944 (1983); Bardakci, K., Berkeley preprint (1983); Gervais, J.L. and Sakita, B., Phys. Rev. Lett. 52, 87 (1984) and Phys. Rev. D30, 1795 (1984).
12. Adkins, G.S., Nappi, C.R. and Witten, E., Nucl. Phys. B228, 552 (1983).
13. Adkins, G.S. and Nappi, C.R., Nucl. Phys. B233, 109 (1984); Breit, J.D. and Nappi, C.R., Princeton preprint (1984); Dey, J. and Le Tourneux, J., Montreal preprint (1984); Gajduk, Ch. and Schwesinger, B., Stony Brook preprint (1984); Hayashi, A. and Holzwarth, G., Siegen preprint (1984).
14. Rho, M., Goldhaber, A.S. and Brown, G.E., Phys. Rev. Lett. 51, 747 (1983); Jackson, A.D. and Rho, M., Phys. Rev. Lett. 51, 751 (1983); Jackson, A., Jackson, A.D. and Pasquier, V., Stony Brook preprint (1984) and references therein; Goldstone, J. and Jaffe, R.L., Phys. Rev. Lett. 51, 1518 (1983); Biedenharn, L.C., Dothan, Y. and Stern, A., Austin preprint (1984).
15. Chodos, A. and Thorn, C.B., Phys. Rev. 12, 2733 (1985); Inoue, T. and Maskawa, T., Prog. Theor. Phys. 54, 1833 (1975).
16. Nair, V.P. and Rodgers, V.G.J., Princeton preprint (1985).
17. Bhattacharya, G. and Rajeev, S.G., Syracuse University preprint (1983); Rubakov, V.A., Moscow preprint (1984); Sriram, M.S., Mani, H.S. and Ramachandran, R., Phys. Rev. D30, 1141 (1984).

18. Balachandran, A.P., Barducci, A., Lizzi, F., Rodgers, V.G.J. and Stern, A., Syracuse preprint (1985).

19. Jaffe, R.L., Phys. Rev. Lett. 38, 195 (1977); Aerts, A.T.M. and Dover, C.M., Phys. Rev. D29, 433 (1984) and references therein.

20. E.g Birge, M. and Banerjee, M., Phys. Lett. 136B, 284 (1984); Kahana, S., Ripka, G. and Soni, V., Phys. Rev. Letters 52, 1743 (1984); Soni, V., Nucl. Phys. A415, March (1984); Callan, C.G. and Witten, Princeton preprint (1983).

21. Braaten, E. and Ralston, J.P., Argonne preprint (1984); Bander, M. and Hayot, F., Saclay preprint (1984).

WHAT IS THE NEXT STEP AFTER ELECTROWEAK UNIFICATION?

G.Rajasekaran*
Institute of Mathematical Sciences,
Madras 600 113, INDIA

ABSTRACT

After a compact description of the standard model of electroweak and strong interactions based on the SU(3) x SU(2) x U(1) gauge group, various approaches towards the next step in High Energy Physics are discussed. These are grand unification, preons, induced gravity, supersymmetry and supergravity, higher dimensions and finally superstrings which at the present time seem to be the most promising.

*Also at Department of Theoretical Physics, University of Madras, Guindy Campus, Madras 600 025, INDIA.

1. INTRODUCTION

The various avenues open to high energy physicists after the successful electroweak unification are the following:

a) Grand Unification

b) Preons

c) Induced Gravity

d) Supersymmetry and Supergravity

e) Higher Dimensional Unification

f) Superstrings

Our aim is to give a brief introduction to each of these possibilities. Some of these are better developed than the rest and some are more promising than the others. However, it is correct to say that all of these are speculative ideas; none of them has an iota of experimental support. So, before we start describing these ideas, let us summarize the standard model of electroweak and strong interactions, for which we have good experimental support.

2. THE STANDARD MODEL

The standard model describes <u>all</u> of presently known High Energy Physics. It is based on the gauge group $SU_3 \times SU_2 \times U_1$. While SU_3 leads to Quantum Chromodynamics (QCD) and describes strong interactions, $SU_2 \times U_1$ leads to Quantum Flavour Dynamics (QFD) and describes electroweak interactions. The gauge bosons of QCD are the 8 gluons G^i_μ ($i = 1....8$). The gauge bosons of QFD are W^a_μ ($a = 1,2,3$) and B_μ or, the physical weak bosons $W^\pm_\mu$ and Z_μ and the photon A_μ.

The physical low-energy separation between weak and electro-magnetic interactions is achieved by breaking the $SU_2 \times U_1$ electroweak symmetry at the scale $M_L \sim 100$ Gev. The symmetry breaking is supposedly induced by the nonvanishing vacuum expectation value of the Higgs scalar ϕ which is chosen to be a doublet under SU_2. The colour symmetry SU_3 is supposed to be unbroken.

The particle sector comprising leptons and quarks is taken to be the 3 generations of fermions:

$$\underbrace{\begin{pmatrix} \nu_e \\ e^- \end{pmatrix} \begin{pmatrix} u_\alpha \\ d_\alpha \end{pmatrix}}_{1} \quad \underbrace{\begin{pmatrix} \nu_\mu \\ \mu^- \end{pmatrix} \begin{pmatrix} c_\alpha \\ s_\alpha \end{pmatrix}}_{2} \quad \underbrace{\begin{pmatrix} \nu_\tau \\ \tau^- \end{pmatrix} \begin{pmatrix} t_\alpha \\ b_\alpha \end{pmatrix}}_{3}$$

In the above, α denotes the colour index of the quarks. Among these fermions, the top quark t which is the heaviest, had remained elusive, but as announced recently, the UAI experiment at the $p\bar{p}$ collider seems to have obtained evidence for its existence around a mass of about 40 Gev.

Since the weak interaction is helicity-dependent, it is necessary to separate the helicities of fermions. Counting the L and R helicities as distinct particles, we have 15 particles for each generation. For the first generation, they are

$$\begin{pmatrix} \nu_e \\ e^- \end{pmatrix}_L , \begin{pmatrix} u_\alpha \\ d_\alpha \end{pmatrix}_L , \; e^-_R , \; u_{\alpha R} , \; d_{\alpha R}$$

where the doublets and singlets under SU_2 are explicitly indicated.

What is the the present experimental status of the standard model?

<u>QFD:</u> All neutral current sectors are in beautiful agreement with $\mathrm{Sin}^2 \theta_W \approx 0.21$ where θ_W is the weak mixing angle defined by the ratio of the U(1) and SU(2) coupling constants:

$$\tan \theta_W = \frac{g_1}{g_2} .$$

The next crucial prediction was the existence of the W and Z bosons with masses given by

$$m_W = \frac{1}{\sin\theta_W} \left(\frac{\pi\alpha}{\sqrt{2}\, G_F} \right)^{1/2} = \frac{37.4 \text{ Gev}}{\sin\theta_W} \approx 84 \text{ Gev} \qquad (1)$$

$$m_Z = \frac{m_W}{\cos \theta_W} \approx 94 \text{ Gev} . \qquad (2)$$

The electroweak theory passed this crucial test in 1983 when the W and Z bosons were discovered at the $p\bar{p}$ collider at CERN [1)].

QCD: There is intense activity at an unprecedented level to fit all hadronic and even nuclear phenomena to QCD. Although there appears to be general confirmation, there is no direct verification of QCD. This is in contrast to the case of QFD. In fact, because of the dogma of colour confinement, QCD is doomed to indirect verification only.

Thus, among the two pillars on which the standard model rests, the strength of one can not be directly measured.

Now let us write down the Lagrangian of the standard model.

$$
\begin{aligned}
\mathcal{L} = & -\frac{1}{4} G^i_{\mu\nu} G^i_{\mu\nu} - \frac{1}{4} W^a_{\mu\nu} W^a_{\mu\nu} - \frac{1}{4} B_{\mu\nu} B_{\mu\nu} \\
& - \sum_n \bar{q}_L^{(n)} \gamma^\mu \left(\partial_\mu + i g_3 \frac{\lambda^i}{2} G^i_\mu + i g_2 \frac{\tau^a}{2} W^a_\mu \right. \\
& \left. + i \frac{g_1}{6} B_\mu \right) q_L^{(n)} \\
& - \sum_n \bar{u}_R^{(n)} \gamma^\mu \left(\partial_\mu + i g_3 \frac{\lambda^i}{2} G^i_\mu + i \frac{2}{3} g_1 B_\mu \right) u_R^{(n)} \\
& - \sum_n \bar{d}_R^{(n)} \gamma^\mu \left(\partial_\mu + i g_3 \frac{\lambda^i}{2} G^i_\mu - i \frac{g_1}{3} B_\mu \right) d_R^{(n)} \\
& - \sum_n \bar{\ell}_L^{(n)} \gamma^\mu \left(\partial_\mu + i g_2 \frac{\tau^a}{2} W^a_\mu - i \frac{g_1}{2} B_\mu \right) \ell_L^{(n)} \\
& - \sum_n \bar{e}_R^{(n)} \gamma^\mu \left(\partial_\mu - i g_1 B_\mu \right) e_R^{(n)} \\
& - \left| \left(\partial_\mu + i g_2 \frac{\tau^a}{2} W^a_\mu + i \frac{g_1}{2} B_\mu \right) \phi \right|^2 + \mu^2 \phi^\dagger \phi - \frac{\lambda}{2} (\phi^\dagger \phi)^2 \\
& - \sum_{m,n} \left(\Gamma^u_{mn} \bar{q}_{mL} \phi^c u_{nR} + \Gamma^d_{mn} \bar{q}_{mL} \phi \, d_{nR} \right. \\
& \left. + \Gamma^e_{mn} \bar{\ell}_{mL} \phi \, e_{nR} + h.c \right) \qquad (3)
\end{aligned}
$$

The first three terms describe the pure gauge field part of the $SU_3 \times SU_2 \times U_1$ non abelian gauge theory. The next group of terms describes the fermions. In these fermionic terms, the SU_3 colour and the SU_2 flavour indices are suppressed and the index n is used to denote the generation number. The left-handed SU(2) doublet quark of the nth generation is denoted by $q_L^{(n)}$ and the corresponding right handed SU(2) singlets are denoted by $u_R^{(n)}$ and $d_R^{(n)}$. For the leptons, $\ell_L^{(n)}$ is the doublet while $e_R^{(n)}$ is the singlet.

The last group of terms describes the Higgs field and its interactions with itself, with the gauge bosons and with the fermions which are respectively responsible for spontaneous symmetry breaking, generation of W and Z masses and generation of the masses for the quarks and the leptons.

As already mentioned, this Lagrangian is supposed to describe <u>all</u> that is known in High Energy Physics. This is the achievement of the past two decades. Dirac is supposed to have said, referring to his relativistic wave equation of the electron, that it describes all of Chemistry and almost all of Physics. Referring to the standard model Lagrangian, we are tempted to say that it describes all of Physics except gravity. However, note the contrast in complexity. Whereas the Dirac equation can be written down on one single line, the standard model Lagrangian requires almost a whole page. That supplies the chief motivation for going beyond the standard model. The model lacks the simplicity which is the hall mark of any truly fundamental theory.

A few words on the generation of the masses of the quarks and leptons in the standard model are in order. These arise from the Yukawa couplings of ϕ given in the last part of the Lagrangian in eq.(3), as can be seen by replacing ϕ by its vacuum expectation value. In contrast to the rest of the Lagrangian, the Yukawa coupling constants Γ^u_{mn}, Γ^d_{mn} and Γ^e_{mn} mix the generations. In fact, the mass terms are non diagonal with respect to parity as well as flavour quantum numbers such as strangeness, charm etc. These mass matrices can be diagonalized, but then the mixing between the generations enters through the charged-current weak interactions (mediated by the $W^{\pm}$ bosons). The mixing is described by the unitary Cabibbo-Kobayashi-Maskawa matrix:

$$U = \begin{pmatrix} c_1 & -s_1c_3 & -s_1s_3 \\ s_1c_2 & c_1c_2c_3 - s_2s_3e^{i\delta} & c_1c_2s_3 + s_2c_3e^{i\delta} \\ s_1s_2 & c_1s_2c_3 + c_2s_3e^{i\delta} & c_1s_2s_3 - c_2c_3e^{i\delta} \end{pmatrix}$$

where $c_1 = \cos\theta_1$, $s_1 = \sin\theta_1$ etc and δ is the CP violating phase. It is important to note that such a CP violating phase exists only if the number of generations is $\geqslant 3$, as was first pointed out by Kobayashi and Maskawa. The charged-current weak interaction is therefore, modified to

$$\bar{u}\ \gamma_\mu\ (1 - \gamma_5)\ U\ d\ W^\mu\ + \text{h.c}$$

where u stands for $(\bar{u}\ \bar{c}\ \bar{t})$ and d stands for

$$\begin{pmatrix} d \\ s \\ b \end{pmatrix}.$$

A similar mixing matrix will exist in the leptonic sector also, if neutrinos have masses. For massless left-handed neutrinos, suitable redefinition of the neutrino fields removes leptonic mixing.

The neutral-current weak interaction on the other hand is not modified and remains diagonal in the generation space, because of the unitarity of the mixing matrix. This is a generalization of the original Glashow-Illiopoulos-Maiani mechanism which achieved the cancellation of the strangeness-changing neutral current.

The diagonal mass matrices are

$$\begin{pmatrix} m_u & & \\ & m_c & \\ & & m_t \end{pmatrix} \approx \begin{pmatrix} 4 & & \\ & 1500 & \\ & & 40{,}000(?) \end{pmatrix} \text{Mev}$$

$$\begin{pmatrix} m_d & & \\ & m_s & \\ & & m_b \end{pmatrix} \approx \begin{pmatrix} 8 & & \\ & 200 & \\ & & 5000 \end{pmatrix} \text{Mev}$$

$$\begin{pmatrix} m_e & & \\ & m_\mu & \\ & & m_\tau \end{pmatrix} = \begin{pmatrix} 0.5 & & \\ & 103 & \\ & & 1784 \end{pmatrix} \text{Mev}$$

The non-diagonal mixing matrix in the quark sector is [2)]

$$U = \begin{array}{c} \\ u \\ c \\ t \end{array} \begin{array}{c} \begin{array}{ccc} \quad d \quad & \quad s \quad & \quad b \quad \end{array} \\ \begin{pmatrix} 0.9730 \pm 0.0007 & 0.231 \pm 0.003 & 0.003 \pm 0.002 \\ 0.231 \pm 0.003 & 0.9715 \pm 0.0011 & 0.050 \pm 0.006 \\ 0.009 \pm 0.006 & 0.049 \pm 0.006 & 0.9984 \pm 0.0007 \end{pmatrix} \end{array}$$

where the absolute values alone are given. The small values of the $b \to u$ and $b \to c$ matrix elements are consequences of the recently discovered long lifetime of the b-flavoured hadrons[3)].

We see that the masses increase rapidly as we go up in the generation number and also that the mixing between generations is weak, especially for the heavier generation. In fact, as a crude approximation, the mixing matrix can be written as [4)]

$$U \approx \begin{pmatrix} 1 & s & s^3 \\ s & 1 & s^2 \\ s^3 & s^2 & 1 \end{pmatrix}$$

where $s \approx 0.23$.

Let us now collect together the standard reasons usually given for attempting to go beyond the standard model.

1. Too Many Parameters:

Counting the coupling constants, the boson masses, the various quark and lepton masses and the quark-mixing parameters the total number of independent parameters in the standard model is about twenty. They are all empirically determined and there is no fundamental theoretical understanding of these numbers. This is one of the weakest points of the standard model and the strongest motivation for considering the next step.

2. Generation Puzzle:

The standard model contains no clue explaining the existence of several generations of quarks and leptons.

3. Pattern Within One Generation:

The model does not explain why quark and lepton charges are quantized in a related way: why does integer charge come with colour singlets and noninteger charge with colour triplets? Also the model does not explain the apparent quark-lepton universality: why do the quarks and leptons possess identical SU(2) properties? Since the same pattern repeats at least 3 times (for the 3 generations), there must be a particularly good reason for this pattern!

4. Unification:

Theoretical Physicists have an innate urge for unification. The three interactions must be unified in some manner so that the 3 coupling constants g_1, g_2 and g_3 are replaced by a single coupling constant.

5. Inclusion of Gravity:

Unification with gravitational interactions is of course an important aim of physics, especially if the other three interactions are already unified and if the (grand) unification scale is already close to the Planck scale.

6. Hierarchy problem:

Assuming that there is an important energy scale beyond the standard model (it may be the Planck mass ~ 10^{19} Gev if gravity is included in unification or somewhat lower energy scale such as the grand unification scale ~ 10^{14} Gev), it is difficult to understand how particles with masses corresponding to the low energy scales of the standard model can survive enormous self-energy corrections. Vector bosons and fermions may be protected from such corrections by gauge symmetry or a chiral symmetry, respectively. Scalars and their vacuum expectation values are not protected, in general.

Grand Unification solves problems 3 and 4 mentioned above. Preons do not solve any of the above problems (nevertheless they may turn out to be the correct next step!). Supersymmetry and supergravity solves 3, 4, 5 and 6. Supersymmetric strings in higher dimensions is expected to solve all the problems.

3. GRAND UNIFICATION

Before we describe models based on particular unifying groups, let us discuss some general results [5)] which are independent of the particular unifying group.

3.1 Coupling Constant Ratios In The Unifying Limit

Let $G_o = G_A \times G_B \times G_C \cdots$ (where G_A, G_B, G_C ... are simple) be the observed group of strong, weak and electromagnetic interactions with corresponding coupling constants g_A, g_B, g_C ... If G_o is embedded in a unifying group G with a single coupling constant g_G, then the various observed coupling constants g_A, g_B ... get related to g_G. There are many choices possible for the unifying group G. But, the trick that facilitates a general analysis is to consider a maximal unifying group $G_{max} = SU(2N_f)$, where N_f is the total number of fermions in the theory and $2N_f$ is the total number of left and right-handed components. The algebraic relations among the coupling constants are derived from G_{max}, but are automatically valid for any unifying group G which is a subgroup of G_{max}.

It is easy to see the validity of the relations:

$$g_G \; T_G^{(\alpha)} = g_A \; T_A^{(i)}$$

$$g_G T_G^{(\beta)} = g_B T_B^{(j)} \tag{5}$$

....

where the T's are all $2N_f \times 2N_f$ dimensional matrices, $T_G^{(\alpha)}$ and $T_A^{(i)}$ are the αth generator matrix for the group G and the corresponding i th generator matrix for the factor group G_A and similarly for $T_G^{(\beta)}$ and $T_B^{(j)}$. Further, $T_G^{(\alpha)}$ and $T_A^{(i)}$ are normalized such that $\mathrm{Tr}\,(T_G^{(\alpha)}\ T_G^{(\alpha)})$ is independent of α and $\mathrm{Tr}\,(T_A^{(i)}\ T_A^{(i)})$ is independent of i, so that

$$\mathrm{Tr}\ (T_G^{(\alpha)}\ T_G^{(\beta)}) = \delta^{\alpha\beta}\ \mathrm{Tr}\ (T_G)^2$$

$$\mathrm{Tr}\ (T_A^{(i)}\ T_A^{(j)}) = \delta^{ij}\ \mathrm{Tr}\ (T_A)^2 , \tag{6}$$

Thus,

$$g_G^2\ \mathrm{Tr}\ (T_G)^2 = g_A^2\ \mathrm{Tr}\ (T_A)^2 = g_B^2\ \mathrm{Tr}\ (T_B)^2 = \ldots . \tag{7}$$

Hence, for the ratios of the strong, electric and weak coupling constants, we get

$$\frac{\alpha}{\alpha_s} = \frac{\mathrm{Tr}\ T_C^2}{\mathrm{Tr}\ Q^2}\ ;\ \sin^2\theta_W = \frac{e^2}{g_L^2} = \frac{\mathrm{Tr}\ T_L^2}{\mathrm{Tr}\ Q^2} \tag{8}$$

where Q, T_L and T_C are the $2N_f \times 2N_f$ dimensional matrices corresponding to the electric charge, the third component of the weak SU_{2L} generators and the third component of the colour generators respectively. In this section, we are using

$$\alpha = \frac{e^2}{4\pi}\ ;\quad \alpha_S = \frac{g_S^2}{4\pi}\ ;\quad \alpha_L = \frac{g_L^2}{4\pi}\ ;\quad \alpha_1 = \frac{g_1^2}{4\pi}$$

where g_S, g_L and g_1 are the coupling constants corresponding respectively to the SU_3, SU_2 and U_1 groups of the standard model. The coupling constant relations (8) are valid only in the unification limit. Let us now give a few examples.

a) <u>Doublets of fermions:</u>

$$\begin{pmatrix} u^\alpha \\ d^\alpha \end{pmatrix} \quad \begin{pmatrix} \nu_e \\ e^- \end{pmatrix}$$

These can be repeated any number of times, corresponding to the number of generations. The generator matrices will, however, be restricted to one generation and they will be of dimension 16 :-

$$\begin{array}{llllll} & & u_L & d_L & \nu_e & e_L^- \\ Q & = \text{diag.} & \Big(\frac{2}{3}, \frac{2}{3}, \frac{2}{3}, & -\frac{1}{3}, -\frac{1}{3}, -\frac{1}{3}, & 0, & -1, \text{ same for R}\Big) \\ T_L^{(3)} & = \text{diag.} & \Big(\frac{1}{2}, \frac{1}{2}, \frac{1}{2}, & -\frac{1}{2}, -\frac{1}{2}, -\frac{1}{2}, & \frac{1}{2}, & -\frac{1}{2}, \text{ zero for R}\Big) \\ T_C^{(3)} & = \text{diag.} & \Big(\frac{1}{2}, \frac{-1}{2}, 0, & \frac{1}{2}, -\frac{1}{2}, 0, & 0, & 0, \text{ same for R}\Big). \end{array}$$

In these diagonal matrices, the first three elements refer to the three colours of u_L, the next three elements refer to the three colours of d_L and the next two elements refer to ν_e and e_L. The other eight elements refer to the corresponding right-handed fermions. We therefore get

$$\text{Tr } Q^2 = \frac{16}{3}, \quad \text{Tr } (T_L^{(3)})^2 = 2, \quad \text{Tr } T_C^2 = 2 \tag{9}$$

and hence

$$\frac{\alpha}{\alpha_S} = \frac{\text{Tr } T_C^2}{\text{Tr } Q^2} = \frac{3}{8} \tag{10}$$

$$\sin^2 \theta_W = \frac{e^2}{g_L^2} = \frac{\text{Tr } T_L^2}{\text{Tr } Q^2} = \frac{3}{8}. \tag{11}$$

The above results are independent of the actual unifying group as long as it is a subgroup of G_{max} which is in this case is SU(16) and as long as the doublet scheme is maintained.

b) Triplets of fermions

$$\begin{pmatrix} u^\alpha \\ d^\alpha \\ x^\alpha \end{pmatrix} \qquad \begin{pmatrix} \nu_e \\ e^- \\ E \end{pmatrix}$$

Again these can be repeated any number of times. Now, we have 24 dimensional matrices:

$$
\begin{array}{llllll}
 & u_L & d_L & x_L & \nu_L \quad e_L & E_L \\
Q = \text{diag.}\Big(\frac{2}{3},\frac{2}{3},\frac{2}{3}, & \frac{-1}{3},\frac{-1}{3},\frac{-1}{3}, & \frac{-1}{3},\frac{-1}{3},\frac{-1}{3}, & 0, \quad -1, & +1, & \text{same for R}\Big) \\
T_L^{(3)} = \text{diag.}\Big(\frac{1}{2},\frac{1}{2},\frac{1}{2}, & \frac{-1}{2},\frac{-1}{2},\frac{-1}{2}, & 0,0,0, & \frac{1}{2}, \quad \frac{-1}{2}, & 0, & \text{zero for R}\Big) \\
T_C^{(3)} = \text{diag.}\Big(\frac{1}{2},\frac{-1}{2},0, & \frac{1}{2},\frac{-1}{2},0, & \frac{1}{2},\frac{-1}{2},0, & 0, \quad 0, & 0, & \text{same for R}\Big)
\end{array}
$$

$$\text{Tr } Q^2 = 8 \;; \quad \text{Tr } T_L^2 = 2 \;; \quad \text{Tr } T_C^2 = 3. \tag{12}$$

The results in this triplet scheme are

$$\frac{\alpha}{\alpha_S} = \frac{\text{Tr } T_C^2}{\text{Tr } Q^2} = \frac{3}{8} \tag{13}$$

$$\sin^2\theta_W = \frac{e^2}{g_L^2} = \frac{\text{Tr } T_L^2}{\text{Tr } Q^2} = \frac{1}{4} \,. \tag{14}$$

This is given only for illustration. Hereafter, we shall assume the doublet scheme only.

3.2 Renormalization Effects

As already mentioned, the coupling constant ratios derived above are valid only in the unification limit. Once the unified symmetry is broken, the coupling constants undergo different renormalizations and hence their ratios relevant (at low energies) after symmetry breaking are different. The energy-dependence of the coupling constants is calculated [5,6] using renormalization group.

Let us assume that the unification group G is broken at the energy scale M_U in a <u>single step</u> into the standard model group $SU_3 \times SU_2 \times U_1$.

Then the variation of the gauge coupling constants g_S, g_L and g_1 as a function of Q^2 for $Q^2 < M_U{}^2$ is given by

$$\frac{1}{\alpha_S(Q^2)} = \frac{1}{\alpha_S(M_U^2)} - \beta_S \ln \frac{Q^2}{M_U^2} \tag{15}$$

$$\frac{1}{\alpha_L(Q^2)} = \frac{1}{\alpha_L(M_U^2)} - \beta_L \ln \frac{Q^2}{M_U^2} \tag{16}$$

$$\frac{3}{5}\frac{1}{\alpha_1(Q^2)} = \frac{3}{5}\frac{1}{\alpha_1(M_U^2)} - \frac{3}{5}\beta_1 \ln \frac{Q^2}{M_U^2} \tag{17}$$

where

$$\beta_S = -\frac{11}{16\pi^2} + F$$

$$\beta_L = -\frac{11}{16\pi^2} \times \frac{2}{3} + F$$

$$\frac{3}{5}\beta_1 = 0 + F\,. \tag{18}$$

The above results are obtained from the one loop contributions of the type given in Fig.1. Decoupling theorem is invoked in this calculation so that we may ignore the contribution of heavy particles. Thus, although unifcation will involve further gauge bosons, they are all assumed to be sufficiently heavy so as not to contribute to these one loop equations for $Q^2 << M^2{}_U$. The non abelian gauge-boson contributions lead to the negative numbers given for β_S and β_L, whereas there is no such contribution for the abelian case β_1. (This is the famous asymptotic freedom of nonabelian gauge theory which plays a crucial role in QCD). The fermion contribution to the β coefficient is denoted F and the contribution is the same for all the three graphs, assuming that the complete fermion multiplet is light. Such an assumption will not be valid for the Higgs scalars and hence they will contribute differently to the three coupling constants. However, we shall ignore the Higgs contribution.

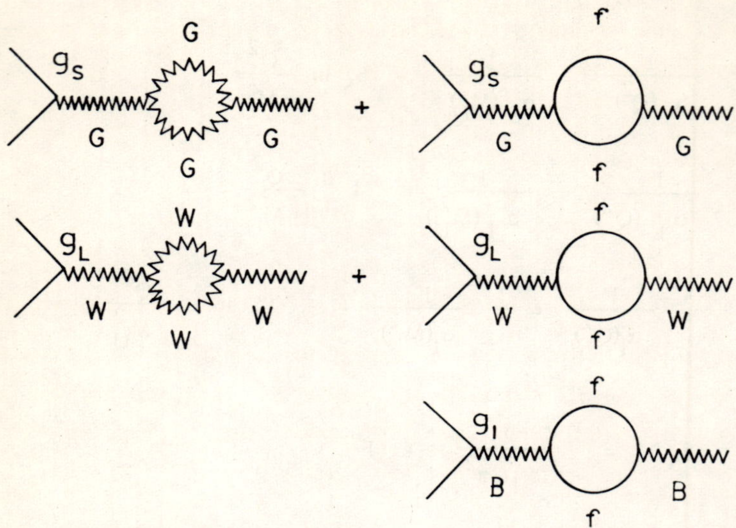

FIG.1 RENORMALIZATION EFFECTS ON THE COUPLING CONSTANTS

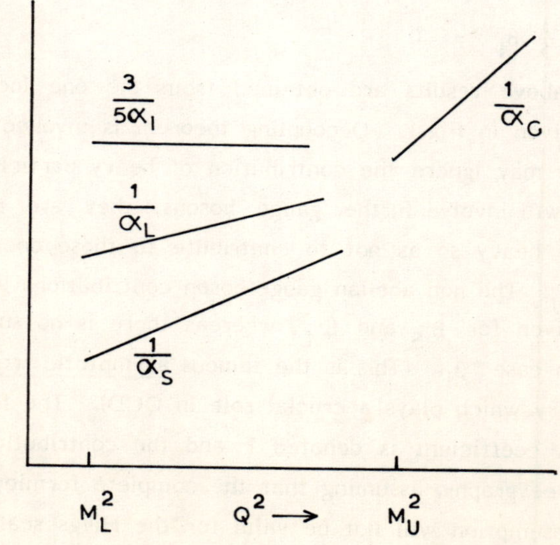

FIG.2 DEPENDENCE OF THE COUPLING CONSTANTS ON Q^2

We should also remark that the factor 3/5 occurring in the above equations is the value of $Tr(T_G)^2/Tr(T_l)^2$ for the doublet scheme of fermions which has been assumed. (vide eqs.(7-11)). Also, we have assumed the usual normalization for the nonabelian generators:

$$Tr\,(T_G)^2 = Tr\,(T_C)^2 = Tr\,(T_L)^2. \tag{19}$$

In the unification limit, we have:

$$\frac{1}{\alpha_S(M_U^2)} = \frac{1}{\alpha_L(M_U^2)} = \frac{3}{5}\,\frac{1}{\alpha_l(M_U^2)} \equiv \frac{1}{\alpha_G(M_U^2)}\,. \tag{20}$$

The behaviour of the coupling constants is schematically plotted in Fig.2

By substraction, we can eliminate the unknown constants from the three equations (15-17) and write the results for the ratios of the coupling constants in the low-energy region $Q \approx M_L \approx 100$ Gev in the form:

$$\frac{\alpha}{\alpha_S} = \frac{3}{8} - \frac{33}{8\pi}\,\alpha\,\ln\frac{M_U}{M_L} \tag{21}$$

$$\sin^2\theta_W = \frac{3}{8} - \frac{55}{24\pi}\,\alpha\,\ln\frac{M_U}{M_L} \tag{22}$$

where α is the fine structure constant of electromagnetism and θ_W is the electroweak mixing angle.

Eq.(21) can be rewritten as

$$\frac{3}{8}\,\frac{1}{\alpha} - \frac{1}{\alpha_S} = \frac{33}{8\,\pi}\,\ln\frac{M_U}{M_L}\,.$$

On the left-hand side we substitute the empirical values of the coupling constants for QCD and QED at the 'low' energy scale M_L:

$$\frac{1}{\alpha_S} \approx 5\;;\quad \frac{3}{8}\,\frac{1}{\alpha} \approx 50\;;\quad \frac{3}{8}\,\frac{1}{\alpha} - \frac{1}{\alpha_S} \approx 45\,.$$

This leads to the unification scale:

$$M_U \approx 100\text{ Gev} \times \exp\frac{45 \times 8\,\pi}{33} \approx 10^{15}\text{ Gev}\,. \tag{23}$$

Thus, because of the slow (logarithmic) variation of the coupling constants, the unification of the strong and electromagnetic coupling constants occurs only at superhigh energy scale.

Substitution of this value of M_U in eq.(22) gives

$$\sin^2 \theta_W \approx 0.21 \qquad (24)$$

which is in very good agreement with the current experimental value.

Further, the above unification scale leads to another important consequence. Noting that the unifying group G, in general, contains B (Baryon-number) - violating processes, small values of M_U would have been catastrophic for the stability of the proton. Assuming a single exchange of the superheavy gauge boson of mass ~ M_U in the amplitude for proton decay, the life-time of proton can be estimated as

$$\tau_p \sim \frac{M_U^4}{m_p^5} \sim 10^{31} \text{ years} \qquad (25)$$

which is long enough to avoid the catastrophe.

Thus, not only the correct value of the electroweak mixing angle θ_W, but also the longevity of the proton is explained. That is the spectacular success of the grand unification scheme.

Since the proton does decay in the grand unified models, its detection in the various underground experiments including the one at Kolar Gold Fields will offer a crucial test of the models.

3.3 Popular Models For Unification

Here we provide only the bare bones of the three popular grand unifcation models. For more details, see the review by Langacker [7].

The group SU_5

The gauge bosons belong to the adjoint representation 24 and their decomposition under $SU_3 \times SU_2$ are as follows:

$$\underline{24} = \underset{G^\alpha_\beta}{(8,1)} + \underset{(W^\pm,\, W^o)}{(1,3)} + \underset{B}{(1,1)} + \underset{(X_\alpha,\, Y_\alpha)}{(3,2)} + \underset{(\bar{X}^\alpha,\, \bar{Y}^\alpha)}{(3^*,2)}$$

The 24 gauge bosons are comprised of the octet of colour gluons G^α_β, the 4 electroweak bosons ($W^\pm$,W^o, B) and the 12 leptoquarks (X_α, Y_α,

$\bar{X}^{\alpha}$, $\bar{Y}^{\alpha}$), where α is the colour index.

The fermions of one family belong to $\underline{5}^*$ + $\underline{10}$ and they are respectively

$$\begin{pmatrix} \nu_e \\ & d^c_\alpha \\ e^- \end{pmatrix}_L \quad \text{and} \quad \begin{pmatrix} & u^\alpha & \\ e^+ & & u^c_\alpha \\ & d^\alpha & \end{pmatrix}_L$$

where the superscript c denotes charge-conjugate. This is repeated for μ - c - s family and the τ - t - b family.

The symmetry-breaking occurs in two stages at two mass scales M_U and M_L :

$$SU_5 \underset{M_U}{\rightarrow} SU_3 \times SU_2 \times U_1 \underset{M_L}{\rightarrow} SU_3 \times U_1^{em}$$

This can be achieved by choosing two Higgs multiplets $\underline{24}$ and $\underline{5}$ with vacuum expectation values equal to M_U and M_L respectively.

<u>The group SO_{10}</u>

The gauge bosons belong to $\underline{45}$ and their decomposition according to $SU_3 \times SU_{2L} \times SU_{2R}$ is given by

$$\begin{array}{ccccccccc} \underline{45} = & (8,1,1) & + & (1,3,1) & + & (1,1,3) & + & (1,1,1) \\ & G^\alpha_\beta & & W^{\pm,0}_L & & W^{\pm,0}_R & & B \\ + & (3^*,2,2) & + & (3,2,2) & + & (3,1,1) & + & (3^*,1,1) \\ & \begin{pmatrix} X & \bar{Y}' \\ Y & \bar{X}' \end{pmatrix} & & \begin{pmatrix} X' & \bar{Y} \\ Y' & \bar{X} \end{pmatrix} & & X_S & & \bar{X}_S \end{array}$$

Now, there are weak bosons corresponding to SU_{2L} as well as SU_{2R} and the lepto-quarks X , Y are also larger in number.

All the fermions of one family can be assigned to a single representation $\underline{16}$ consisting of

$$\begin{pmatrix} u^{\alpha} \\ d^{\alpha} \end{pmatrix}_L \quad \begin{pmatrix} \nu \\ e^- \end{pmatrix}_L \quad \begin{pmatrix} d^c_{\alpha} \\ -u^c_{\alpha} \end{pmatrix}_L \quad \begin{pmatrix} e^+ \\ -\nu^c \end{pmatrix}_L$$

Note that a right-handed neutrino (or a left-handed antineutrino) has been added.

In contrast to SU_5, many patterns of symmetry-breaking are possible for SO_{10}, leading to quite a rich intermediate - energy physics.

The group E_6

The gauge bosons belong to 78 :

$$\underline{78} = (8,1,1) + (1,8,1) + (1,1,8) + (3,3^*,3^*) + (3^*,3,3)$$
$$= 45 + 16 + 16^* + 1$$

where the decomposition given in the first and second lines correspond to representations of $SU_3 \times SU_{3L} \times SU_{3R}$ and SO_{10} respectively.

The fermions can be taken to be in 27 and two types of models are possible, depending on whether $SU_3 \times SU_{3L} \times SU_{3R}$ or SO_{10} is chosen as the intermediate symmetry group.

(a) $SU_3 \times SU_{3L} \times SU_{3R}$ type :

$$\underline{27} = (1,3^*,3) + (3,3,1) + (3^*,1,3^*)$$

$$\begin{pmatrix} N_1 & E^- & e^- \\ E^+ & N_2 & \nu_e \\ e^+ & \nu^c_e & N_3 \end{pmatrix} \qquad \begin{pmatrix} u \\ d \\ h \end{pmatrix} \qquad (u^c \ d^c \ h^c)$$

This can be repeated three times for the three generations. A new type of quark h with electric charge - 1/3 and new type of leptons N_1, N_2, N_3 and E^- and corresponding objects for higher generations are needed.

(b) SO_{10} type :

In this type of model, each of the 3 families is a 27 with the following

SO_{10} decomposition:

$$\underline{27} = \underline{16} + \underline{10} + \underline{1}$$

where the $\underline{16}$ is the standard SO_{10} multiplet given earlier, the $\underline{10}$ is comprised of new fermions belonging to $\underline{5}$ and $\underline{5}^*$ of SU_5 :

$$\begin{pmatrix} & N & \\ D^c_\alpha & & \\ & E & \end{pmatrix}_L + \begin{pmatrix} & N^c & \\ D_\alpha & & \\ & E^c & \end{pmatrix}_L$$

and the $\underline{1}$ is a neutral fermion N_{OL} . Here D, N, E and N_o and the corresponding repetitions for the higher generations are the new objects needed.

3.4 Comparison And Comments

A comparison between the good and bad points of the three unification groups SU_5, SO_{10} and E_6 is offered in Table 1. SU_5 is the minimal group

Table 1. A comparison between SU_5, SO_{10} and E_6 as grand unified groups.

	SU_5	SO_{10}	E_6
Minimal group	✓	✗	✗
Almost unique prediction	✓	✗	✗
SU_5: desert (✗); SO_{10}, E_6: fertile valleys (✓)	✗	✓	✓
Single fermionic representation (per family)	✗	✓	✓
L-R symmetry	✗	✓	✓
q-ℓ symmetry	✗	✓	✗

with no more than the four commuting generators of the standard $SU_3 \times SU_2 \times U_1$ model, whereas SO_{10} and E_6 have one and two additional commuting generators respectively. Consequently, SU_5 has only a single step of symmetry-breaking, leading to the 'almost unique' predictions of $\sin^2 \theta_W$ and τ_p already mentioned in Sec.2. In contrast, no unique prediction is possible in SO_{10} and E_6. This is the good point of SU_5. On the other hand, this is bought at the expense of the sterile desert which exists in SU_5 extending from 100 GeV to 10^{14} GeV. Because of the many interesting patterns of symmetry breaking possible in SO_{10} and E_6, these groups can lead to much richer physics.

All the fermions of one family can be put in a single irreducible representation of SO_{10} and E_6; because of anomalies, this is not possible in SU_5. Another good point of SO_{10} and E_6 is their left-right symmetry. Can a fundamental unified theory be as asymmetric as SU_5? Finally, if one regards quark-lepton symmetry as a desirable feature, then only SO_{10} scores a point.

4. PREONS

A brief look at the history of atoms, nuclei and then quarks would suggest that preons must be the next natural step after the quarks and leptons. There may exist in Nature a never-ending layered structure.

In the hey-days of S matrix and bootstrap philosophy (early 60's) it was even proposed by some physicists that perhaps the end of the road was in sight and that no more constituent structure beyond the hadrons was possible. But the subsequent development of physics seems to have shown the fallacy of this argument. It is now believed by almost every physicist that hadrons are made of quarks. Are quarks, inturn, made of preons?

4.1 Examples of Preonic Models

The fermions of the standard model can be denoted by the compact notation f_{ijk} where $i = 1,2$ (SU(2) index), $j = 1,2,3,4$ (1,2,3 denoting SU(3) colour and the fourth colour 4 denoting the colour singlet lepton) and $k = 1,2,3$ .... (generation index). Factorizing the three types of labels into three types of spin-1/2 preons, we write f_{ijk} as a three-particle composite of these preons, one from each type:

$$f_{ijk} \sim \omega_i c_j h_k$$

where ω_i is an SU(2) doublet, c_j is a SU(3) triplet for j = 1,2,3 or a singlet for j = 4 while h_k is a singlet under both SU(2) and SU(3). Such models have been constructured by Pati, Salam and Strathdee [8] as well as by Terazawa [9].

A more economical model was sought by Harari [10] and Shupe [11]. It was claimed that just two spin 1/2 - preons T and V with electrical charges 1/3 and 0 could be used to construct all the quarks and leptons of one generation in the following way:

$$e^+ \sim TTT \quad ; \quad u \sim TTV \ , \quad TVT \ , \quad VTT$$

$$\nu \sim VVV \quad ; \quad \bar{d} \sim TVV \ , \quad VTV \ , \quad VVT$$

It was suggested by Harari and Shupe that the three colours of u may arise from the three possibilities of rearrangement shown above and similarly for $\bar{d}$. At the present time, nobody believes that colour could be explained by this novel trick.

Later, Harari and Seiberg [12] proceeded to endow this V-T model with more conventional properties. They assigned the usual colour SU(3) quantum numbers to T and V and further invoked hypercolour SU(3) for the binding mechanism. (More on the hypercolour idea later). Under $SU_{3C} \times SU_{3H}$, (C = colour ; H = hypercolour), the assignment are

$$T \sim (3, 3)$$

$$V \sim (3^*, 3)$$

It can be easily checked that for the composite structure of e^+, ν , u and d given above, the leptons can be made singlets under colour and hypercolour, while the quarks can be singlets under hypercolour, but triplets under colour.

Many more examples of preonic models exist in the literature and they are reviewed in the recent reports [13-16].

4.2 The Sizes Of Quarks And Leptons

So far, there is no experimental evidence for any composite structure for leptons and quarks. All that are available are upper limits on their sizes R, or lower limits on the energy scale of compositeness Λ_H which is equal to R^{-1}. It is useful to note

$$1 \text{ Gev} = (2 \times 10^{-14} \text{ cm})^{-1}$$

The most reliable limits on the sizes of the leptons come from the measured values of their (anomalous) magnetic moments. Composite structure at a scale Λ_H leads to deviations from the QED predictions for the value of (g-2) of a charged lepton of mass m_ℓ :

$$\begin{aligned} \delta\,(g-2) &\sim O\left(\frac{m_\ell}{\Lambda_H}\right) \text{; if there is no symmetry ,} \\ &\sim O\left(\frac{m_\ell}{\Lambda_H}\right)^2 \text{; if the model has chiral symmetry.} \end{aligned} \tag{26}$$

Using present values of (g-2) of μ , we get

$$\Lambda_H > 500 \text{ Gev} \tag{27}$$

which is the weaker bound based on chiral symmetry. This is the most model-independent bound available.

Model-independent bounds are also obtainable from the Bhabha-type scatterings:

$$\begin{aligned} e^+e^- &\to e^+e^- \\ &\to \mu^+\mu^- \\ &\to q^+q^- \end{aligned}$$

Any possible deviation of the experimental values of the cross-sections for these processes from the prediction of quantum electrodynamics can be described by effective interactions:

$$\Lambda_H^{-2}\ \bar{e}\,e\ \ \bar{e}\,e\ , \quad \Lambda_H^{-2}\ \bar{e}\,e\ \ \bar{\mu}\mu\ , \quad \Lambda_H^{-2}\ \bar{e}\,e\ \ \bar{q}\,q\ ,$$

Of course, there can be other interpretations for the deviations; nevertheless it is obvious that this method provides a lower limit on Λ_H. Recent experimental data give

$$\Lambda_H > 700 \text{ Gev} . \tag{28}$$

There are other processes which can give more stringent limits on Λ_H, but these are model-dependent. The absence of the processes

$$\begin{aligned} \mu &\to e\gamma \\ &\to 3e \\ K &\to \mu e \\ &\to \pi\mu e \\ \mu N &\to eN \end{aligned}$$

provide such model-dependent bounds on compositeness. Such transitions can be described by the effective interactions:

$$\Lambda_H^{-2} \ \bar{\mu} e \bar{e} e \ , \quad \Lambda_H^{-2} \ \bar{s} d \bar{e} \mu \ .$$

The experimental limits on the above processes lead to the very stringlent bounds:

$$\Lambda_H \gtrsim 10^4 - 10^5 \text{ Gev} . \tag{29}$$

However, it should be emphasized that these bounds (29) are on an entirely different footing as compared to the earlier ones (27) and (28). No selection rule can forbid an $\bar{e}\ e\ \bar{e}\ e$ effective interaction, but selection rules can certainly forbid the conversion of e to μ in which case, these stringent bounds (29) do not exist. Even if there is no absolute selection rule, there can exist strong suppression factors for these conversions, rendering the bounds (29) useless. A very stringent bound following from the well-measured $K_S - K_L$ mass-difference also suffers from the same defect.

Finally, proton decay describable by the effective interaction

$$\Lambda_H^{-2} \ u u d \bar{e}$$

can lead to a fantastic bound:

$$\Lambda_H > 10^{15} \text{ Gev} , \quad (30)$$

since

$$\tau_p \sim \frac{\Lambda_H^4}{m_p^5} > 10^{31} \text{ years} . \quad (31)$$

But, this is terribly model-dependent. Proton decay is forbidden in some preonic models. In some models, it is allowed in second order, in which case

$$\Lambda_H > 10^7 \text{ Gev} . \quad (32)$$

In some models, it is allowed in third order so that

$$\Lambda_H > 10^5 \text{ Gev} . \quad (33)$$

All the above experimental information on the compositeness scale are collectively depicted in the Harari-diagram [14] (Fig.3), which also gives scales of other theories for comparison.

Let us next compare the values of the parameter mR (where m is the mass of the composite and R is the size of the composite) for various known composites. This is given in Table 2.

Table 2. Values of mR for different levels of substructure

Particle	m(Gev)	R (cm)	mR
H atom	1	10^{-8}	5×10^5
Positronium	10^{-3}	10^{-8}	10^2
Deuteron	2	10^{-13}	10
Nucleus A	A	$A^{1/3} 10^{-13}$	$5\ A^{4/3}$
Proton	1	10^{-13}	5
Quarks and leptons	10^{-8} - 10	$< 10^{-16}$cm	$< 10^{-2}$

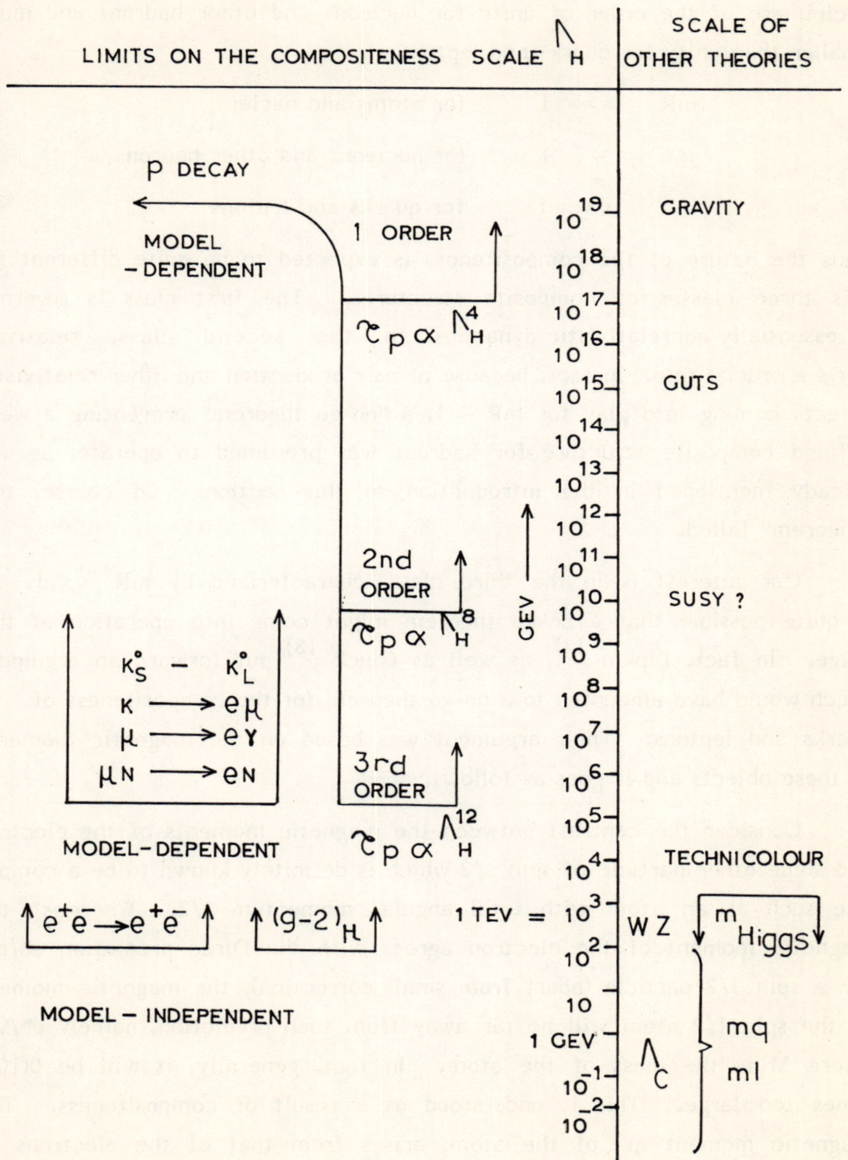

FIG.3 EXPERIMENTAL LIMITS ON THE COMPOSITENESS SCALE Λ_H

The values of mR are much larger than unity for molecules,atoms and nuclei, are of the order of unity for nucleons and other hadrons and much smaller than unity for quarks and leptons.

$$mR \gg 1 \quad \text{for atoms and nuclei}$$

$$\sim 1 \quad \text{for nucleons and other hadrons}$$

$$\ll 1 \quad \text{for quarks and leptons}$$

Thus the nature of the compositeness is expected to be quite different for this three classes of composite structures. The first class is governed by essentially nonrelativistic dynamics. In the second class, relativity plays a crucial role. In fact, because of pair production and other relativistic effects coming into play for $mR \sim 1$, a "no-go theorem" preventing a well-defined composite structure for hadrons was presumed to operate, as was already mentioned in the introduction to this section. Of course, this "theorem" failed.

Our interest is in the third class, characterized by $mR \ll 1$. It is quite possible that a no-go theorem might come into operation at this stage. In fact, Lipkin [17)], as well as Gluck [18)] put forward an argument which would have amounted to a no-go theorem for the compositeness of quarks and leptons. Their argument was based on the magnetic moments of these objects and it goes as follows.

Consider the contrast between the magnetic moments of the electron and some other particle of spin 1/2 which is definitely known to be a composite such as an atom with total angular momentum 1/2. Whereas the magnetic moment of the electron agrees with the Dirac prediction $e\hbar/mc$ for a spin 1/2 particle (apart from small correction), the magnetic moment of the spin 1/2 atom will be far away from such prediction, namely $e\hbar/Mc$ where M is the mass of the atom. In fact, generally, it will be $0(10^3)$ times too large. This is understood as a result of compositeness. The magnetic moment μ of the atom, arises from that of the electrons as well as that of the nucleus:

$$\vec{\mu}_{atom} = \vec{\mu}_e + \vec{\mu}_N .$$

Since $\mu_e >> \mu_N$, $\mu_{atom} \approx \mu_e$ and this explains the above fact. Now, the argument of Lipkin and Gluck is that such a deviation from Dirac prediction is not observed for e and μ and hence these particles should be elementary.

However, the above argument is not valid for those composite systems for which $mR << 1$ as was first pointed out by Brodsky and Drell [19] as well as Shaw, Silverman and Slansky [20]. For such composite systems, it can be shown that the deviation from the Dirac prediction is only of the order of mR. In fact, we already used this information to set limits on the size of the muon (vide eq.(26)).

Thus, since leptons and quarks satisfy $mR << 1$, the closeness of their magnetic moments to the Dirac prediction cannot be used to rule out their compositeness.

4.3 Steps Towards Preonic Dynamics

Regarded as composite systems, quarks and leptons are unusually light. In fact, with respect to their compositeness scale Λ_H, they are essentially massless. As we have already seen, this is what distinguishes quarks and leptons from earlier composite systems. Such masslessness must emerge from a symmetry principle. The simplest symmetry which may prevent a fermion from acquiring a mass is chiral symmetry. Such a chiral symmetry would rise naturally if the preons were massless.

Hence, the logical sequence of ideas is as follows:-

1) The fundamental Lagrangian contains massless preons and therefore possesses a chiral symmetry.
2) The full chiral symmetry, or at least, a chiral subsymmetry remains unbroken at the composite level.
3) The chiral symmetry of the effective Lagrangian containing the composite fermions, prevents the latter from gaining a mass.
4) We thus end up with composite massless fermions to be identified with quarks and leptons.

The following three questions arise:-

a) If the preons are massless, why don't we observe them?

b) What is the interaction which binds the preons inside the composite quark or lepton?

c) If both the preons and the composites are massless, what provides us with the necessary 'compositeness scale' Λ_H ?

All three questions can be immediately answered by one postulate, if we assume a new colour-like force called hypercolour with a scale parameter Λ_H. All preons carry hypercolour. They are confined by hypercolour forces of characteristic scale Λ_H into hypercolour-singlet composite fermions with an effective radius Λ_H^{-1}. The confined preons cannot be experimentally observed. Whereas the QCD scale parameter Λ_C is of the order of 100 Mev the hypercolour scale parameter Λ_H is expected to be greater than 500 Gev. This scenerio provides an attractive framework for the construction of a composite model.

Let us then assume that we have constructed a composite model of quarks and leptons based on an $SU(n)_H$ hypercolour gauge group and containing massless preons belonging to the fundamental representation $\underline{n}$ of $SU(n)_H$. Let there be N number of such objects. Then, the underlying Lagrangian automatically possesses a global chiral $SU(N)_L \times SU(N)_R \times U(1)$ symmetry, the U(1) factor being a vector charge which counts the number of preons.

Such a model contains "flavour" triangle anomalies corresponding to products of the SU(N) currents and the U(1) current. Such anomalies are perfectly legitimate and do not affect the validity of the original $SU(n)_H$ gauge theory since the $SU(N)_L \times SU(N)_R \times U(1)$ symmetry is not gauged. However, these lead to non-trivial constraints on thedynamics. These anomalies can be calculated both at the preonic level and at the level of the low-energy effective theory containing the massless composite fermions and either calculation must produce the same result. This is t'Hooft's anomaly matching condition[21] and provides a severe constraint on the spectrum of massless composite fermions.

In fact, t'Hooft's condition is often powerful enough to tell us what the spectrum of massless composite fermions is in a given hypercolour theory and this is the way the condition has been used in the considerable amount of work done on composite-model building[22].

However, there may be difficulties with this programme. How do we know that chiral symmetry is not spontaneously broken? In fact, in QCD, chiral symmetry is known to be spontaneously broken. Hence, if we want chiral symmetry to remain unbroken in hypercolour dynamics, it is clear that preonic hypercolour dynamics is going to be different from QCD. One possibility is that preonic dynamics may be supersymmetric and supersymmetry may supply the required difference *. In supersymmetric models perhaps it is possible to keep the chiral symmetry unbroken[23]. A number of supersymmetric preon models satisfying t'Hooft's condition has been constructed in the recent literature[24].

Once supersymmetry is admitted into the game, other ways of solving the problem of massless fermions suggest themselves. For instance, we may imagine a preonic theory which is supersymmetric and which has a global symmetry group. If this global symmetry is spontaneously broken, then there will be Nambu - Goldstone bosons. However, because of supersymmetry, these bosons will be accompanied by fermionic partners-which have been called quasi Nambu - Goldstone fermions. One suggestion is that quarks and leptons are the quasi Nambu - Goldstone fermions of a confining supersymmetric theory and models based on this idea also have been constructed[25].

To sum up:

i) A large number of preonic models are available. The review by Lyons[15] written two years ago already contained more than 30 models.

ii) No model has been worked out to its logical limit; this may not be possible at the present level of our understanding of composite particles in

---

* Discussions with A.S.Joshipura on this as well as other aspects of preonic models are gratefully acknowledged.

quantum field theory.

iii) The internal consistency of no model has been established.

iv) No model is elegant enough.

v) No model has the ring of Truth about it.

vi) No model is as yet required by experimental data.

vii) Nevertheless, Nature might have already chosen one preonic model !

5. INDUCED GRAVITY

Induced gravity may also be called the elasticity of the vacuum. Let us start with the analogy of ordinary elasticity of a solid body. The macroscopic theory of the elasticity of a solid body can be worked out in terms of the stress and strain tensors and the symmetry properties. However, we know that this is not a fundamental theory, for the elastic properties of the solid have to be explained in terms of the atomic structure of the solid. It is the atomic forces that give rise to the observed macroscopic elastic properties. In an analogous way, what is observed as gravitation in the macroscopic Universe may not be a fundamental property of space-time, but may be a consequence of the existence in vacuum of the quantized fields of the strong and electroweak forces of nature. These fields give rise to generalized forces which oppose the curving of space-time. So, there is a "metrical elasticity of space-time", which manifests itself as gravitation in the large.

The idea of induced gravity was originally proposed by Sakharov [26)] in 1968, but was resurrected in the modern gauge-theoretic context by Adler [27)] in 1980. Related contributions were made by Zee and others [28-30)].

If we write the total action including gravity as

$$\int d^4x \sqrt{-g}\,\mathcal{L} + \frac{1}{16\pi G_N}\int d^4x \sqrt{-g}\,R \qquad (34)$$

where the first term is the matter part of the action and the second term is the Einstein-Hilbert action for gravity, the contention is that the second term is induced by the first term.

So, let us start with matter action in curved space-time

$$S\,[\varphi\,,\,g_{\mu\nu}\,] \;=\; \int d^4 x\, \sqrt{-g}\, \mathcal{L}\,(\varphi\,,\,g_{\mu\nu}) \qquad (35)$$

which describes the matter fields (collectively denoted by φ) dynamically evolving in a given background matric $g_{\mu\nu}$. To start with there are no derivative terms for $g_{\mu\nu}$ and so the metric is not yet a dynamical field. They are to be generated by quantum fluctuations.

We define $S_{eff}\,[g_{\mu\nu}]$ as the effective action obtained by integrating over the matter fields:

$$\exp\; i\; S_{eff}[g_{\mu\nu}] \;=\; \int D\,\varphi \;\;\exp\; i\; \int d^4\, x\; \sqrt{-g}\,\mathcal{L}\,(\varphi\,,\; g_{\mu\nu})\,. \qquad (36)$$

The resulting $S_{eff}\,[g_{\mu\nu}]$ is expanded in terms of the background metric $g_{\mu\nu}$ and its derivatives:

$$S_{eff}\,[g_{\mu\nu}] \;=\; \int d^4\, x\, \sqrt{-g}\, \{\, <\mathcal{L}_0>_0 \,+\, <\mathcal{L}_2>_0 R \,+\, <\mathcal{L}_4>_0 R^2 \ldots\ldots \} \qquad (37)$$

where R is the curvature scalar obtained in terms of the Riemann curvature tensor:

$$R \;=\; g^{\mu\nu}\;\; R_{\mu\nu} \;=\; g^{\mu\nu}\;\; R^{\lambda}_{\;\mu\,\lambda\,\nu}$$

and R^2 stands for R^2 as well as terms such as $R^2_{\mu\nu}$, $R^2_{\mu\nu\alpha\beta}$ etc.

The first term in the expansion $<\mathcal{L}_0>_0$ can be regarded as the "induced cosmological constant". It is the second term which is the induced gravity and Sakharov's idea is to regard this as the entire gravitational action so that

$$<\mathcal{L}_2>_0 \;=\; -\,\frac{1}{16\,\pi G_N}\quad .$$

However, in general, the vacuum expectation values $<\mathcal{L}_0>_0$, $<\mathcal{L}_2>_0$ are divergent quantities and are to be absorbed into the counterterms contained in the original $\mathcal{L}$ and so nothing can be calculated. In 1980 Adler pointed out that in unified gauge theory without fundamental scalar fields, the situation is more favourable.

The mass-dimension of R is 2, since it contains two derivatives of $g_{\mu\nu}$ and so the dimensions of $<\mathcal{L}_0>_0$, $<\mathcal{L}_2>_0$, $<\mathcal{L}_4>_0$ are 4, 2 and 0 respectively. In unified gauge theories containing only the gauge fields A_μ and spinor fields ψ, but no scalar fields ϕ, a local gauge invariant operator $\mathcal{L}_2$ of dimension 2 does not exist; for, $A^a_\mu A^a_\mu$ is not gauge-invariant. Consequently, no counter term in $\mathcal{L}$ is possible to absorb the divergence in $<\mathcal{L}_2>_0$ and hence $<\mathcal{L}_2>_0$ is finite and calculable. This is Adler's observation.

Let $\mathcal{L}(\psi, A_\mu, g_{\mu\nu})$ be the Lagrangian of the gauge theory in a background metric $g_{\mu\nu}$, but without any sclar fields. Then, we define $S_{eff}[g_{\mu\nu}]$ by integrating over the ψ and A_μ fields:

$$\exp i\, S_{eff}[g_{\mu\nu}] \sim \int D\psi\, D\bar{\psi}\, DA_\mu \exp i \int d^4x \sqrt{-g}\, \mathcal{L}(\psi, A_\mu, g_{\mu\nu}) . \qquad (39)$$

Diagramatically, this is equivalent to obtaining $S_{eff}[g_{\mu\nu}]$ by doing a one-loop integration over the ψ and A_μ fields with given external $g_{\mu\nu}$.

$S_{eff}[g_{\mu\nu}] \sim$ + + + ...

where the solid lines denote ψ and A_μ while the dotted lines denote $g_{\mu\nu}$. Thus, in this theory, gravitons should be regarded as composites of ψ and A_μ.

A remarkable fall-out of the theory of induced gravity is a formula for Newton's gravitational constant:

$$\frac{1}{16\pi G_N} = \frac{i}{96} \int d^4x\, x^2 < 0|\, T(T^\lambda_\lambda(x)\; T^\mu_\mu(0))|\, 0 > \qquad (40)$$

where T^μ_ν is the energy-momentum tensor of the matter fields (i.e. ψ and A_μ). The right-hand side of the above equation should yield a finite number, in view of Adler's argument. But so far nobody has succeeded in calculating it. Even the sign is not known.

In Einstein's equation

$$R_{\mu\nu} - 1/2\ g_{\mu\nu} R = -8\pi G_N T_{\mu\nu} \tag{41}$$

the left-hand side comes from geometry which Einstein regarded as beautiful while the right-hand side is the ugly junk coming from the matter-part of the action. Einstein's fond hope was to generate the effects of the ugly matter-part somehow from the beautiful geometry-part. In induced gravity one hopes to achieve just the reverse. Finally we may also mention that if induced gravity is a correct idea, there is no need of bothering about renormalization of quantum gravity, since it is only an effective interaction.

6. SUPERSYMMETRY AND SUPERGRAVITY

Supersymmetry [31-34] transformations convert bosons into fermions and vice versa. Supersymmetry offers a revolutionary quantum jump in our conceptual framework; however, it is not yet organically integrated into physics. Supersymmetry involves an enlargement of ordinary space-time into superspace which includes Grassmann coordinates in addition to ordinary space-time coordinates. There must be some direct physical manifestation of superspace. This must be identified and looked for experimentally. However this is for the future.

A practical motivation for the immediate introduction of supersymmetry into high energy physics came in the wake of the hierarchy problem. The light Higgs boson of mass ~ 100 Gev picks up a large self-energy contribution of order 10^{14} Gev because of its coupling to the heavy Higgs boson of mass $\sim 10^{14}$ Gev and hence the hierarchy of masses with the large ratio of 10^{12} cannot be maintained. This is the hierarchy problem. Supersymmetry can solve this problem [35] because of the cancellation between the contributions to the self-energy of the boson coming from the bosonic internal line and the corresponding fermionic internal line:

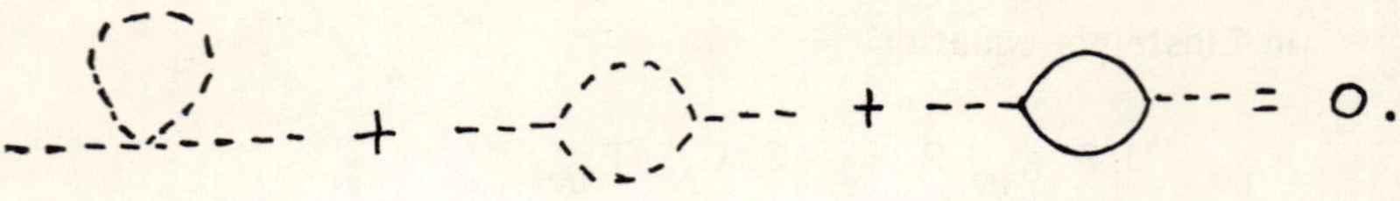

$$\text{(tadpole)} + \text{(boson loop)} + \text{(fermion loop)} = 0.$$

More generally, supersymmetric quantum field theories have better ultraviolet behaviour because of the cancellation of divergences between the contributions from the fermionic and bosonic members of the supermultiplet.

Since we already have both fermions (quarks and leptons) as well as bosons (gauge bosons and Higgs bosons) in our description of Nature, it would have been nice if supersymmetry could connect these fermions and bosons. Unfortunately that turns out to be impossible. Instead, supersymmetry requires so-far unobserved superpartners to be postulated for every known fermion and boson. We need spin-0 superpartners to quarks and leptons named squarks and sleptons respectively as well as spin-1/2 super-partners to gauge bosons and Higgs bosons named gauginos (gluino, Wino, Zino etc) and Higgsinos. Thus, supersymmetry seems to lead to a further proliferation in the population of particles.

Where are these superpartners? Obviously supersymmetry must be broken so that the unseen members of the supermultiplet can be pushed up to the heavier mass-region which is not yet so well probed by experiments. The current belief of the optimists [36)] is that the superpartners have masses in the region of several tens or hundreds of Gev so that they will burst forth in the experiments of the Tev-region. In fact, the peculiar monojet events [37)] apparently seen in the 540 Gev $p\bar{p}$ collider data (if they survive more rigorous analysis) may be the signature of the superpartners such as gluinos and squarks. At least this is the claim of the optimists in this field [38-40)].

It is quite possible that supersymmetry is a correct idea, but perhaps we are applying it in the wrong way. May be, we should not apply it linearly. Nonlinear realization [41)] of supersymmetry is one approach which might save us from the disease of proliferation, but this has not been pushed far enough.

Another remarkable facet of supersymmetry is that it provides us with a natural stepping stone into quantum gravity. Consider two successive supersymmetric transformations applied on a boson. The first one will transform the boson into a fermion while the second one will transform the fermion into a boson, but it turns out that the final boson obtained is not at the same space-time point as the initial boson we started with, but it is at a displaced point. This is a reflection of the fact that the anticommutation algebra of the supersymmetric generators Q_α and $\bar{Q}_\beta$ involves the momentum operator P_μ which is the displacement operator in space-time:

$$\{ Q_\alpha , \bar{Q}_\beta \} = 2\gamma^\mu_{\alpha\beta} P_\mu \tag{42}$$

where γ^μ is the Dirac matrix (with α and β as Dirac indices). Hence, if we now make supersymmetry into a local or gauged symmetry, then that gauges the generators P_μ also and thus gravitation results. Thus

Local supersymmetry = Supergravity.

Of course, this is a supersymmetric version of gravity which is a generalization of Einstein's gravity and includes the gravitino as the spin-3/2 supersymmetric partner to the spin-2 graviton.

Again, supergravity has better ultraviolet behaviour than ordinary gravity. It was hoped in the beginning that supergravity might even be a finite theory; but that hope has proved false, divergences remain at the level of 3 or higher member of loops.

A further generalization is extended supersymmetry. This is described by the supersymmetric generators Q^i_α (i = 1...N) where N denotes the number of supersymmetries in the extended supersymmetric theory. N=1 corresponds to simple (unextended) supersymmetry. The maximal extension considered so far is N=8; N is connected to the maximum spin J in the theory through the relation

$$J_{max} = \frac{N}{4} \text{ or } \frac{N+1}{4}, \text{ for N even or odd.} \tag{43}$$

It is believed that spins greater than 2 cannot be described by a consistant quantum field theory and hence the restriction to $N < 8$. Some of the maximally supersymmetric theories of great beauty are given in Table 3.

Table 3. Supersymmetric Theories

Helicity	N=1 Wess-Zumino theory	N=4 Yang-Mills theory	N=8 supergravity
2			1
$\frac{3}{2}$			8
1		1	28
$\frac{1}{2}$	1	4	56
0	2	6	70
$-\frac{1}{2}$	1	4	56
-1		1	28
$-\frac{3}{2}$			8
-2			1

The above mentioned rule on J_{max} can be seen to be satisfied in the three theories enumerated in the Table. The equality of the total number of bosonic components to the fermionic components is also manifest in each theory.

The N=4 Yang-Mills theory has been proved to be a finite theory [42). So, for the first time in history we have a quantum field theory in 4 dimensional space-time which is free of ultraviolet divergences. The consequences of this remarkable achievement have yet to be exploited in model-construction.

The N=8 supergravity is the maximally extended theory including gravity and it might even describe the whole world, especially after Cremmer and Julia [43)] discovered a hidden SU(8) in the theory. This hidden SU(8) can incorporate the standard SU(3) x SU(2) x U(1) model or a grand unified SU(5) model [44]. However, it is worthwhile to note that in this theory, the known leptons and quarks as well as the known gauge bosons are no longer elementary, but have to arise as composites of preons. In any case, it is possible to envisage a superunified theory including gravity although the details of such a theory have not been completely worked out.

7. HIGHER DIMENSIONAL UNIFICATION

General coordinate invariance of the four-dimensional world leads to Einstein's theory of gravity while local internal symmetry leads to the gauge theory of strong and electroweak forces. If the number of space-time dimensions is greater than four, gauge invariance with respect to internal quantum numbers becomes a part of the general coordinate invariance of a higher-dimensional world. This is the higher-dimensional unification of gravity with the other forces of nature [45-48]. The mechanism by which this is achieved is so elegant that Nature would have surely used it at some level. The question is: Is that level already accessible to present-day Theoretical Physics?

Let us first discuss the original 5 dimensional Kaluza-Klein theory [45,46]. We have

$$x^A = (x^\mu , x^5) \tag{44}$$

where $A = 1 \ldots . 5$, $\mu = 1 \ldots . 4$, but the 5th coordinate is curled up into a tiny circle of radius 'a' so that

$$x^5 = \theta a \tag{45}$$

and θ goes from 0 to 2π. Consider the metric tensor g_{AB}. Under the usual 4 dimensional transformation, $g_{\mu 5}$ transforms as a 4-vector. Consider the following transformation:

$$\begin{aligned} x^\mu &= x'^\mu \\ x^5 &= x'^5 + a\,\lambda(x'^\nu) \\ &\text{or} \\ \theta &= \theta' + \lambda(x'^\nu) \end{aligned} \tag{46}$$

i.e. the circle at each point is rotated by a space-time dependent amount. Under this transformation, $g_{\mu 5}$ transforms as follows:

$$g_{\mu 5} \rightarrow g_{AB} \frac{\partial x^A}{\partial x'^{\mu}} \frac{\partial x^B}{\partial x'^5}$$

$$= g_{\sigma\rho} \frac{\partial x^{\sigma}}{\partial x'^{\mu}} \frac{\partial x^{\rho}}{\partial x'^5} + g_{\sigma 5} \frac{\partial x^{\sigma}}{\partial x'^{\mu}} \frac{\partial x^5}{\partial x'^5}$$

$$+ g_{5\rho} \frac{\partial x^5}{\partial x'^{\mu}} \frac{\partial x^{\rho}}{\partial x'^5} + g_{55} \frac{\partial x^5}{\partial x'^{\mu}} \frac{\partial x^5}{\partial x'^5}$$

$$= g_{\mu 5} + a \partial_{\mu} \lambda \tag{47}$$

where we have put $g_{55} = 1$. We see that the transformation of $g_{\mu 5}$ is identical to the gauge transformation of the vector potential in electromagnetism:

$$A_{\mu} \rightarrow A_{\mu} + \partial_{\mu} \lambda . \tag{48}$$

Hence, we make the identification:

$$g_{\mu 5} = a A_{\mu} . \tag{49}$$

Thus, we see that gauge transformations are contained in the coordinate transformations of the 5-dimensional theory.

We partition the 5-dimensional metric tensor and write the ansatz :

$$g_{AB} (x^{\alpha}, x^5) = \left(\begin{array}{c|c} g_{\mu\nu} (x^{\alpha}) & a A_{\mu} (x^{\alpha}) \\ \hline a A_{\nu} (x^{\alpha}) & g_{55} (x^{\alpha}) \end{array} \right) . \tag{50}$$

We assume that g_{AB} depends on x^{α} only. This is the ansatz which breaks the original 5-dimensional symmetry and the 5-dimensional manifold M^5 splits into the 4-dimensional manifold M^4 and the circle S^1.

$$M^5 \rightarrow M^4 \times S^1 \tag{51}$$

Substituting the above ansatz into the 5-dimensional Einstein-Hilbert action we get the 4-dimensional Einstein-Hilbert action plus the Maxwell action:

$$\int d^5 x \ \sqrt{-g^{(5)}} \ R^{(5)} \rightarrow \int d^4 x \ \sqrt{-g^{(4)}} \ \{ a^{-2} \ R^{(4)} + F_{\mu\nu} \ F^{\mu\nu} \} \quad (52)$$

where $R^{(d)}$ refers to the curvature scalar in d-dimensions and $F^{\mu\nu}$ is the usual Maxwell field. Thus, the strength of the gravitational interaction is fixed by the radius 'a'. Identifying with the Newtonian constant, one finds that 'a' is essentially the Planck length, which is the inverse of the Planck mass M_{Pl}:

$$a \sim \frac{1}{M_{Pl}} \sim (G_N)^{1/2} \sim 10^{-33} \text{cm} \ . \quad (53)$$

Thus, we understand why the experimental physicists have not yet discovered the existence of these tiny circles.

Nevertheless, it may be worthwhile to ask whether there are direct physical manifestations of these extra dimensions. In this connection, the following remark may be in order. Dimensional regularization [49] which has been found to be one of the best methods of regularization of divergent Feynman integrals, plays a very important role in present-day quantum-field-theoretic calculations. Is the success of dimensional regularization connected in some way to the physical reality of extra dimensions? Or, should regularization be regarded merely as a mathematical technique devoid of any physical meaning? We shall not pursue this speculation here.

Coming back to eq.(50), we should replace this ansatz on $g_{AB}(x^{\alpha}, x^5)$ by a more general expression containing dependence on the 5th coordinate. The 5th dimension being a circle, the dependence must be of the form:

$$\exp \ i \ \frac{2\pi \ n \ x^5}{a} \ .$$

Since

$$\frac{\partial^2}{\partial x_A \ \partial x^A} = \frac{\partial^2}{\partial x_\mu \ \partial x^\mu} - \frac{4\pi^2 n^2}{a^2} \ , \quad (54)$$

it is clear that the masses of these excitations are

$$m_n \sim \frac{n}{a} \sim nM_{Pl} \quad . \tag{55}$$

Hence, the x_5 dependent excitations have masses of the order of Planck-mass or larger and our original x_5-independent ansatz corresponds to a low-energy approximation in which these supermassive excitations are ignored.

The 5-dimensional theory unifies electromagnetism with gravity. What about unification of electroweak and strong forces with gravity? That requires $d > 5$.

Let us take $d = 4 + n$. The $4 + n$ dimensional theory can be worked out in analogy to the 5-dimensional case. The manifold M^{4+n} splits into $M^4 \times B$ where B is a compact manifold of dimension n. This is implemented by the ansatz:

$$g_{AB}(x^\alpha, \theta^k) = \left(\begin{array}{c|c} g_{\mu\nu}(x^\alpha) & a \sum_c A^c_\nu (x^\alpha)\, K_i^{\ c}(\theta^k) \\ \hline a \sum_c A^c_\mu (x^\alpha) K^c_j(\theta^k) & \gamma_{ij}(\theta^k) \end{array} \right) \tag{56}$$

where θ^k ($k = 1 \ldots n$) are the coordinates for the internal space of B, T^c ($c = 1 \ldots N$) are the generators of the symmetry group G of B, these generators T^c acting on θ^i produce $\theta^i + K^c_i(\theta)$, and $K^c_i(\theta)$ are the Killing vectors. Substitution of this ansatz into the $4 + n$ dimensional action produces 4-dimensional gravity along with the gauge theory of the Group G:

$$\int d^{4+n}x \sqrt{-g^{(4+n)}}\, R^{(4+n)} \to \int d^4x \sqrt{-g^{(4)}} \{ a^{-2} R^{(4)} + \sum_c F^c_{\mu\nu} F^c_{\mu\nu} \} \tag{57}$$

One can also verify that for transformations of the type

$$(x^\alpha, \theta^i) \to (x^\alpha, \theta^i + \sum_c \varepsilon^c (x^\alpha) K^c_i (\theta) \tag{58}$$

We get

$$A^a_\mu(x) \to A^a_\mu(x) + f^{abc} \varepsilon^b A^c_\mu + \partial_\mu \varepsilon^a \tag{59}$$

where f^{abc} are the structure constants of the group G. Thus, the non-abelian gauge transformations are contained in the coordinate transformations.

The important question now is: What is the dimension n of the internal manifold B? Witten [48)] has shown that for the standard model group

$$G_{std} = SU_3 \times SU_2 \times U_1 ,$$

the minimal manifold is $CP^2 \times S^2 \times S^1$ with dimension $n = 4 + 2 + 1 = 7$. Hence, adding the 4 of space-time, the minimum number of dimensions needed to describe the world is $d = 4 + 7 = 11$.

This same 11-dimensional world turns up, from an entirely different direction also. It is well known that $d = 11$ is the maximal number of dimensions for supergravity theory. For, 11-dimensional N=1 supergravity is known to be equivalent to the 4-dimensional N=8 extended supergravity and as already mentioned in the last section, 8 is the maximal extension possible if spins >2 are to be avoided. This remarkable coincidence between the minimal number of dimensions for Kaluza-Klein type of unification and the maximal number of dimensions for supersymmetric unification suggests the physical reality of $d = 11$.

However, there is a serious problem with the original programme of higher-dimensional unification. This is the problem of the chiral fermions. The Dirac equation in $4 + n$ dimensions (for ψ of definite helicity) is

$$\sum_{A=1}^{4+n} \gamma^A \partial_A \psi = 0$$

which can be re-written as

$$\sum_{\mu=1}^{4} \gamma^\mu \partial_\mu \psi = - \sum_{i=5}^{4+n} \gamma^i \partial_i \psi$$

The operator on the right-side is a differential operator in the internal compact space B and the nonvanishing eigenvalues of this operator (to be called D) will be of the order of Planck mass or higher, as we have already seen. So, we need eigenstates of D with zero eigenvalues to be identified with the observed leptons and quarks. It can be shown [50] that such eigenstates with the required chiral properties (forming complex representations of the internal symmetry group) do not exist for the operator D.

There are two ways of solving the above problem :

(i) In $d > 4$, gravitation is perhaps described by a theory which does not obey equivalence principle. In other words, go to a non-Riemannian manifold to describe the higher-dimensional world. For this approach, see Weinberg [51].

(ii) Add extra gauge fields in the higher-dimensional world, and compactify the extra dimensions with these gauge fields having a topologically nontrivial vacuum configuration in the compact space. For this apprach, see Frampton et al [52].

The second approach (ii) certainly goes against the spirit of the original Kaluza-Klein programme of unification of gravity and gauge fields since the extra gauge fields cannot be derived from the higher dimensional gravity and their introduction is adhoc. Remarkably enough, in the superstring theory to be discussed in the next section, these extra gauge fields are not adhoc, but are already present. This is only one of the many merits of the superstrings.

8. SUPERSTRINGS

Strings entered into High Energy Physics many years ago as a possible description of hadrons [53-58]. Starting with Veneziano's discovery of a formula for the scattering amplitude which had both Regge behaviour and crossing symmetry the S-matrix approach reached its pinnacle in the construction of the dual model. However, ironically enough, just at that juncture, Nambu identified the spectrum of the dual model with the vibrations of a string and the S-matrix approach was soon abandoned in favour of the canonical dynamical formalism of the string. The dynamics was reformulated in terms of the Nambu-Goto action which is equal to $(2\pi\alpha')^{-1}$ times the area swept by the string during its time-evolution. Here $(2\pi\alpha')^{-1}$ is the string tension T and α' is the slope of the Regge trajectory.

Remarkable discoveries soon followed:-

(1) Supersymmetry (in 2 dimensions) was discovered in the attempt to include fermionic excitations in the string spectrum.

(2) A consistent relativistic quantum dynamics could be formulated only if the number of space-time dimensions were equal to 26 for the bosonic string and 10 for the supersymmetric string.

(3) The zero slope limit (i.e.$\alpha' \to 0$) of the open string theory was found to yield Yang-Mills gauge theory, while the zero-slope limit of the closed string theory yielded gravitation.

In retrospect, we can see that these are precisely the ideas which animate present-day theoretical physics. These ideas, namely, supersymmetry, higher-dimensions, gauge theory and quantum theory of gravitation had really emerged out of strings. However, these were not viewed in this light in the early 70's when they were discovered. This was because of the preoccupation with the hadrons. In the hadronic context, these properties (especially the unwanted extra dimensions) were regarded as bizarre and efforts were made to get rid of them. The remarkable discoveries made in the zero-slope limit which showed the intimate connection of the strings to gauge theory and gravitation were apparently seen as a mere curiosity.

The following deserves special emphasis. Whereas one can construct the theory of a point particle in any number of space-time dimensions, the relativistic quantum mechanics of a string requires a specific number of space-time dimensions (26 or 10) for consistency. Further, the bosonic string in 26 dimensions has a tachyonic ground state, which arises from the zero-point energy of the vibrations of the string. In the supersymmetric string (in 10 dimensions) the fermionic contribution to the zero-point energy cancels the bosonic contribution and hence there is no tachyon. Thus, in contrast to the quantum mechanics of a point particle, the quantum mechanics of a string requires both higher dimensions and supersymmetry. Thus, string theory supplies a strong motivation for studying both higher dimensions and supersymmetry.

Pursuing the historical thread again, because of the 'troubles' of the string theory, the popularity of the strings waned after about 1974. An additional reason for the loss of interest on strings as a description of hadrons was the discovery of QCD which soon developed into the basic theory of the hadrons.

Meanwhile, motivated by Scherk's original discovery of the connection of the strings to gauge and gravitational fields in the zero-slope limit, Green and Schwarz [59)] decided to liberate the string theory from the hadronic world and that liberated the string theory from all the 'troubles'. All they had to do was to change the slope parameter α' from $(1 \text{ Gev})^{-2}$ to $(10^{19} \text{ Gev})^{-2}$. Thus was born the Superstring theory. It is the theory of supersymmetric strings in 10 dimensions.

The Regge trajectory for both the values of $\alpha' (= d\,\alpha/dm^2)$ are given in Fig.4. For $\alpha' = (1 \text{ Gev})^{-2}$, the spectrum of the vibrations of the strings are spaced by about 1 Gev and these were supposed to describe the hadrons. For $\alpha' = (10^{19} \text{ Gev})^{-2}$ the scale is given by the Planck mass 10^{19} Gev and now the theory describes the primary objects including gravity and gauge fields. In the limit $\alpha' \to 0$, one can see that all the finite masses are chased away to infinity and only the $m^2 = 0$ state survives which is the graviton if the string is closed, or Yang-Mills quantum if the string is open. In the supersymmetric theory the graviton and the Yang-Mills quantum are also accompanied by their superpartners. Thus, the supersymmetric string theory in 10 dimensions with $\alpha' = (10^{19} \text{ Gev})^{-2}$ emerges as a theory of the primary objects and this is the superstring theory. In this context, the higher dimension is no longer a difficulty, but infact an advantage, since it offers the possibility of unifying all the interactions via the Kaluza-Klein route.

An important discovery which in some sense ran counter to the above development of the superstring theory in dimension 10 was made by Polyakov [60)]. He showed that a proper quantization (taking account of the quantum anomalies) required the addition of the Liuville action in the string theory. He also showed that the coefficient in front of this additional piece in the action was (d-26) for the bosonic string and (d-10)

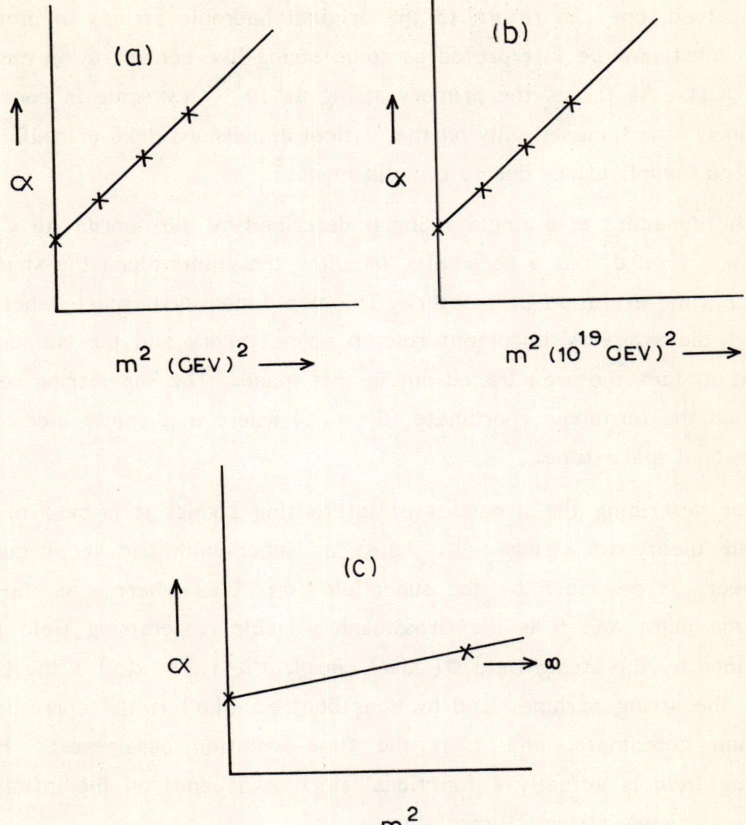

FIG. 4 REGGE TRAJECTORIES FOR (a) THE HADRONIC STRING WITH $\alpha' = (1\,\mathrm{GEV})^{-2}$ (b) THE PRIMARY STRING WITH $\alpha' = (10^{19}\,\mathrm{GEV})^{-2}$ AND (c) THE LIMIT $\alpha' \to 0$

for the supersymmetric string, thus explaining why the strings could be quantized only in these dimensions. With these additional terms, quantization can in principle be done in any number of dimensions including 4. However, quantization of the Liuville theory is still an unsolved problem. When this is solved, one can return to the original hadronic strings in dimension 4 which must now be interpreted as some string-like configuration emerging out of QCD. As far as the primary string at 10^{19} Gev scale is concerned, attention is now focussed only on the critical dimensions d=26 or d=10, where there is no complications due to Liuville modes*.

The dynamics of a single string is described by the coordinate $x_\mu(\sigma,\tau)$ where $\mu = 1 \ldots d$, σ is a parameter labelling the points along the string and τ is the time-evolution parameter. The two-dimensional space labelled by σ and τ plays a very important role in string theory and the Nambu-Goto action is in fact the area traced out in this space. The superstring requires in addition the fermionic coordinate $\theta_a(\sigma,\tau)$ where a is spinor index in the d-dimensional space-time.

For describing the dynamics of interacting strings it is best to go to the field theory of strings [61)]. Just as supersymmetric local quantum field theory is described by the superfield $\Phi(x, \theta)$ where x is the usual space-time point and θ is the Grassmann variable, superstring field theory is described by the string field $\Phi[x(\sigma), \theta(\sigma), \tau]$. Here, $x(\sigma)$ is the coordinate on the string parametrized by σ as beofre, $\theta(\sigma)$ is the corresponding Grassmann coordinate, and τ is the time-evolution parameter. Hence, the string field is actually a functional since it depends on the infinite set of points along the string. Pictorially,

Local quantum field : $\Phi(\,.\,)$

String quantum field : $\Phi(\sim)$

---

*See however Jain, S., Shankar, R. and Wadia, S.R. "Conformal invariance and String theory in compact space: Bosons", Tata Institute preprint TIFR/TH/85-3 (1985).

Thus, the string field theory is actually a nonlocal quantum field. Construction of a consistent nonlocal quantum field theory had long remained an unsolved problem. Superstring field theory is therefore an important fundamental advance.

Another point is worthy of note. In section 6, we had mentioned the importance of looking for physical manifestation of the Grassmann coordinate occuring in supersymmetric theories of nature. Actually because of the finite number of Grassmann variables θ, the local superfield $\Phi(x, \theta)$, is only polynomial in θ and hence θ could be regarded merely as a book-keeping device in local quantum field theory. In contrast, in superstring field theory, there are an infinite number of Grassmann variables $\theta(\sigma)$ and the superstring field $\Phi[x(\sigma), \theta(\sigma)]$ is no longer a polynomial in θ. Thus, the Grassmann coordinates have acquired a deeper significance as a new type of dimension.

Type I superstring theory describes open and closed strings both of which are unoriented, and they live in 10 dimensions and have one supersymmetry. Further, the open string is characterized by a Yang-Mills symmetry group SO(n), U Sp(2n) or a direct product of these. The open string field Φ contains the massless super Yang-Mills field (gauge field + gaugino) plus massive excited 'string' modes, whereas the closed string field Ψ contains the supergravity multiplet plus massive excited modes. The superfield action contains the following 7 types of interaction terms:

$$g\,\Phi^3,\ g\,\Phi\,\Psi,\ \kappa\,\Psi^3,\ \kappa\,\Phi^2\Psi,\ \kappa\,\Phi^4,\ \kappa\,\Phi^2 \text{ and } \kappa\,\Psi^2$$

where $g\Phi^3$ contains the usual cubic super Yang-Mills couplings plus couplings among the massive particles and $\kappa\Psi^3$ contains the usual supergravity couplings plus others. The Yang-Mills coupling constant g and the gravitational coupling constant κ get related to the string tension T by the relation:

$$\kappa \sim g^2 T$$

Type II theory contains only closed oriented strings living in 10 dimensions and having two supersymmetries (N=2). Since ψ is the only field in this theory, the interaction contains the single term $\kappa\,\psi^3$ only and so it describes supergravity plus the couplings among the massive modes.

The seven types of string interactions allowed in type I theory are indicated in Fig.5. The first interaction $g\,\Phi^3$ which is cubic in the open string field describes two open strings joining to give a single open string or a single open string splitting into two open strings. The second interaction $g\,\Phi\Psi$ describes a open string closing up and becoming a closed string or the reverse process. The third interaction $\kappa\Psi^3$ describes two closed strings touching at a single point and joining into a single closed string. Since the strings are unoriented in type I theory, the two initial strings can join in either of the configurations shown in the figure. The next two interactions describe $\kappa\Phi^2\Psi$ and $\kappa\Phi^4$. The last two interaction terms $\kappa\Psi^2$ and $\kappa\Phi^2$ arise only because of the absence of orientation in the strings.

Type II theory contains the single $\kappa\Psi^3$ interactions shown in Fig.5. Since the strings are oriented in this theory (as indicated by the arrow) the final configuration now depends on the relative direction of the arrows of the two approaching strings.

The most remarkable thing about the string theory is that no more interactions beyond those described above seem to be allowed. As a consequence, string theory is polynomial, whereas the effective supergravity theory (obtained in the limit $\alpha' \to 0$) is nonpolynomial with vertices occuring at arbitrarily high order. Higher-order contact terms of supergravity are replaced by effects arising from exchange of the infinite spectrum of the massive string excitations. This is analogous to the way massive vector exchanges of the electroweak theory of the standard model replace the old four fermi coupling of the weak interaction. So, the hope is that, in the same way as in the case of weak interaction theory the ultra-violet divergences of gravity theory are ameliorated so as to yield a renormalizable or even a finite theory of gravity.

Let us now consider the limits of the superstring theories when the compactification length R and the Regge slope α' go to zero. The following are the results:

Superstring theory I (N=1 , d=10) $\xrightarrow[R \to 0]{}$ Superstring theory (N=4, d=4)

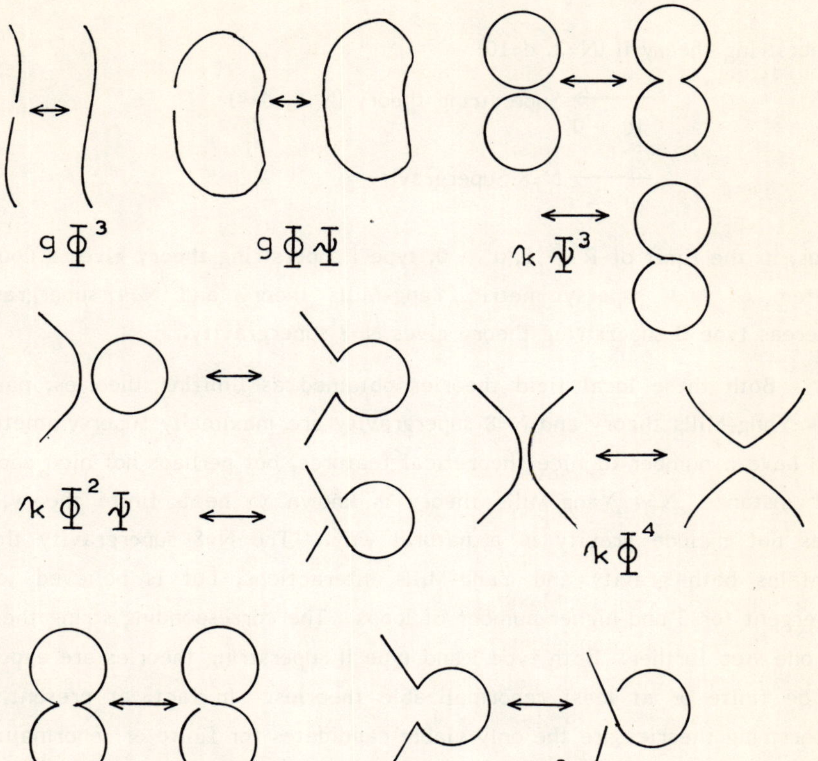

$g\Phi^3$ $g\Phi\Psi$ $\kappa\Psi^3$

$\kappa\Phi^2\Psi$ $\kappa\Phi^4$

$\kappa\Psi^2$ $\kappa\Phi^2$

FIG.5a INTERACTIONS IN TYPE I STRING THEORY

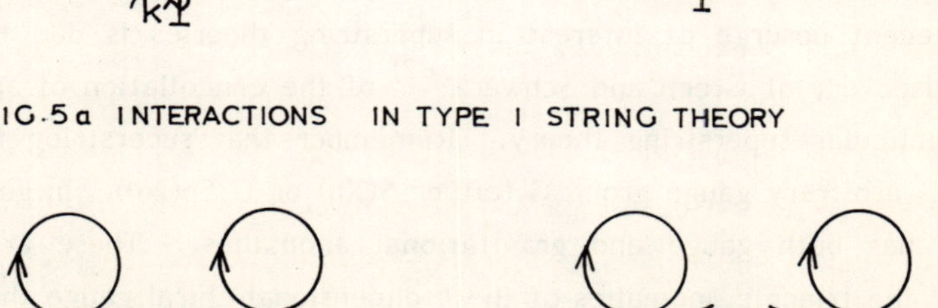

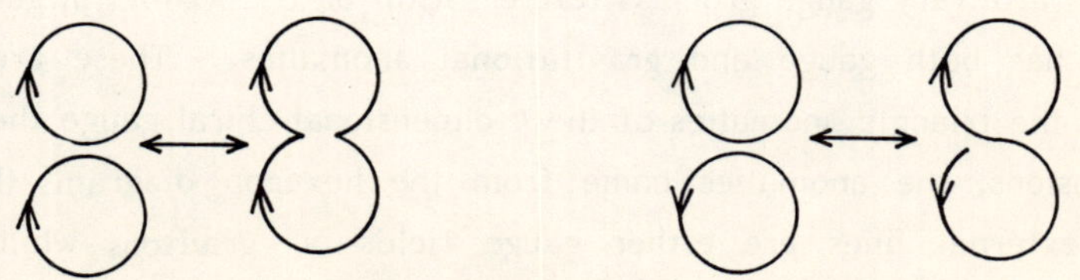

FIG.5b THE UNIQUE $\kappa\Psi^3$ INTERACTION IN TYPE II STRING THEORY

$$\xrightarrow[\alpha' \to 0]{} \text{N=4 SUSY Y.M. theory + N=4 Supergravity}$$

Superstring theory II (N=2, d=10)

$$\xrightarrow[R \to 0]{} \text{Superstring theory (N=8, d=4)}$$

$$\xrightarrow[\alpha' \to 0]{} \text{N=8 Supergravity}$$

Thus, in the limit of $R \to 0, \alpha' \to 0$, type I superstring theory gives a coupled system of N=4 supersymmetric Yang-Mills theory and N=4 supergravity, whereas type II superstring theory gives N=8 supergravity.

Both these local field theories obtained as limiting theories, namely N=4 Yang-Mills theory and N=8 supergravity are maximally supersymmetrical and have a number of nice theoretical features, but perhaps not nice enough. For instance, N=4 Yang-Mills theory is known to be a finite theory, but does not include gravity in a natural way. The N=8 supergravity theory contains both gravity and Yang-Mills interactions, but is believed to be divergent for 3 and higher number of loops. The corresponding string theories go one step further. Both type I and type II superstring theories are expected to be finite or at least renormalizable theories. In fact, at present, the superstring theories are the only viable candidates for finite or renormalizable theories of quantum gravity.

The recent upsurge of interest in superstring theories is due to the "stunning" discovery of Green and Schwarz [62)] of the cancellation of anomalies in a particular superstring theory. Remember that superstring theory of type I has arbitrary gauge group G (either SO(n) or U Sp(2n)). In general, this theory has both gauge and gravitational anomalies. These are the analogues of the triangle anomalies of the 4 dimensional chiral gauge theories. In 10 dimensions, the anomalies come from the hexagon diagrams (Fig.6) where the external lines are either gauge fields or gravitons while the internal lines are spin 1/2, spin 3/2 and (curiously) bosons of an antisymmetric tensor field of rank 3. However, the remarkable thing is that these anomalies cancel if and only if

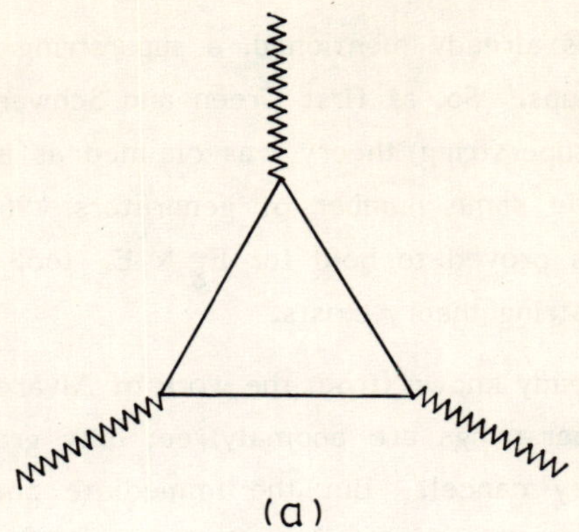

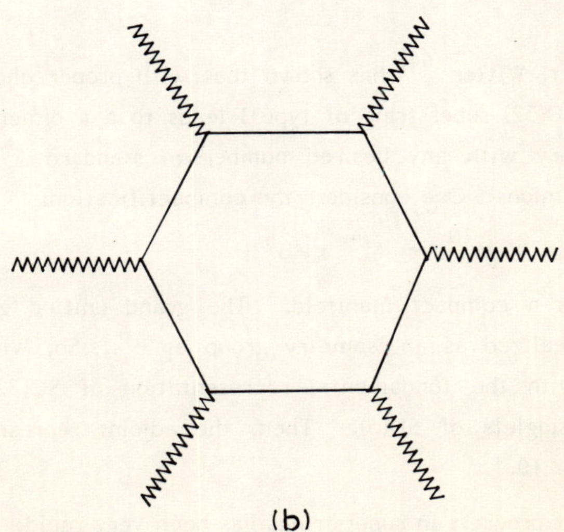

FIG. 6a THE TRIANGLE DIAGRAM FOR THE CHIRAL ANOMALY IN 4 DIMENSIONS

6b THE HEXAGON DIAGRAM FOR THE GAUGE AND GRAVITATIONAL ANOMALIES IN 10 DIMENSIONS

$$G = SO(32) \text{ or } E_8 \times E_8$$

Actually, as was already mentioned, a superstring theory allows only SO(n) or U Sp(2n) groups. So, at first Green and Schwarz considered only SO(32) and so SO(32) superstring theory was claimed as a unique possibility. But $E_8 \times E_8$ has the same number of generators 496 as SO(32) and anomaly cancellation was proved to hold for $E_8 \times E_8$ too. Thus, the possibility of a $E_8 \times E_8$ superstring theory exists.

It was already known (from the work of Alvarez-Gaume and Witten)[63] that type II superstrings are anomalyfree; only gravitational anomalies are possible and they cancel. But the immediate phenomenological prospects of the anomaly-free type I superstring theory are considered to be much brighter.

In the last section, mention was made of the need for extra gauge fields to solve the problem of chiral fermions. The SO(32) or $E_8 \times E_8$ gauge fields of the superstring theory are precisely the needed extra gauge fields.

In fact, Witten [64] has shown that with proper choice of the ground state, the SO(32) superstring of type I leads to a 4 dimensional SU(5) grand unified theory with any desired number of standard $5^* + 10$ generations of chiral fermions. One considers the compactification:

$$M^{10} \to M^4 \times B^6$$

where B^6 is a compact manifold. The grand unified group SU(5) is too big to be realized as an isometry group of B^6. So, Witten embeds SU(5) in SO(32) with the fundamental representation of SO(32) transforming as $5 + 5^*$ + singlets of SU(5). Then, the adjoint representation of SO(32) contains $5^* + 10$.

Recent progress in superstrings has been very rapid. Even the discarded d=26 bosonic strings have been combined with the d=10 superstrings to form what are called "heterotic strings" [65,66]. String theory with $E_8 \times E_8$ has emerged as a better candidate [67]; this can incorporate the old E_6 grand unified theory of section 3.

9. CONCLUSION

At the present time, superstring theory seems to be the most promising avenue for further development. However, as we have tried to show, this is not independent of the other approaches. In fact, superstring theory is only a further development along the road of grand unification, supersymmetry, higher-dimensions etc. and is built on these ideas plus something more .... .

Finally, it is sobering to ask where all this is leading to. Where is the experimental support for all this? For, after all, physics is an experimental science. Will it ever be possible to do experiments bearing on the energy scale 10^{15} - 10^{19} Gev? Or, are we doomed for ever to "verify" our speculative theories by comparing them with other speculations on the origin of the Universe?

REFERENCES

1. UA1 - Collaboration, Arnison, G. et al, Phys. Lett. 122B, 103 (1983); UA2 - Collaboration, Banner, M. et al, Phys. Lett. 122B, 476 (1983).
2. Pakvasa, S. XXI International Conference on High Energy Physics, Paris, (1982) 631.
3. Lee - Franzini, J. Europhysics Topical Conference on Flavour Mixing in Weak Interactions, Erice, 1984; Kleinknecht, K. Ibid.
4. Wolfenstein, L. Ibid.
5. Bajaj, J.K., and Rajasekaran, G. Pramana 14, 395 and 411 (1980).
6. Georgi, H. Quinn, H. and Weinberg, S. Phys. Rev. Lett. 33, 451 (1974).
7. Langacker, P. Physics Reports 72C, 185 (1981).
8. Pati, J., Salam, A. and Strathdee, J. Phys. Lett. 59B, 265 (1975).
9. Terazawa, H. Phys. Rev. D22, 184 (1980).
10. Harari, H. Phys. Lett. 86B, 83 (1979).
11. Shupe, M.A. Phys. Lett. 86B, 87 (1979).
12. Harari, H. and Seiberg, N. Phys. Lett. 96B, 269 (1981), Nucl. Phys. B204, 141 (1982).
13. Gupta, V. "If Preons are there ... " Talk at the Winter School on 'Experimental methods in High Energy Physics', Mahabaleswar, January 1985.

14. Harari, H. Phys. Reports 104C, 159 (1984).
15. Lyons, L., "An Introduction to the Possible Substructure of Quarks and Leptons", Prog. in Particle and Nuclear Physics 10, 227 (1983).
16. Ramachandran, R. "Composite Models", Talk at the High Energy Physics Symposium, Jammu, 1984.
17. Lipkin, H.J. Phys. Lett. 89B, 358 (1980).
18. Gluck, M. Phys. Lett. 87B, 247 (1979).
19. Brodsky, S.J. and Drell, S.D. Phys. Rev. D22, 2236 (1980).
20. Shaw, G.L., Silverman, D. and Slansky, R. Phys. Lett. 94B, 57 (1980).
21. t'Hooft, G. Lectures at the Cargese Summer Institute (1979).
22. Barbieri, R., Maiani, L. and Petronzio, R. Phys. Lett. 96B, 63 (1980); Banks and Yankielowicz Phys. Lett. 101B, 159 (1981); Bars, I. Phys. Lett. 109B, 73 (1982); Frishman, Y. et al Nucl. Phys. B177, 157 (1981).
23. Sharatchandra, H.S. Phys. Lett. 139B, 301 (1984) and Max Planck Institute preprints MPI-PAE/Pth 85/83, 91/83, 8/84.
24. Gerard, J.M. et al Phys. Lett. 128B, 65 (1983); Gupta, V., Joshipura, A.S., Mani, H.S. and Ramachandran, R. Phys. Rev. D30, 1982 (1984); Pati, J. Phys. Rev. D30, 1144 (1984).
25. Buchmuller, W., Love, S.T., Peccei, R.D. and Yanagida, T. Phys. Lett. 115B, 233 (1982); Buchmuller, W., Peccei, R.D. and Yanagida, T. Phys. Lett. 124B, 67 (1983), Nucl. Phys. B227, 503 (1983); Nucl. Phys. B231, 53 (1984); Barbieri, R., Masiero, A. and Veneziano, G. Phys. Lett. 128B, 179 (1983).
26. Sakharov, A.D. Doklady Akad. Nauk, S.S.S.R. 117, 67 (1967) Sov. Phys. Doklady 12, 1040 (1968).
27. Adler, S.L. Phys. Rev. Lett. 44, 1567 (1980); Phys. Lett. 95B, 241 (1980); Rev. Mod. Phys. 54, 729 (1982).
28. Zee, A. Phys. Rev. Lett. 42, 417 (1979); Phys. Rev. D23, 858 (1981).
29. Smolin, L. Nucl. Phys. B160, 253 (1979).
30. Minkowski, P. Phys. Lett. 71B, 419 (1977).
31. Fayet, P. and Ferrara, S. Phys. Rep. 32C, 249 (1977).
32. Van Nieuwenhuizen, Phys. Rep. 68C, 189 (1981).
33. Wess, J. and Bagger, J., Supersymmetry and Supergravity (Princeton

University Press, Princeton, N.J. 1983).

34. Gates, J., Grisaru, M., Rocek, M. and Siegel, W., Superspace or 1001 Lectures in Supersymmetry (Benjamin Reading, Mass., 1983).

35. Kaul, R.K. Phys. Lett. 109B, 19 (1982); Kaul, R.K. and Majumdar, P. Nucl. Phys. B199, 36 (1982); Sakai, N. Z. Phys. C 11, 153 (1981).

36. Haber, H.E. and Kane, G.L. The search for supersymmetry: Probing physics beyond the Standard Model, Phys. Reports 117C, 76 (1985).

37. Arnison, G. et al., Phys. Lett. 139B, 115 (1984).

38. Reya, E. and Roy, D.P., Phys. Rev. Lett. 51, 867 1307 E (1983) and 53, 881 (1984), and Phys. Lett. 141B, 442 (1984).

39. Barger, V. et al., Phys. Rev. Lett. 53, 641 (1984), and Phys. Lett. 145B, 147 (1984).

40. Ellis, J. and Kowalski, H. Phys. Lett. 142B, 441 (1984) and Nucl. Phys. B246, 189 (1984).

41. Samuel, S. and Wess, J. Columbia University preprint CU-TP-264, June, 1983.

42. Mandelstam, S. Nucl. Phys. B213, 149 (1983).

43. Cremmer, E. and Julia, B. Phys. Lett. 80B, 48 (1978); Nucl. Phys. B159, 141 (1979).

44. Ellis, J., Gaillard, M.K. and Zumino, B. Phys. Lett. 94B, 343 (1980).

45. Kaluza, Th. Sitz. preuss. Acad. Wiss. Phys. Math. Kl. 966 (1921).

46. Klein, O. Z. Phys. 37, 895 (1926).

47. Zee, A. Grand Unified Theories and Related Topics (4th Kyoto Summer Institute), World Scientific (1981), p.143.

48. Witten, E. Nucl. Phys. B186, 412 (1981).

49. Leibbrandt, G. Rev. Mod. Phys. 47, 849 (1975).

50. Witten, E., Fermion Numbers in Kaluza-Klein theory, Princeton University preprint (October 1983).

51. Weinberg, S., Phys. Lett. 138B, 47 (1984).

52. Frampton, P.H. and Yamamoto, K. Phys. Rev. Lett. 52, 2016 (1984); Frampton, P.H. and Kephart, T.W. Phys. Rev. Lett. 53, 867 (1984).

53. Schwarz, J.H. Dual Resonance Theory, Phys. Rep. 8C, 269 (1973).

54. Veneziano, G. Dual Models, Phys. Rep. 9C, 199 (1974).

55. Rebbi, C., Dual Models and Strings, Phys. Rep. 12C, 1 (1974).

56. Mandelstam, S. Dual Resonance Models, Phys. Rep. 13C, 25 (1974).

57. Scherk, J., Dual Models and Strings, Rev. Mod. Phys. 47, 123 (1975).

58. Frampton, P.H. Dual Resonance Models, Benjamin (1974).

59. Schwarz, J.H. Superstring Theory, Phys. Reports 89C, 223 (1982).

60. Polyakov, A. M. Phys. Lett. 103B, 207, 211 (1981).

61. Green, M. B., and Schwarz, J.H., Nucl. Phys. B243, 475 (1984).

62. Green, M.B. and Schwarz, J.H., Phys. Lett. 149B, 117 (1984).

63. Alvarez-Gaume, L. and Witten, E. Nucl. Phys. B234, 269 (1983).

64. Witten, E., Phys. Lett. 149B, 351 (1984).

65. Gross, D.G. et al, Phys. Rev. Lett. 54, 502 (1985); and Princeton preprint, January, 1985.

66. Freund, P.G.O., Phys. Lett. 151B, 387 (1985).

67. Candelas, P. et al, Vacuum Configurations for superstrings, Santa Barbara preprint, NSF - ITP - 84 - 170 (1984).

KALUZA-KLEIN THEORIES - A MODERN PERSPECTIVE

A. Maheshwari
Department of Physics
Regional College of Education
Mysore 570 006, India

1. INTRODUCTION

Einstein's idea which gave rise to general relativity was the proposal that the space-time is non-trivially curved and the curvature is responsible for the gravity. The Kaluza-Klein[1)] idea takes this one step further, proposing that there are more than one time and three space dimensions and that the curvature of the higher dimensional space-time in the low energy approximation is perceived in the effective ordinary 4-dimensions as unified theory of gravity and gauge fields. In 1921 Kaluza had suggested that gravitation and electromagnetism could be unified in a five-dimensional theory of gravity. In the classic Kaluza-Klein theory the fifth coordinate was made invisible through a process of dimensional reduction. This idea has been revived several times, but recently it has gained popularity because of the introduction of the concept of spontaneous compactification[2)] of extra dimensions. The starting point of all Kaluza-Klein theories is a theory of gravitation coupled to some matter fields in D = 4 + K dimensions. The classical field equation of the metric and the matter fields induce compactification of the extra space dimensions such that the 4 + K dimensional background manifold, $M_1 \times M_2$, emerges out as a product of a compact space of K dimensions, M_2, and an ordinary 4-dimensional space-time, M_1. This is called spontaneous compactification of space-time. The next step in the Kaluza-Klein programme is to expand the matrix and matter fields around this particular solution using `harmonic` functions for the invariance group G of the compact space of K dimensions. In general a finite number of massless modes and infinite number of massive modes are obtained using this programme.

Although the Kaluza-Klein programme in $D > 5$ dimensions has not so far provided an answer to a physical theory, yet the question of the "true" dimensionality of the world and the possibility of unifications of non-Abelian gauge fields with gravity has made it an extremely interesting theoretical scheme.

Many good review articles on Kaluza-Klein theories are available in the literature[3,4]. Therefore, the scope of this review has been limited to provide for non-experts a brief overview of the well established results such as spontaneous compactification, the identification of massless modes and the symmetries of compactified manifold.

2. SPONTANEOUS COMPACTIFICATION

One starts with gravity, described by the Einstein lagrangian ($\sqrt{g}\, R$), coupled to matter fields in space-time of D dimensions with $D > 4$ and signature $- + + \ldots +$. To distinguish the compact dimensions from the remaining four of ordinary space-time, we choose the following notational conventions. For the coordinates write

$$z^M = \{x^m, y^\mu \text{ ; } m, n \ldots \text{ taking the values } 0,1,2,3; \text{ and}$$

$$\mu, \nu, \ldots \text{ taking the values } 1,2 \ldots K \ (4 + K = D)\}.$$

If the field equations have to induce compactification, we look for solutions such that the ground state of the D-dimensional space-time becomes the product of a 4-dimensional Minkowskian space-time and a compact K-dimensional Riemannian manifold; $M = M^4 \times B^K$. This can be achieved if the solution of the classical equation for the metric has the structure

$$g_{MN} = \begin{bmatrix} g_{mn}(x) & 0 \\ 0 & g_{\mu\nu}(y) \end{bmatrix} , \qquad (1)$$

and the field equation for the metric decouples into equations of the type

$$R_{mn} = \gamma_4 \; g_{mn} ,$$

$$R_{\mu\nu} = \gamma_K \; g_{\mu\nu} , \qquad (2)$$

$$R_{\mu m} = 0.$$

In our notation[5] a compact Riemann manifold has negative scalar curvature. Thus γ_4 should be negative and γ_K of opposite sign and of a magnitude such that the compactification size is of the order of Planck length.

We shall consider the cases of pure gravity and gravity with cosmological term before discussing the model of Freund and Rubin[6] as an example of spontaneous compactification.

2.1 Pure Gravity

The field equation for the metric

$$R_{MN} = 0 \qquad (3)$$

obviously possesses a solution so that the background manifold is Minkowskian space-time in D dimensions, M^D. There is also a solution with the background manifold being given by M^4 X T^K (S^1 X S^1...X S^1). The choice between these solutions is not obvious since the energy is not well defined. The original Kaluza-Klein unified theory of gravitation and electromagnetism belongs to this category. It can be understood as a theory of spontaneous symmetry breaking in which the group of general coordinate transformations in five dimensions is spontaneously broken to the symmetries of the ground state $M^4 \times S^1$, which are the four-dimensional Poincare symmetries acting on M^4, and a U(1) group of rotation of the circle S^1. These symmetries would be observed as local or gauge symmetries in the apparent four-dimensional world because the whole theory started with the Einstein action which is generally covariant. This case is not very interesting.

2.2 Pure Gravity with Cosmological Constant

The field equation for the metric in this case is

$$R_{MN} - 1/2\, g_{MN}\, R + \Lambda\, g_{MN} = 0 \tag{4}$$

This gives

$$R_{MN} = 2\,\Lambda /D-2 \cdot g_{MN} \tag{5}$$

A metric of the type given in Eq. (1) will decouple Eq. (5) into equations of the type

$$R_{mn} = \gamma g_{mn} \tag{6}$$

$$R_{\mu\nu} = \gamma g_{\mu\nu} \tag{7}$$

We want $\gamma_4 \neq \gamma_7$, therefore gravity has to be coupled to matter inorder to obtain realistic spontaneous compactification. In the Cremmer-Scherk[2)] model gravity is coupled to Yang-Mills fields in D = 4 + K dimensions. This would defeat the purpose of Kaluza-Klein because we want Yang-Mills fields to emerge out as components of the higher-dimensional metric. We discuss next the Freund-Rubin[6)] model of spontaneous compactification.

2.3 Freund-Rubin Spontaneous Compactification

In the Freund-Rubin model matter field is introduced as a completely antisymmetric tensor $F_{M \ldots M}$ of rank-s. This field is the curl of a rank-(s-1) antisymmetric tensor potential $A_{M_1 \cdots M_{s-1}}$,

$$F_{M_1 M_2 \cdots M_s} = \partial_{[M_1} A_{M_2 \cdots M_s]}. \tag{8}$$

The coupling of the tensor field with the metric in D dimensions is taken to be of the form

$$S = \int d^D Z \sqrt{-g} \{1/C_1 R + 1/C_2 F_{M_1 \cdots M_s} F^{M_1 \cdots M_s}\}, \tag{9}$$

where C_1 and C_2 are two dimensional constants. The metric g_{MN} has the signature $(- + + \ldots +)$ and $-g = |\det g_{MN}|$. The field equations for the metric g_{MN} and the vector potential A_{NPQ} are easily found to be

$$R_{MN} - 1/2\, g_{MN} R = - C_1 s/C_2 \,(F_{MM_1 \cdots M_{s-1}} F_N{}^{M_1 \cdots M_{s-1}} - (1/2s) F_{M_1 \cdots M_s} F^{M_1 \cdots M_s} g_{MN}), \tag{10}$$

$$(\sqrt{-g}\, F^{MM_1 \cdots M_{s-1}}) = 0 \tag{11}$$

$C_1 s/C_2$ can be made equal to $8\pi G_N$, if we wish to write Eq. (10) in the usual Einstein form

$$R_{MN} - 1/2\, g_{MN} R = - 8\pi G_N T_{MN} \tag{12}$$

A trivial solution of Eq. (11) is $F^{M_1 \cdots M_s} = 0$. Then Eq. (10) reduces to an equation for pure gravity in D dimensions. This case we have already studied and discarded because it does not lead to interesting physics in 4 dimensions. On the other hand, Eq. (11) also admits a solution with a non-trivial vacuum,

$$F^{m_1 m_2 \cdots m_s} = (1/\sqrt{s!|g_s|})\, \varepsilon_{(s)}{}^{m_1 m_2} f, \tag{13}$$

where $\varepsilon_{(s)}{}^{m_1 m_2 \cdots m_s}$ is the Levi-Civita tensor in s-dimensions, and f is a constant. To evaluate the right-hand-side of Eq. (10) we use the relation

$$\varepsilon_{m_1 \cdots m_n p_{n+1} \cdots p_s} \varepsilon^{n_1 \cdots n_n p_{n+1} \cdots p_s}$$

$$\equiv n!\,(s-n)!\, g_s\, \delta_{[m_1 \cdots m_n]}{}^{n_1 \cdots n_n} \tag{14}$$

where

$$\delta_{pq \cdots}{}^{mn \cdots} = \delta_p{}^m \delta_q{}^n$$

Using Eqs. (13) and (14) it can be easily seen that

$$F_{M_1 \cdots M_s} F^{M_1 \cdots M_s} = f^2 \eta_s , \tag{15}$$

and

$$F_{mm_2 \cdots m_s} F_n{}^{m_2 \cdots m_s} = g_{mn} f^2 \eta_s / s , \tag{16}$$

where the sign $\eta_s = g_s / |g_s|$. Equation (10) breaks up into the following equations.

$$R_{mn} = -\lambda \frac{D-s-1}{D-2} g_{mn} , \tag{17}$$

$$R_{\mu\nu} = \lambda \frac{s-1}{D-2} g_{\mu\nu} , \tag{18}$$

$$\lambda = 8 \pi G_N f^2 \,\text{sign}\,(g_s) \tag{19}$$

We note that the signs of the curvature scalars of the two product spaces are opposite (D > s+1), and are determined by the sign of λ. Therefore for D > s+1 either s or D-s space-like dimensions compactify. The background manifold is either $M^s \times B^{D-s}$ or $B^s \times M^{D-s}$.

In d=11 supergravity, supersymmetry forces the introduction of an anti-symmetric tensor field with s=4. The Cremmer-Julia-Scherk[7] action for 11 - dimension supergravity is

$$S = \int d^{11}z \sqrt{-g} \{1/4 R + 1/48 F_{MNPQ} F^{MNPQ}$$

$$- 2/(12)^4 1/\sqrt{-g} \varepsilon^{MNPQRSTUVWX} F_{MNPQ} F_{RSTU} A_{VWX} + \ldots\} \quad (20)$$

The form of action given in Eq. (20) is similar to the Freund-Rubin action, Eq. (9). The Freund-Rubin type analysis goes through with minor changes. The 11-dimensional manifold has a ground state that leads to the preferential compactification of 7 or 4 space-like dimensions, so that physical space-time must be either 4 or 7 dimensional. Equations (17) and (18) correspond to Einstein spaces and possess large number of solutions. In a maximally symmetric solution, M^4 can be the anti-de Sitter space-time and B^7 can be the 7-sphere, S^7. This case has been studied in detail by Duff and Pope[8]. As mentioned earlier, because the compactification size being of the order of Planck length the ordinary space-time gets its own cosmological constant of opposite sign but again of order G_N^{-1}. This is immense, so how does it get removed ? The cosmological constant of the space-time can be made equal to zero by introducing a bare cosmological constant. The argument against this procedure is that besides being ad hoc the fine-tuning will be spoiled by the renormalisation effects.

3. FOURIER EXPANSION

The physical spectrum is obtained by studying small oscillations about the ground state. Fourier expansion of all the fields, including the gravity, is performed about the ground state. Denoting these fields generically by Φ (x,y) where x^m (m = 0,1,2,3) are the coordinates on M_1, and y^n (m = 1,2 ... K, 4 + K = D) are the coordinates on M_2, then symbolically

$$\Phi (x,y) = \sum_n \Phi_n(x) Y_n(y) , \quad (21)$$

$$M^2 Y_n(y) = m_n^2 Y_n(y) . \quad (22)$$

The expansion coefficients $\phi_n(x)$ will be interpreted as physical fields in d=4 and the Y_n's are eigen functions of the mass matrix M^2.

This mass matrix will be given by some appropriate differential operator on M_2 determined by expanding the field equations to first order in fluctuations about the ground state.

As an illustrative example[4)], we consider a real scalar field on the product manifold $M^4 \times B^K$ (B^K is some compact space of K dimensions.) Let the metric of $M^4 \times B^K$ be of the form

$$g_{MN} = \begin{bmatrix} g_{mn}(x) & 0 \\ 0 & g_{\mu\nu}(y) \end{bmatrix}, \tag{23}$$

Let the term in the action containing the scalar field be of the form

$$S_\Phi = \int \sqrt{|\det g_{MN}|}\, d^DZ(-1/2\, \Phi \Box_D \Phi + 1/2(m^2 + \xi R_D)\Phi^2), \tag{24}$$

$$\xi = \text{constant},$$

and $$\Phi = \Phi(x,y)$$

From Eq. (23) we get

$$|\det g_{MN}|\, d^DZ = |\det g_{mn}|\, d^4x\, |\det g_{\mu\nu}| d^Ky ,$$

$$\Box_D = \Box_4 + \Box_K ,$$

and $R_D = R_4 + R_K$. Now let $\{ Y_n(y) \}$ be the set

$$(-\Box_K + m^2 + \xi R_K)\, Y_n(y) = \lambda_n Y_n(y) , \tag{25}$$

with

$$\int |\det g_{\mu\nu}(y)|^{1/2}\, d^Ky\;\; Y^*_n(y)\, Y_m(y) = \delta_{nm} . \tag{26}$$

In terms of these functions the harmonic expansion for $\Phi(x,y)$ is

$$\Phi(x,y) = \sum_n \phi_n(x)\, Y_n(y) . \tag{27}$$

Inserting this expansion in the action functional, Eq. (24), and integrating over y^μ then yields the action functional of an effective four-dimensional theory on M_1 ;

$$S_\Phi = -1/2 \sum_n \int |\det g_{mn}|^{1/2} d^4x\, \phi^*_n(x)\{\Box_4 - \lambda_n - \xi R_4\}\, \phi_n(x). \quad (28)$$

In general we obtain an infinite tower of fields with masses $\sqrt{\lambda_n}$. In all Kaluza-Klein theories the metric field carries an infinite number of new degrees of freedom corresponding to the propagation of excitations on the compactified dimensions. The masses are of the order of inverse size of the compact dimensions. If the size of the compact space is of the order of Planck length the masses are in Planck unit $\sim 10^{19}$ GeV. As a rule these excitations are ignored because of their large masses. The original Kaluza-Klein theory in addition to containing a massless graviton, a massless photon and a massless Brans-Dicke scalar has in it an infinite tower of massive spin-2 fields.

3.1 Zero Mass States

Kaluza-Klein theories also contain finite number of massless modes, which presumably would constitute the low-energy physics. It has already been mentioned that the massless modes of the 5-dimensional Kaluza-Klein theory are a spin-two graviton and a spin-one photon, which are gauge particles of the symmetries of $M^4 \times S^1$, and a Brans-Dicke scalar. The ansatz which exhibits the massless modes is the following:

$$g_{MN}(x,y) = \begin{bmatrix} g_{mn}(x) + \kappa^2\phi(x)A_m(x)\,A_n(x) & \kappa\phi(x)\,A_m(x) \\ \kappa\phi(x)\,A_n(x) & \phi(x) \end{bmatrix} \quad (29)$$

In the classic work on Kaluza-Klein theory it is shown[9)] that under this ansatz

$$|\det g_{MN}|^{1/2} R^{(5)} d^4x\, dy$$

$$= \int |\det g_{mn}|^{1/2} d^4x (1/\kappa^2 R^{(4)} + 1/4 F_{nm} F^{nm}$$

$$+ 1/2 g^{mn} \partial_m \phi \partial_n \phi). \quad (30)$$

U(1) gauge transformations emerge as a special case of coordinate transformations of the five-dimensional theory. Under the transformations

$$(x^m, y) \to (x'^m, y') = (x^m, y + \kappa\varepsilon(x)), \quad (31)$$

$$g_{MN}(Z) \to g'_{M'N'}(Z') : g_{MN}(Z) = \frac{\partial Z'^{M'}}{\partial Z^M} \frac{\partial Z'^{N'}}{\partial Z^N} g'_{M'N'}(Z') \quad (32)$$

The transformation of the metric component

$$g_{m5}(x) \to \frac{\partial Z^M}{\partial x'^m} \frac{\partial Z^N}{\partial y'} g_{MN}(Z)$$

gives

$$A_m(x) \to A_m(x) - \frac{\partial \varepsilon}{\partial x^m} \quad (33)$$

This allows $A_m(x)$ to be identified as the electromagnetic potential.

The ansatz which generalizes the 5-dimensional case is the following[10]:

$$g_{MN} = \begin{bmatrix} g_{mn} + g_{\mu\nu} V^\mu_m V^\nu_n & V^\mu_m g_{\mu\nu} \\ g_{\mu\nu} V^\nu_n & g_{\mu\nu} \end{bmatrix} \quad (34)$$

Under the coordinate transformations

$$\delta g_{MN} = - \nabla_M \zeta_N - \nabla_N \zeta_M ,$$

$$\text{with} \quad \zeta_M = (\zeta_n(x), 0) , \quad (35)$$

$g_{mn}(x,y)$ transform as a tensor of spin-2,

$V_m(x,y)$ transform as vectors of spin-1, and

$g_{\mu\nu}(x,y)$ transform as scalars of spin-0. (36)

We impose the following y-dependence,

$$g_{mn}(x,y) = g_{mn}(x) , \tag{37}$$

$$V_m^\mu(x,y) = A_m^{\hat\alpha}(x)\, K_{\hat\alpha}^\mu(y) , \tag{38}$$

and $g_{\mu\nu}(x,y) = g_{\mu\nu}(y)$ (metric on M_2) (39)

$K_{\hat\alpha\mu}$, the Killing vectors on M_2, are the solution of the Killing equation

$$\nabla_\mu K_{\hat\alpha\nu}(y) + \nabla_\nu K_{\hat\alpha\mu}(y) = 0 .$$

Index $\hat\alpha$ runs over the number of independent solutions of the Killing equation. The vector fields

$$K_{\hat\alpha} \equiv K_{\hat\alpha}^\mu \frac{\partial}{\partial y^\mu} \tag{40}$$

determine the isometry group G of the manifold M_2. The structure constants $f_{\hat\alpha\hat\beta}{}^{\hat\gamma}$ of the group G are determined by the Lie-brackets,

$$[K_{\hat\alpha}, K_{\hat\beta}] = f_{\hat\alpha\hat\beta}{}^{\hat\gamma} K_{\hat\gamma} \tag{41}$$

From the form of the metric given in Eq. (34) with the y-dependence as given in Eqs. (37)-(39), the identification of $A_m^{\hat\alpha}(x)$ as the Yang-Mills fields can be seen as a special case of general coordinate transformaations of the D-dimensional metric. Under the transformation

$$\delta Z^M = \zeta^M , \quad \delta g_{MN} = - \nabla_M \zeta_N - \nabla_N \zeta_M . \tag{42}$$

If we restrict ζ_M to the form

$$\zeta_M = (0 \ , \quad \varepsilon_{\hat{\alpha}}(x)\, K^{\hat{\alpha}}_{\mu}\ (y)) \ , \tag{43}$$

Eq. (42) gives

$$\delta\, g_{mn} = 0 \tag{44}$$

and

$$\delta\, A^{\hat{\alpha}}_m(x) = \partial_m\, \varepsilon^{\hat{\alpha}}(x) - f_{\hat{\gamma}\hat{\beta}}{}^{\hat{\alpha}}\ A^{\hat{\gamma}}_m(x)\ \ \varepsilon^{\hat{\beta}}(x) \tag{45}$$

$A^{\hat{\alpha}}_m(x)$ are the Yang-Mills gauge fields appropriate to G and must be massless. As already stated the gauge group G is the isometry group of the manifold with compactified dimensions, similarly $g_{mn}(x)$ transform like a 4-dimensional metric tensor and corresponds to a massless graviton.

4. COSET SPACES

If the internal space for explaining the internal symmetries belongs to the most symmetrical category-the coset spaces-then the Fourier expansions and the study of massless modes can be carried out by an ingenious procedure due to Salam and Strathdee[3]. If H is a subgroup of the compact continuous group G, then the space of left cosets, denoted by G/H, is invariant under the action of G. We shall suppose that the internal space M_2, obtained by solving the 4+K - dimensional equations is such a quotient space and its dimensions K = dim G - dim H.

The eleven-dimensional supergravity compactifies to a ground state manifold M = Ads X S^7. The 7-sphere is isomorphic to the coset space SO(8)/SO(7). A detailed spectrum analysis for this case has been done by Duff and Pope[8]. The massless states in this case are[8)]

spin	2	3/2	1	1/2	0
number	1	8	28	56	35

.

This is not a physical model because the symmetry group SO(8) of S^7 does not contain SU(3) X SU(2) X U(1). The largest dimensional non-trivial subgroup of SU(3) X SU(2) X U(1) is SU(2) X U(1) X U(1). Therefore the coset spaces with SU(3) X SU(2) X U(1) isometry can be[10])

$$M^7 = \frac{SU(3) \times SU(2) \times U(1)}{SU(2) \times U(1) \times U(1)}$$

If the gauge fields of SU(3) X SU(2) X U(1) are to arise as components of gravity in higher dimensions by the Kaluza-Klein mechanism, an eleven-dimensional world is suggested. Even if a suitable starting action in 11 dimensions with appropriate matter fields can be found so that the ground state is compactified to

$$M^4 \times \frac{SU(3) \times SU(2) \times U(1)}{SU(2) \times U(1) \times U(1)} ,$$

it may not turn out to be a physical theory because of the problem of chiral fermions. In the minimal GUT at least 45 fermion degrees of freedom of a given helicity are required for placing zero mass quarks and leptons in appropriate representations of the gauge group. The observed quarks and leptons of left-handed helicity transform differently from those of right-hand helicity. This is the reason that in the standard model quarks and leptons do not have bare masses but receive mass from the Higgs-mechanism. If left-handed and right-handed fermions transfor the same way under the gauge group arbitrarily large bare masses would develop. Within the framework of Kaluza-Klein theories it is not difficult to have sufficient number of massless fermion modes, but it is difficult to perceive how only chiral fermions can be made to arise as excitations over the ground state.

The gauge coupling constants are obtained by substituting the harmonic expansion of the metric and the matter fields in the action functional and integrating over the compactified dimensions. This can easily be performed in the 5-dimensional Kaluza-Klein giving the fine-structure constant $e^2 \sim G_N/L^2$, where L is the size of the compactified dimension. This fixes L to be of the order of 10^{-33} cm. The size of the compactified dimensions in $D > 5$ Kaluza-Klein theories will also be of the order of Planck length and therefore such theories will contain in addition to finite number of zero mass modes infinite towers of massive states corresponding to the propagation of excitations in the compactified manifold. These massive modes cannot be ignored in quantum theory, their physical interpretation will have to be found.

Inspite of these difficulties the Kaluza-Klein unification is a beautiful idea. It is a rapidly developing field still very much in its infancy. Recently important advances have been made in topics such as Kaluza-Klein cosmology, quantum theory and Kaluza-Klein monopole. Readers interested in seriously learning this field may like to consult Ref. 11.

ACKNOWLEDGEMENT

The author had his first introduction to the Kaluza-Klein theories in the 1983 Trieste Summer Workshop on High Energy Physics. He wishes to thank Mike Duff for a stimulating course of lectures given by him in that workshop. He is thankful to Mr. S. Manjunath for assistance in the preparation of the manuscript.

REFERENCES

1. Kaluza, Th., Preuss. Akad. Wiss. Berlin, Phys. Math. Kl, 966 (1921); Klein, O., Phys. 37, 895 (1926).
2. Cremmer, E. and Scherk, J., Nucl. Phys. B108, 409 (1976) : B118, 61 (1977); Luciani, J.F., Nucl. Phys. B135, 111 (1978).
3. Salam, A. and Strathdee, J., Annals of Physics 141, 316 (1982).
4. Mecklenburg, W., 'The Kaluza-Klein Idea : Status and Prospects', ICTP Preprint IC/83/32 (1983).
5. Weinberg, S., "Gravitation and Cosmology" (Wiley, New York, 1972).

6. Freund, P.G.O. and Rubin, M.A., Phys. Lett. B97, 233 (1980).
7. Cremmer, E., Julia, B. and Scherk, J., Phys. Lett. B76, 409 (1978).
8. Duff, M.J. and Pope, C.N., in Supersymmetry and Supergravity'82, edited by Ferrara, S., Taylor, J.G. and Van Nieuwenhuizen, P., (World Scientific, Singapore, 1983).
9. Lord, E.A., "Tensors Relativity and Cosmology" (Tata McGraw Hill Publishing Co. Ltd., New Delhi, 1976).
10. Witten, E., Nucl. Phys. B186, 412 (1981).
11. Lee, H.C., "An Introduction to Kaluza-Klein Theories" (World Scientific, Singapore, 1984).

BOUNDARY CONDITIONS, FRACTIONAL FERMION NUMBER AND THETA VACUA*

S.M. ROY

Tata Institute of Fundamental Research
Homi Bhabha Road, Colaba, Bombay 400 005
INDIA

ABSTRACT

We discuss the determination of allowed boundary conditions in quantum theory using the criterion of self-adjointness of the Hamiltonian. Recent applications to the phenomena of fractional fermion number and theta - vacua in quantum field theory are summarized.

CONTENTS

I. ROLE OF BOUNDARY CONDITIONS IN QUANTUM THEORY

I.1 Is there a path integral description?

The boundary conditions we discuss here should be carefully distinguished from initial conditions. E.g., for a free particle, classically the equation of motion $m\ddot{x} = 0$ has to be supplemented by initial conditions on $x(0)$, $\dot{x}(0)$; quantum mechanically the eqn.

$$H_o\psi = i\frac{\partial\psi}{\partial t}, \quad H_o = -\frac{1}{2m}\frac{d^2}{dx^2} \tag{1.1}$$

has to be supplemented by the initial condition $\psi(x,t = 0)$. Neither classically, nor quantum mechanically do we need any spatial boundary condition. The next simple example of the quantum oscillator,

$$H_{osc} = H_o + \frac{1}{2}m\omega^2 x^2 \tag{1.2}$$

also needs no boundary conditions to specify the dynamics.

Consider now a particle in a box described by $m\ddot{x} = 0$ classically, and by (1.1) quantum mechanically, except that

$$x \in [0,L]. \tag{1.3}$$

Both classically and quantum mechanically boundary conditions are needed to specify the dynamics. Classically we have to specify perfectly reflecting walls, sticky walls etc. Here we encounter the difficulty we wish to focus on: even in this simple problem there seems to be no systematic classification of the boundary conditions allowed classically! On the other hand, quantum mechanically, the requirement that the Hamiltonian H_o be a self-adjoint operator gives the following precise answer. There is a four parameter family of allowed boundary conditions:

$$\begin{bmatrix} \psi(L) + (i/m)\psi'(L) \\ \psi(0) - (i/m)\psi'(0) \end{bmatrix} = U \begin{bmatrix} \psi(L) - (i/m)\psi'(L) \\ \psi(0) + (i/m)\psi'(0) \end{bmatrix} \tag{1.4}$$

where $\psi'(x)$ denotes $d\psi(x)/dx$, and U is a 2 x 2 unitary matrix specified by four real parameters. The more familiar boundary conditions, Dirichlet, Neumann and periodic correspond to special choices of U:

$$U = -1 \Rightarrow \psi(0) = \psi(L) = 0 \quad \text{(Dirichlet)}$$
$$U = +1 \Rightarrow \psi'(0) = \psi'(L) = 0 \quad \text{(Neumann)} \qquad (1.5)$$
$$U = \sigma_1 \Rightarrow \psi(L) \Rightarrow \psi(0),\ \psi'(L) = \psi'(0) \ \text{(Periodic)}$$

(Here σ_1 denotes a Pauli matrix). But any unitary U is a valid specification of boundary conditions.

This situation leads us to ask the provocative question: is there a path integral description of quantum theory? In the path integral description

$$L_{classical} \xrightarrow{\text{(leads to)}} \text{Quantum theory.} \qquad (1.6)$$

With $L_{class} = \frac{1}{2}m\dot{x}^2$, but no classification of classically allowed boundary conditions, and given a four parameter family of quantum theories, the statement (1.6) does not make sense. Quantum theory with what boundary conditions - must be specified, and in order to do that must know the path integral prescription corresponding to each unitary matrix in eq. (4).

The path integral prescription which corresponds to a general U is unknown. We must hasten to add that there are many specific choices of U for which a path integral prescription has been found (G1,G2,S1,S2,R1,R2), and shown to entail a modification of the naive Feynman formula. Let us mention here a curious and instructive fact. With the same generalized path integral description (Rothe & Swieca[1]) which gives

$$L_{cl} = \frac{1}{2} m\dot{x}^2 \implies U = \sigma_1, \text{ (periodic)}, \qquad (1.7)$$

We also obtain,

$$L_{cl} = \frac{1}{2} m\dot{x}^2 - \frac{\theta\hbar}{L}\dot{x} \Rightarrow U = \begin{pmatrix} 0 & e^{i\theta} \\ e^{-i\theta} & 0 \end{pmatrix}, \text{ (Quasi-periodic).} \tag{1.8}$$

For a given time-slicing, the generalization consists in having to sum over all paths ($x(t_i) \equiv x_i$) from x_i to $x_{i+1} + nL$, n = 0, 1, 2, ..., instead of x_i to x_{i+1}, i.e., over all winding numbers, if one imagines x = 0 and x = L to be identified. The striking fact is that the two classical Lagrangians in (1.7) and (1.8) which yield the same classical equation of motion result nevertheless in different quantum theories. This fact is intimately related to the theta vacua occurring in gauge theories (J1,C1,R2). We may therefore expect that an attempt to obtain a path integral description for general U might lead to a better understanding of theta vauca.

I.2 Physical effects related to unusual boundary conditions

If one considers space to be infinite and a 'box' to be a mere idealization can we dispense with the worry about boundary conditions? The answer is no. Boundary conditions may be needed even in infinite space. Here are some examples.

(i) The non-relativistic Schrödinger Eqn. with attractive $1/r^2$ potential needs an unusual boundary condition at $r = 0$ if (C2)

$$[\ell(\ell+1) + 2m \lim_{r\to 0} r^2 V(r)] < -1/4, \tag{1.9}$$

where ℓ = orbital angular momentum, m = mass of the particle.

(ii) A Dirac particle in a Coulomb potential described by

$$H = \vec{\alpha}.\vec{p} + \beta m - Ze^2/r, \tag{1.10}$$

and usual boundary conditions at r = 0, yields the ground state energy

$$E(1S) = m\sqrt{1 - (Ze^2/(\hbar c))^2} \tag{1.11}$$

Clearly this formula gives an imaginary energy for $Z > 137$ and must break down there. Case (C2) discussed the remedy in the form of unusual boundary conditions at $r = 0$ and others (e.g. Z1) in the form of replacing the Coulomb potential $-Ze^2/r$ by that due to a nucleus of finite size.

(iii) The fermion monopole system described by

$$H = \vec{\alpha}.(\vec{p} - e\vec{A}_{monopole}) + \beta m, \tag{1.12}$$

where $\vec{A}_{monopole}$ is the vector potential due to a monopole of strength g at $r = 0$ needs an unusual boundary condition at the origin for the lowest angular momentum $(|eg| - 1/2)$ state (K1,C3,G3). The nonexistence of a self-adjoint helicity operator and appearance of a fermion monopole bound state (G3), the phenomenon of the monopole becoming a dyon of fractional charge (W1,G4,Y1), and the possibility of monopoles catalyzing proton decay (R3,C4) are related to this boundary condition.

(iv) Even in infinite space, the finite angular range (e.g. $0 \leqslant \theta < \pi$, $0 \leqslant \phi < 2\pi$) may lead to unconventional boundary conditions. Such boundary conditions have recently been discussed in connection with fractional statistics and quantized Hall effect (W2,W3,T1,A2), topological supersymmetry breaking (A1), Aharonov-Bohm plus monopole potential (R4), Klein Gordon theta vacua (R5).

I.3 Boundary conditions and Nonuniqueness of infinite volume limit

Finite volume systems are considered either as representing the true physical situation or as mathematical artifice in regularizing field theory. A qualitative result arising in certain cases, e.g. in the discussion of fractional fermion number (J2,G5,R6,R7,S3) needs further investigation. In infinite volume the Hamiltonian is well defined without boundary conditions and leads to (J2) fractional fermion number of solitons. In finite volume one obtains a boundary condition dependent but volume independent fermion number (R7). Hence volume $\rightarrow \infty$ limit of the fermion number, being boundary condition dependent, is nonunique. The infinite volume result obtained directly agrees with the finite volume result with a particular boundary condition. What is the role of the other boundary conditions? Perhaps they model other infinite volume Hamiltonians.

II. ALLOWED BOUNDARY CONDITIONS

II.1 Physical Principles

The "Superposition principle" motivates the identification,

$$\text{observable} \equiv \text{self-adjoint operator.} \tag{2.1}$$

E.g., if A is an observable, an arbitrary state ψ is written as a superposition of the eigenstates ψ_n of A with eigenvalues a_n, i.e., (if the eigenvalues are discrete)

$$\psi = \sum_n C_n \psi_n, \; A\psi_n = a_n \psi_n, \tag{2.2}$$

and the probability of finding the value a_n in a measurement is identified with $|C_n|^2$. Since the spectrum of a self-adjoint operator can always be associated with a resolution of the identity, the expansion (2.2) generalized to include the case of continuum spectrum, and the associated physical interpretation are justified if we identify an observable with a self-adjoint operator. We are making an important distinction between hermitian and self-adjoint operators to be explained shortly.

Not every operator has to be self-adjoint. The Hamiltonian operator H which defines the time development,

$$\psi(t) = U(t)\,\psi(0), \; U(t) = \exp(-iHt), \tag{2.3}$$

must be: that will guarantee existence, continuity and unitarity of U(t) and hence probability conservation. A self-adjoint Hamiltonian specifies a unique, acceptable dynamics.

II.2 Self-adjointness, deficiency indices, Von Neumann's theorem on self-adjoint extensions

We begin with a review of Hilbert space concepts. This presentation is based mainly on references W4,A3,N1,R8. The definitions adopted are those of ref. N1.

A Hilbert space $\mathcal{H}$ is a complete inner product space. Completeness means that every Cauchy sequence of vectors (or points) $f_n \in H$ (i.e., vectors such that $||f_n - f_m|| \to 0$ for $m,n \to \infty$) converges to some $f \in H$ (i.e., $\exists\, f \in H$ s.t. $||f - f_n|| \to 0$). An f in H is called a limit point of a subset S of H, if $\forall\, \varepsilon > 0$ there exist an infinite number of $g \in S$ s.t. $||g - f|| < \varepsilon$. The set consisting of S and its limit points is called the closure of S and denoted $\bar{S}$. A set S is closed if $S = \bar{S}$. A set S in H is called dense in H if $\bar{S} = H$. If S is dense in H, it follows that for every $f \in H$ and $\forall\, \varepsilon > 0$ there is a $g \in S$ s.t. $||g - f|| < \varepsilon$. Further, S is dense in H iff for $f \in H$, $f \perp S \Rightarrow f = 0$, where $f \perp S$ means that f is orthogonal to all vectors in S. A Hilbert space H is called separable if it contains a countable dense set. A hilbert space H is separable if and only if there exists a complete orthonormal sequence of vectors ϕ_n s.t. any $f \in H$ can be expressed as

$$f = \sum_n \phi_n \, (\phi_n, f), \tag{2.4}$$

where (ϕ_n, f) denotes the inner product in H.

In quantum theory we assume the states to be represented by vectors in a separable Hilbert space. We therefore assemble here some results which apply to any abstract separable Hilbert space. It might be useful to keep in mind the following concrete examples occurring in quantum mechanics and field theory. The separable Hilbert space

$$L^2(a,b) = \{f(x) | \int_a^b |f(x)|^2 \, dx < \infty\} \tag{2.5}$$

consists of all complex-valued functions f(x) which are measurable and quadratically summable in the Lebesgue sense over a finite or infinite interval (a,b).

Another separable Hilbert space is a Fock space occurring in the theory of neutral mesons. It consists of vectors of the form

$$\Phi = \left\{\phi_o + \sum_{n=1}^{\infty} \int d^3k_1 \ldots . d^3k_n \frac{\phi_n(\vec{k}_1, \ldots \vec{k}_n)}{\sqrt{n!}} a^\dagger(\vec{k}_1) \ldots a^\dagger(\vec{k}_n)\right\} \Phi_o, \tag{2.6}$$

where Φ_o denotes the vacuum state, ϕ_o a complex number, ϕ_n a complex function symmetric under interchange of any two arguments $\vec{k}_i, \vec{k}_j$, and such that,

$$(\Phi, \Phi) < \infty, \tag{2.7}$$

where the scalar product of two vectors is

$$(\psi,\Phi) = \psi_o^\star \phi_o + \sum_{n=1}^{\infty} \int d^3k_1 \ldots d^3k_n \, \psi_n^\star(\vec{k}_1 \ldots \vec{k}_n) \phi_n(\vec{k}_1 \ldots \vec{k}_n) \tag{2.8}$$

1. Domain and Range

If an operator A is defined on a subset D(A) of H such that for $f \in D(A)$, $Af \in H$, and $R(A) = \{Af |\ f \in D(A)\}$, then the subsets D(A) and R(A) = AD(A) of H are called respectively the domain and range of A.

2. Direct sum

Given two Hilbert spaces $H_1 = \{f_1\}$, $H_2 = \{f_2\}$, their direct sum is $H_1 \oplus H_2 = \{f_1, f_2\}$, with addition of vectors, multiplication by scalars and scalar product of vectors in $H_1 \oplus H_2$ being defined by

$$\{f_1,f_2\} + \{g_1,g_2\} = \{f_1+g_1, f_2+g_2\}; \lambda\{f_1,f_2\} = \{\lambda f_1, \lambda f_2\} \tag{2.9}$$

$$(\{f_1,f_2\}, \{g_1,g_2\}) = (f_1,g_1) + (f_2,g_2)$$

3. Graph

For $f \in D(A)$, f and $Af \in H$. Hence $\{f, Af\} \in H \oplus H$. The set of these pairs is called the graph of the operator A and denoted G_A:

$$G_A = \{ \{f, Af\} | \ f \in D(A)\} \subset H \oplus H \tag{2.10}$$

4. Closed operator

An operator is called closed if its graph is closed, i.e., A is closed if for $f_n \in D(A)$

$$\{f_n, Af_n\} \to \{f,g\} \Longrightarrow f \in D(A) \text{ and } g = Af. \tag{2.11}$$

Now, given f, Af is unique. Hence, if there are two sequences $\{f_n, Af_n\}$, $\{f_n', Af_n'\}$ which tend to $\{f, g\}$ and $\{f', g'\}$ with $f' = f$, then A closed implies $g = g'$. Note that a closed operator need not be continuous: $f_n \to f$ need not imply that Af_n has a limit.

If A is not closed, is it closable by means of extension of D(A) with inclusion of limits of sequences f_n for which $Af_n \in H$? For this to be possible Eq. (2.11) must hold with A replaced by some operator $\bar{A}$. Hence, A is closable if

$$\{f_n, Af_n\} \to \{f,g\} \text{ and } \{f_n', Af_n'\} \to \{f,g'\} \Rightarrow g' = g. \tag{2.12}$$

5. Closure

If A is closable, its closure $\bar{A}$ is defined by $G_{\bar{A}} = \bar{G}_A$, i.e.,

$$D(\bar{A}) = \{f | f = \lim f_n, f_n \in D(A), \lim Af_n = g \in H\} \tag{2.13}$$

and

$$\bar{A}f \equiv g, \text{ i.e., } \bar{A} \lim f_n \equiv \lim Af_n \tag{2.14}$$

6. Extension

If $D(B) \supset D(A)$ and $Bf = Af \ \forall f \in D(A)$, then we say that B is an extension of A, and write $B \supset A$. Thus $\bar{A} \supset A$.

7. Hermitian operator

An operator A is called Hermitian if

$$(g, Af) = (Ag, f), \forall f, g \in D(A) \qquad (2.15)$$

8. Symmetric operator

A hermitian operator with dense domain is called a symmetric operator.

9. Adjoint

Let A be a linear operator with dense domain. Its adjoint $A^\dagger$ is defined to have the domain

$$D(A^\dagger) = \{g \in H | (g, Af) = (h,f) \ \forall f \in D(A), \text{ for some } h \in H\}, \qquad (2.16)$$

and for $g \in D(A^\dagger)$, $A^\dagger g$ is defined by

$$(g, Af) = (A^\dagger g, f), \forall f \in D(A). \qquad (2.17)$$

Eq.(2.17) defines $A^\dagger g$ uniquely, since D(A) dense, and $(h,f) = (h',f)$ $\forall f \in D(A)$ imply that $h = h'$. We see that the adjoint $A^\dagger$ may be so defined whenever D(A) is dense.

10. Isometric and unitary operators

An operator U is called isometric if $(Uf, Ug) = (f,g) \ \forall f,g \in D(U)$. An isometric U is called unitary if its domain D(U) and range R(U) coincide with the entire Hilbert space H.

11. $A^\dagger$ is closed

The domain $D(A^\dagger)$ consists of those $g \in H$ for which there exists $h \in H$ such that for all $f \in D(A)$

$$(g, Af) = (h,f), \text{ i.e., } (\{g,h\}, \{iAf, -if\}) = 0 \qquad (2.18)$$

In terms of an unitary operator U defined on vectors in $H + H$ by

$$U\{f_1,f_2\} = \{if_2, -if_1\}, \qquad (2.19)$$

the condition (2.18) reads

$$(\{g,h\}, U\{f, Af\}) = 0. \tag{2.20}$$

Thus, the graph of $A^\dagger$ consists of those $\{g,h\}$ in $H \oplus H$ which are orthogonal to $U\, G_A$, i.e. ,

$$G_{A^\dagger} = (H \oplus H) - U\, G_A. \tag{2.21}$$

Consider a sequence $\{g_n, h_n\} \in G_{A^\dagger}$ which has a limit in $H \oplus H$. From Schwarz inequality we deduce that

$$\{g_n, h_n\} \perp U\, G_A \Longrightarrow \lim \{g_n, h_n\} \perp U\, G_A. \tag{2.22}$$

Hence the graph $G_{A^\dagger}$ is closed, i.e., $A^\dagger$ is closed. This is a special case of the general result that every orthogonal complement is a closed subspace. Eq.(2.21) also implies the basic fact that

$$B \supset A \Rightarrow A^\dagger \supset B^\dagger, \tag{2.23}$$

i.e., when the domain of an operator is expanded the domain of the adjoint shrinks.

12. $\underline{A^{\dagger\dagger} = \bar{A}}$

If $D(A)$ and $D(A^\dagger)$ are dense, then $A^\dagger$ and $A^{\dagger\dagger}$ are defined, with

$$G_{A^\dagger} = (H \oplus H) - U\, G_A,\ G_{A^{\dagger\dagger}} = (H \oplus H) - U\, G_{A^\dagger}. \tag{2.24}$$

Now, using $U^2 = 1$ and $U\,(H + H) = (H \oplus H)$,

$$U\, G_{A^\dagger} = (H \oplus H) - G_A. \tag{2.25}$$

Hence

$$\bar{G}_A = (H \oplus H) - U\, G_{A^\dagger} = G_{A^{\dagger\dagger}}. \tag{2.26}$$

This means that the closure $\bar{A}$ exists and

$$\bar{A} = A^{\dagger\dagger}. \tag{2.27}$$

13. Self-adjoint

If $A^\dagger = A$, then A is called self-adjoint. This means that $D(A^\dagger) = D(A)$, and $A^\dagger f = Af$ for $f \varepsilon D(A)$.

14. Symmetric versus self-adjoint

An operator A with dense domain is symmetric if and only if $A \subset A^\dagger$. First if $A \subset A^\dagger$, then for all f,g in D(A)

$$(g, Af) = (Ag, f) \tag{2.28}$$

and hence A is symmetric. Conversely if A is symmetric, then for all f,g in D(A) (2.28) holds, and hence $g \varepsilon D(A^\dagger)$ with $A^\dagger g = Ag$, i.e., $A \subset A^\dagger$.

The condition $A \subset A^\dagger$ allows $A = A^\dagger$, but does not imply it. Hence a symmetric operator need not be self-adjoint, but a self-adjoint operator is necessarily symmetric.

15. Essentially self-adjoint

A is called essentially self-adjoint if its closure $\bar{A}$ is self-adjoint. Equivalently A is essentially self-adjoint if $A^\dagger$ is self-adjoint:

$$\bar{A} = (\bar{A})^\dagger \Leftrightarrow A^{\dagger\dagger} = A^{\dagger\dagger\dagger} = \overline{A^\dagger} = A^\dagger \tag{2.29}$$

An essentially self-adjoint A has the unique self-adjoint extension $\bar{A}$. A symmetric A may have one, many or no self-adjoint extension. We illustrate the distinction between symmetric and essentially self-adjoint operators by means of simple quantum mechanical examples.

16. Momentum operator and radial momentum operator

Consider the formal differential operator (formal means: domain unspecified)

$$\ell = -i \frac{d}{dx}, \; x \varepsilon (a,b). \tag{2.30}$$

Corresponding to ℓ the linear operator on $L^2(a,b)$ which has maximal domain is given by,

$$P\psi = \ell\psi \text{ for } \psi \in D(P),\ D(P) = \{\psi | \psi(x) \text{ absolutely continuous},$$

$$\psi \text{ and } P\psi \in L^2(a,b)\}. \tag{2.31}$$

For ϕ and $\psi \in D(P)$, we have

$$(\phi,\ell\psi) - (\ell\phi,\psi) = -i\phi^*(x)\psi(x)\Big|_a^b . \tag{2.32}$$

Consider first (b-a) finite. Since the right-hand side does not vanish for all ϕ,ψ in D(P), P is not symmetric. We may define a symmetric $P_o \subset P$ by setting $P_o\psi = \ell\psi$ for $\psi \in D(P_o)$ with

$$D(P_o) = \{\psi | \psi \in D(P),\ \psi(a) = \psi(b) = 0\}. \tag{2.33}$$

However, neither P_o nor P is self-adjoint. We see easily from (2.32) that

$$P_o^\dagger = P, \quad P^\dagger = P_o \tag{2.34}$$

We also see that for

$$\phi^*(b)\psi(b) = \phi^*(a)\,\psi(a) \tag{2.35}$$

the only boundary condition on ψ which implies the identical condition on ϕ (when $\psi \notin D(P_o)$) is

$$\psi(b) = \psi(a)\, e^{i\theta},\ \theta \text{ real}. \tag{2.36}$$

Hence the only self-adjoint extensions of P_o are $P^{(\theta)}$, with $P^{(\theta)}\psi = \ell\psi$ for $\psi \in D(P^{(\theta)})$, and

$$D(P^{(\theta)}) = \{\psi | \psi \in D(P),\ \psi(b) = \psi(a)e^{i\theta},\ \theta \text{ real}\} \tag{2.37}$$

For each real θ we then have translational invariance in the sense that

$$U(t) \equiv \exp(it\, P^{(\theta)}) = \text{Unitary} \tag{2.38}$$

Next consider the semi-infinite interval case which occurs in quantum mechanics in discussions of the radial momentum operator. Let

$$\ell = -i\frac{d}{dx},\ x \in (0,\infty) \qquad (2.39)$$

The new feature now is that for $\psi \in D(P)$ (the maximal domain) $\psi(\infty)$ must vanish. To see this, consider for $\psi \in D(P)$

$$\int_{x_1}^{x_2} dx\left(\psi^* \frac{d\psi}{dx} + \psi\frac{d\psi^*}{dx}\right) = |\psi(x)|^2\Big|_{x_1}^{x_2}. \qquad (2.40)$$

Since ψ and $d\psi/dx \in L^2(0,\infty)$, it follows from Schwarz inequality that the left-hand side $\to 0$ for x_1 and $x_2 \to \infty$ independently, and so must the right-hand side. Hence lim $|\psi(x)|^2$ exists for $x \to \infty$; further this limit must vanish since $\psi \in L^2$.

Hence, for ϕ and $\psi \in D(P)$,

$$(\phi,\ell\psi) - (\ell\phi,\psi) = i\phi^*(0)\psi(0). \qquad (2.41)$$

Thus P is not symmetric, but $P_0 \subset P$, with

$$D(P_0) = \{\psi |\ \psi \in D(P),\ \psi(0) = 0\} \qquad (2.42)$$

leads to a symmetric P_0. Again $P_0^\dagger = P$, and $P^\dagger = P_0$. <u>The new and striking thing is that now P_0 has no proper symmetric extension and hence no self-adjoint extension</u>. For if $P_1 \supset P_0$ and there is at least one $\psi \in D(P_1)$ such that $\psi \notin D(P_0)$, i.e., $\psi(0) \neq 0$, then

$$(\psi,P_1\psi) - (P_1\psi,\psi) = i|\psi(0)|^2 \neq 0. \qquad (2.43)$$

For quantum mechanics this means that the radial momentum operator P_0 is not an observable, nor can we define an extension of P_0 that would be an observable. Essentially this fact led to the recent conclusion (C3,G3,K1) that in the lowest angular momentum fermion-monopole system helicity is not an observable.

Finally the case

$$\ell = -i\frac{d}{dx},\ x \in (-\infty,\infty) \qquad (2.44)$$

is the simplest. We see that for $\psi \in$ the maximal domain $D(P)$, $\psi(\pm\infty) = 0$, and hence for ϕ and $\psi \in D(P)$,

$$(\phi,P\psi) - (P\phi,\psi) = 0, \qquad (2.45)$$

i.e., P is symmetric. It also follows easily that P is in fact self-adjoint.

17. Deficiency Indices

We begin by considering in more detail the example of the symmetric radial momentum operator P_o in the Hilbert space $L^2(0,\infty)$ associated with the formal operator $\ell = -id/dr$. Since $P_o^\dagger = P$, the maximal operator, the function

$$\psi(r) = \sqrt{2p}\, e^{-pr},\ p > 0 \qquad (2.46)$$

belongs to $D(P_o^\dagger)$. Further,

$$P_o^\dagger\psi = P\psi = \ell\psi = ip\psi, \qquad (2.47)$$

showing that ψ is an eigenvector of $P_o^\dagger$ with imaginary eigenvalue. This is possible because $P_o^\dagger$, unlike P_o, is not a symmetric operator. More generally,

$$P_o^\dagger\psi = \lambda\psi,\ \psi = \text{const}\ \exp(i\lambda r),\ \text{Im}\lambda > 0. \qquad (2.48)$$

Thus the adjoint $P_o^\dagger$ of the symmetric operator P_o has exactly one linearly independent eigenvector with eigenvalue λ if $\text{Im}\lambda > 0$. On the otherhand for an essentially self-adjoint operator A, $A^\dagger$ being self-adjoint can have no complex eigenvalues.

If A is symmetric but not essentially self-adjoint, its departure from essential self-adjointness is measured by two numbers (m,n) called its deficiency indices:

m = number of linearly independent solutions of $A^{\dagger}\psi = i\psi$.

n = " " " " " $A^{\dagger}\psi = -i\psi$.

Thus, for $L^2[0,\infty)$, P_o has deficiency indices (1,0). We already saw that P_o has no self-adjoint extension. This is a special case of the following more general theorem.

Theorem

A symmetric operator has a self-adjoint extension if and only if its deficiency indices are equal.

With the help of this theorem let us study the existence of self-adjoint extensions of a symmetric operator associated with the formal operator

$$\ell = -\frac{d^2}{dx^2}, \; x \in (a,b). \tag{2.49}$$

Define the maximal operator H associated with this formal operator:

$$D_H = \{\psi | \; \psi(x), \; \psi'(x) \text{ absolutely continuous, and } \psi, \; \psi'' \in L^2(a,b)\}, \tag{2.50}$$

and

$$H\psi = \ell\psi, \text{ for } \psi \in D_H \tag{2.51}$$

Then, for $\phi, \psi \in D_H$ we have

$$(\phi, H\psi) - (H\phi, \psi) = (\phi'^{*}\psi - \phi^{*}\psi')\Big|_a^b . \tag{2.52}$$

Clearly, there exist $\phi,\psi \in D_H$ for which the right-hand side does not vanish. Hence H is not symmetric. We define the symmetric H_o,

$$D_{Ho} = \{\psi | \; \psi \in D_H, \; \psi(a) = \psi(b) = \psi'(a) = \psi'(b) = 0\}, \tag{2.53}$$

and

$$Ho\psi = \ell\psi, \text{ for } \psi \in D_{Ho} \tag{2.54}$$

In eq. (2.53) the boundary condition at a boundary point at infinity is to be omitted when a or b becomes infinite. Now $H_o^{\dagger} = H$.

To calculate the deficiency indices of H_o note first the formal equations

$$-\frac{d^2\psi}{dx^2} = i\psi \text{ for } \psi = e^{\lambda x}, \lambda = \pm \frac{1-i}{\sqrt{2}}, \qquad (2.55)$$

$$-\frac{d^2\psi}{dx^2} = -i\psi, \text{ for } \psi = e^{\lambda x}, \lambda = \pm \frac{1+i}{\sqrt{2}}. \qquad (2.56)$$

Noting that when $b = \infty$, $\exp(\lambda x) \in D(H)$ requires $Re\lambda < 0$, and when $a = -\infty$, $\exp(\lambda x) \in D(H)$ requires $Re\lambda > 0$, we deduce the following table of deficiency indices of H_o.

Interval (a,b)	Deficiency indices of H_o
a and b finite	(2,2)
a finite, $b = \infty$	(1,1)
$a = -\infty$, b finite	(1,1)
$a = -\infty$, $b = +\infty$	(0,0)

In each case H_o has one or more self-adjoint extensions because the deficiency indices are euqal. The boundary conditions corresponding to self-adjoint extensions are in each case calculated easily using Eq.(2.52). The result for (b-a) finite has been stated before. For $b = \infty$ and 'a' finite we get a one parameter family of self-adjoint operators corresponding to the boundary condition $\psi'(a) = \lambda\psi(a)$ with λ real. A similar family obtains for $a = -\infty$, b finite. For $a = -\infty$, $b = \infty$, H is self-adjoint without any boundary conditions.

18. Deficiency spaces

Let λ be any complex number with $Im\lambda \neq 0$, and A any symmetric operator. Then

$$R_\lambda \equiv \text{Range of } A-\lambda = (A-\lambda)D_A \tag{2.57}$$

$$R_{\lambda*} \equiv (A-\lambda^*)D_A .$$

The deficiency spaces N_λ and $N_{\lambda*}$ are defined to be the orthogonal complements of the ranges of $A-\lambda$ and $A-\lambda^*$ respectively, i.e.,

$$N_\lambda = \mathcal{H}-R_\lambda,\ N_{\lambda*} = \mathcal{H}-R_{\lambda*}\ . \tag{2.58}$$

We shall prove that N_λ and $N_{\lambda*}$ are equal to the spaces of eigenvectors of $A^\dagger$ with eigenvalues λ^* and λ respectively, i.e.,

$$N_\lambda = \{f|\ A^\dagger f = \lambda^* f\},\ N_{\lambda*} = \{f|\ A^\dagger f = \lambda f\}. \tag{2.59}$$

Now N_λ consists of f such that

$$(f, (A-\lambda)g) = 0,\ \forall g \in D_A, \tag{2.60}$$

$$\text{i.e., } (f,Ag) = (\lambda^* f,g),\ \forall g \in D_A. \tag{2.61}$$

Hence N_λ consists of f which belong to $D_{A^\dagger}$, with $A^\dagger f = \lambda^* f$. $N_{\lambda*}$ similarly consists of $f \in D_{A^\dagger}$ with $A^\dagger f = \lambda f$.

We shall also prove that for a symmetric A, D_A, N_λ and $N_{\lambda*}$ are linearly independent subspaces of $\mathcal{H}$, i.e., if

$$f + g + h = 0,\ f \in D_A,\ g \in N_\lambda,\ h \in N_{\lambda*}, \tag{2.62}$$

then

$$f = g = h = 0. \tag{2.63}$$

Applying $(A^\dagger - \lambda^*)$ on Eqn. (2.62), we have, since $A^\dagger \supset A$,

$$(A-\lambda^*)f + (\lambda-\lambda^*)h = 0 \tag{2.64}$$

The first term belongs to $R_{\lambda*}$ and the second term to its orthogonal complement $N_{\lambda*}$. Hence

$$(A-\lambda^*)f = h = 0 \tag{2.65}$$

Since a symmetric A can have no eigenvector with nonreal eigenvalue, we have

$$f = h = 0,\ g = 0 \tag{2.66}$$

19. Cayley transform

The cayley transform $V(\lambda)$ of a symmetric operator A is the operator defined for a non-real λ by,

$$V(\lambda) = (A-\lambda)(A-\lambda^*)^{-1},\ \text{Im}\lambda \neq 0. \tag{2.67}$$

(Here $\lambda = \lambda\times$identity operator).

Its domain consists of vectors $g \in H$ such that $(A-\lambda^*)^{-1}\ g \in D_A$, i.e., $g \in (A-\lambda^*)D_A = R_{\lambda*}$. Its range therefore consists of vectors $(A-\lambda)(A-\lambda^*)^{-1}(A-\lambda^*)f$, where $f \in D_A$. Hence

$$\text{Domain } (V(\lambda)) = R_{\lambda*},\ \text{Range } (V(\lambda)) = R_\lambda \tag{2.68}$$

For $g = (A-\lambda^*)f$, $V(\lambda)g = (A-\lambda)f$, for $f \in D_A$.

The symmetry of A gives $(f_1,Af_2) = (Af_1,f_2)$ and hence, for $g_1,g_2 \in$ domain of $V(\lambda)$,

$$(g_1,g_2) = (Af_1,Af_2) + |\lambda|^2(f_1,f_2) - (\lambda + \lambda^*)(f_1,Af_2). \tag{2.69}$$

Since $V(\lambda)g$ is obtained from g by $\lambda \to \lambda^*$, under which Eq.(2.69) is invariant,

$$(g_1,g_2) = (V(\lambda)g_1,\ V(\lambda)g_2), \tag{2.70}$$

for all g_1,g_2 in $R_{\lambda*}$. Hence $V(\lambda)$ is an isometric operator.

20. Domain of adjoint of closed symmetric operators

For a closed symmetric A, the domain of the adjoint has the following precise characterization:

$$D_{A^\dagger} = D_A + N_\lambda + N_{\lambda*}\ ,\ \text{Im}\lambda \neq 0. \tag{2.71}$$

Since $A^\dagger \supset A$, $D_{A^\dagger} \supset D_A + N_\lambda + N_{\lambda*}$. It is necessary only to prove that every $u \in D_{A^\dagger}$ also belongs to $D_A + N_\lambda + N_{\lambda*}$. The main important step is to recognise that for a closed symmetric A, the isometry of the Cayley transform $V(\lambda)$ guarantees that $R_{\lambda*}$ is closed. For, consider

$$h_n = (A-\lambda^*)f_n \to h \in H \tag{2.72}$$

Since $V(\lambda)$ is isometric, $V(\lambda)\, h_n$ are a cauchy sequence, and must converge:

$$V(\lambda)h_n = (A-\lambda)f_n \to g \in H \tag{2.73}$$

Hence

$$\begin{aligned} f_n &\to \frac{1}{\lambda-\lambda^*}(h-g) \in H \\ Af_n &\to \frac{1}{\lambda-\lambda^*}(\lambda h-\lambda^* g) \in H \end{aligned} \tag{2.74}$$

Since A is closed,

$$(h-g) \in D_A, \text{ and } A(h-g) = \lambda h-\lambda^* g. \tag{2.75}$$

Hence,

$$h = (A-\lambda^*)(h-g)/(\lambda-\lambda^*) \in R_{\lambda*} \tag{2.76}$$

Hence $R_{\lambda*}$ (and similarly R_λ) are closed. Hence the relation $N_\lambda = H - R_\lambda$ takes the stronger form,

$$H = N_\lambda + R_\lambda \tag{2.77}$$

Now, consider $u \in D_{A^\dagger}$. Since $(A^\dagger-\lambda)u \in H$, there exist $g \in N_\lambda$ and $f \in D_A$ such that

$$(A^\dagger-\lambda)u = (\lambda^*-\lambda)g + (A-\lambda)f,\ g \in N_\lambda, f \in D_A. \tag{2.78}$$

Further $A^\dagger g = \lambda^* g$, and $A^\dagger f = Af$ yield,

$$(A^\dagger-\lambda)(u - g - f) = 0, \tag{2.79}$$

i.e., $u - g - f \in N_{\lambda*}$. Hence there exist f,g,h such that

$$u = f + g + h, \ f \in D_A, \ g \in N_\lambda, \ h \in N_{\lambda*}, \tag{2.80}$$

for every $u \in D_{A^\dagger}$. This proves (2.71).

21. Von Neumann's theorem on self-adjoint extensions of a closed symmetric operator

The deficiency indices of a symmetric operator A are (m,n), where

$$m = \dim N_{-i}, \ n = \dim N_i, \tag{2.81}$$

with $\dim N_\lambda$ denoting the number of linearly independent eigenvectors of $A^\dagger$ with eigenvalue λ^*. We state without proof the following special case of Von Neumann's theorem.

Let A be a closed symmetric operator in a separable Hilbert space , with finite and equal deficiency indices (n,n). Let

$$A^\dagger \phi_{\pm j} = \pm i \phi_{\pm j}, \ j = 1,2,\dots n, \tag{2.82}$$

where the $\phi_{\pm j}$ are orthonormal eigenvectors of $A^\dagger$. The self-adjoint extensions of A are then characterized by n x n unitary matrices U. For each such U, the operator A_U with domain $D(A_U)$ (subset of $D_{A^\dagger}$)

$$D(A_U) = \{\psi_U | \psi_U = \psi + \sum_{j=1}^{n} C_j (\phi_{+j} - \sum_{k=1}^{n} U_{jk} \ \phi_{-k}),$$

$$\psi \in D_A, \ C_j \text{ complex constants}\}, \tag{2.83}$$

and

$$A_U \ \psi_U = A^\dagger \psi_U, \tag{2.84}$$

is a self-adjoint extension of A. Since an n x n unitary U is characterized by n^2 parameters, there are an n^2 parameter family of self-adjoint extensions of A.

The next task is to connect this elegant theorem with boundary conditions in the cases of differential operators encountered in practice.

II.3. Differential operators

Consider $a < x < b$ and $H = L^2(a,b)$. The first step in defining self-adjoint differential operators in $L^2(a,b)$ is the construction of a formal differential operator ℓ which is formally self-adjoint, i.e., for which $(g,\ell f) - (\ell g,f)$ can be expressed in terms of boundary values of $f(x)$, $g(x)$ and their derivatives. We collect here some results from Ref.N1, R8.

1. Self-adjoint quasidifferential expressions

A self-adjoint quasi-differential expression of order 2n is

$$\ell f = (-1)^n(p_o f^{(n)})^{(n)} + (-1)^{n-1}(p_1 f^{(n-1)})^{(n-1)} + \ldots + p_n f,$$
$$= \sum_{k=0}^{n} (-1)^k(p_k\ f^{(n-k)})^{(n-k)}, \tag{2.85}$$

where $f^{(k)}$ denotes $d^k f/dx^k$, the $p_k(x)$ are real functions such that

$$\frac{1}{p_o(x)},\ p_1(x), p_2(x), \ldots, p_n(x) \tag{2.86}$$

are Lebesgue measurable in (a,b) and are Lebesgue summable in any closed subinterval $[\alpha,\beta]$ of (a,b):

$$\int_\alpha^\beta dx/|p_o(x)| < \infty,\ \int_\alpha^\beta dx\ |p_i(x) < \infty \tag{2.87}$$

It is not necessary in a quasi-differentialexpression that the coefficient functions $p_n(x)$ have derivatives. If, however $p_{n-r}(x)$ has derivatives upto the rth order inclusive then we may rewrite (2.85) as

$$\ell f = \sum_{r=0}^{2n} q_r\ f^{(2n-r)}, \tag{2.88}$$

and the quasi-differential expression becomes a differential expression.

To justify the name 'self-adjoint quasi-differential expression' we use Lagrange's identity:

$$g^*\ell f - (\ell g)^* f = \frac{d}{dx}[f,g], \tag{2.89}$$

where

$$[f,g] = \sum_{k=1}^{n} (f^{[k-1]} g^{*[2n-k]} - g^{*[k-1]} f^{[2n-k]}). \tag{2.90}$$

Here $f^{[k]}$ are 'quasi-derivatives' of f defined by $f^{[0]} = f$; $f^{[k]} = d^k f/dx^k$ for $k = 1,2,...(n-1)$; $f^{[n]} = p_o d^n f/dx^n$;

$$f^{[n+k]} = p_k\, d^{n-k} f/dx^{n-k} - df^{[n+k-1]}/dx \text{ for } k = 1,2,...n. \tag{2.91}$$

Then $\ell f = f^{[2n]}$, and

$$g^*\ell f = g^* f^{[2n]} = g^* p_n f + g^{*[1]} f^{[2n-1]} - \frac{d}{dx}(g^* f^{[2n-1]}).$$

Similarly,

$$g^{*[k]} f^{[2n-k]} = g^{*[K]} p_{n-k} f^{[k]} + g^{*[k+1]} f^{[2n-k-1]} - \frac{d}{dx}[g^{*[k]} f^{2n-k-1}]$$

for $k = 1,...n-2$,

$$g^{*[n-1]} f^{[n+1]} = g^{*[n-1]} p_1 f^{[n-1]} + \frac{1}{p_o} g^{*[n]} f^{[n]} - \frac{d}{dx}[g^{*[n-1]} f^{[n]}].$$

Adding the above n equations and subtracting a similar result for $(\ell g)^* f$ we obtain the Lagrange identity (2.89). Integration gives

$$\int_\alpha^\beta (g^* \ell f - (\ell g)^* f)dx = [f,g]\Big|_\alpha^\beta, \tag{2.92}$$

justifying the name self-adjoint quasi-differential expression for ℓf.

2. Regular and Singular Expressions and End points

The expression ℓf is called regular if (b-a) is finite and the functions $1/p_o, p_1, ...p_n$ are summable in the whole interval (a,b). Otherwise the expression is called singular.

The end point a is called singular if $a = -\infty$, or if $a > -\infty$ but the integrability conditions (2.87) fail for $\alpha = a$. Otherwise a is called a regular end point. Similarly b is called a singular end point if $b = \infty$, or if $b < \infty$ but the integrability conditions (2.87) fail for $\beta = b$. Otherwise b is called a regular end point.

3. Solutions of quasi-differential equations

Consider the equation

$$(\ell - \lambda)f = g \tag{2.93}$$

where λ is a complex constant and g(x) a given function. For $\ell f = f^{[2n]}$ to be defined, all the quasi-derivatives $f^{[i]}$ for i = 1,2,... (2n-1) must be absolutely continuous. A function f(x) with absolutely continuous quasi-derivatives upto order (2n-1) in every finite sub-interval $[\alpha,\beta]$ of (a,b) is called a solution of the quasi-differential eqn. (2.93) if it satisfies the equation almost everywhere in (a,b). On existence of solutions the following theorem is useful.

Theorem: If g(x) is measurable in (a,b) and summable in every finite subinterval $[\alpha,\beta]$ of (a,b), then for arbitrary complex constants C_0, $C_1 \ldots C_{2n-1}$ and for every internal $x_0 \in (a,b)$, the eqn. $(\ell-\lambda)f = g$ has in (a,b) a unique solution satisfying the 2n boundary conditions $f^{[k]}(x_0) = C_k$, k = 0,1,...(2n-1). Further, the theorem remains true if x_0 is a regular end point of (a,b) provided g(x) is also summable in every finite sub-interval $[\alpha,\beta]$ with x_0 as end point.

For explicit solutions it is useful to start from a set of 2n linearly independent solutions $f_1, f_2, \ldots f_{2n}$ of the homogenous eqn. $(\ell-\lambda)f = 0$. Such a system of solutions is called a fundamental system. In obtaining a fundamental system the following theorem is useful.

Theorem: If the solutions $f_1, f_2, \ldots, f_{2n}$ of the homogenous eqn. $(\ell-\lambda)f = 0$ are linearly dependent, then the Wronskian

$$W(f_1, f_2, \ldots f_{2n}) = \begin{vmatrix} f_1 & f_2 & . & . & f_{2n} \\ f_1^{[1]} & f_2^{[1]} & . & . & f_{2n}^{[1]} \\ . & . & . & . & . \\ . & . & . & . & . \\ f_1^{[2n-1]} & f_2^{[2n-1]} & . & . & f_{2n}^{[2n-1]} \end{vmatrix} \qquad (2.94)$$

vanishes identically in the interval (a,b). Conversely, if W vanishes at at least one point in (a,b) then the solutions $f_1, \ldots, f_{2n}$ are linearly dependent.

A sufficient condition for linear independence is given by the following theorem.

<u>Theorem</u>: A system of solutions $f_1, f_2, \ldots f_{2n}$ of the homogenous eqn. $(\ell-\lambda)f = 0$ obeying the boundary conditions

$$f_j^{[k-1]}(x_o) = C_{jk}, \text{ for } j,k = 1,2,\ldots 2n,$$

where x_o is an internal point or a regular end point of (a,b) and C_{jk} are complex constants with det $[C_{jk}] \neq 0$, is a fundamental system.

We now have the useful result:

<u>Theorem</u>: Every solution of the homogenous eqn. $(\ell-\lambda)f = 0$ is a linear combination of a fixed arbitrarily chosen fundamental system. Let $f_1, f_2, \ldots f_{2n}$ be a fundamental system. The general solution of the inhomogenous eqn. $(\ell-\lambda)f = g$ where g is measurable in(a,b) and summable in every finite subinterval $[\alpha, \beta]$ has the form

$$f(x) = \sum_{k=1}^{2n} f_k(x)[a_k + \int_{x_o}^{x} U_k(\xi)\, g(\xi) d\xi], \qquad (2.95)$$

where the a_k are arbitrary complex constants, x_o an internal point and

$$U_k(x) = (-1)^{2n+k} \frac{W(f_1,\ldots,f_{k-1},f_{k+1},\ldots f_{2n})}{W(f_1,\ldots,f_{2n})}, \tag{2.96}$$

$$\text{for } k = 1,\ldots,2n.$$

4. Differential operators in $L^2(a,b)$

Let us define L to be the linear operator associated with the self-adjoint differential expression ℓf with the largest possible domain. Then,

$$D_L = \{f|\ f \in L^2(a,b), f^{[k]}(x) \text{ absolutely continuous in } (a,b) \text{ for } k = 0,1,\ldots,2n-1,\ \ell f \in L^2(a,b)\}, \tag{2.97}$$

and

$$Lf = \ell f \text{ for } f \in D_L. \tag{2.98}$$

We see first that the maximal operator D_L is not hermitian. If f and g $\in D_L$, and $[\alpha,\beta] \subset (a,b)$, Lagrange's identity gives

$$\int_\alpha^\beta dx(g^* \ell f - (\ell g)^* f) = [f,g]\Big|_\alpha^\beta . \tag{2.99}$$

The left-hand side has a limt for $\beta \to b$ or $\alpha \to a$. Hence $[f,g]_a$ and $[f,g]_b$ exist, and we have for $f,g \in D_L$

$$(g,\ell f) - (\ell g,f) = [f,g]\Big|_a^b . \tag{2.100}$$

Since the right-hand side need not vanish for all $f,g \in D_L$, L is not hermitian. Consider a restriction L'_o of the operator L defined by

$$D_{L'_o} = \{f|f \in D_L,\ f(x) = 0 \text{ for x outside finite interval } [\alpha,\beta] \subset (a,b)\}. \tag{2.101}$$

Here the finite interval $[\alpha,\beta]$ may be different for different functions. Continuity of $f^{[k]}(x)$ now imply that for $f \in D_{L'_o}$, $g \in D_L$, the right-hand side of (2.100) vanishes and

$$(g,L_o'f) - (Lg,f) = 0 \tag{2.102}$$

i.e., $g \in D_{L_o}'^{\dagger}$ and $L_o'^{\dagger}g = Lg$, i.e.,

$$L \subset L_o'^{\dagger} \tag{2.103}$$

Since Eq.(2.102) holds in particular for all f,g in D_{L_o}', the operator L_o' is Hermitian. It may also be shown that D_{L_o}' is dense, and hence L_o' is symmetric. Its closure therefore exists, and we define

$$L_o = \overline{L_o'} \tag{2.104}$$

Since $L_o^{\dagger} = L_o'^{\dagger}$, we have

$$L \subset L_o'^{\dagger} = L_o^{\dagger} \tag{2.105}$$

In fact, it may be proved that

$$L = L_o'^{\dagger} = L_o^{\dagger} \tag{2.106}$$

showing that L is a closed operator. Since L_o is a closed symmetric operator, we deduce that

$$L^{\dagger} = L_o^{\dagger\dagger} = \bar{L}_o = L_o, \tag{2.107}$$

and hence L and L_o are adjoints of each other.

What are the deficiency indices of the closed symmetric operator L_o? Consider the eqn.

$$L_o^{\dagger}f = \lambda f, \text{ i.e., } Lf = \lambda f, \tag{2.108}$$

where $\text{Im}\lambda > 0$. Then f^* also belongs to D_L and further, due to the reality of $p_o, p_1, .., p_n$ we have

$$Lf^* = \lambda^* f^*. \tag{2.109}$$

Hence the deficiency indices of L_o have the form (m,m). Since $Lf = \lambda f$ has at most 2n linearly independent solutions, we have the following result.

<u>Theorem</u>: The operator L_o is closed, symmetric with deficiency indices (m,m), where $0 \lesssim m \lesssim 2n$, and obeys $L_o^\dagger = L$, $L^\dagger = L_o$.

The domain of L_o may now be given more explicity using $L_o = L^\dagger$ and $L_o \subset L$. The domain $D_{L^\dagger}$ consists of $g \in D_L$ such that for all f in D_L

$$(g,Lf) = (L^\dagger g,f), \text{ or } (g,\ell f) = (\ell g,f) \qquad (2.110)$$

Hence, using (2.100)

$$D_{L_o} = \{g|g \in D_L,\ [f,g]\Big|_a^b = 0\ \forall f \in D_L\}. \qquad (2.111)$$

We may also derive (i) if both a and b are regular points, then

$$D_{L_o} = \{g|g \in D_L,\ g^{[k]}(a) = g^{[k]}(b) = 0,\ k = 0,1,..,(2n-1)\}, \qquad (2.112)$$

and L_o has deficiency indices (2n,2n);

(ii) if one end point (say a) is regular and the other end point (say b) is singular then

$$D_{L_o} = \{g|g \in D_L,\ g^{[k]}(a) = 0,\ k = 0,1,..,(2n-1);\ [f,g]_b = 0 \text{ for all } f \in D_L\}, \qquad (2.113)$$

and L_o has deficiency indices (m,m) where $n \lesssim m \lesssim 2n$;

(iii) if $c \in (a,b)$ then L_{o-} and L_{o+} the operators L_o associated with the same differential expression on (a,c) and (c,b) respectively have each at most one singular end point, and the deficiency index of L_o is given by

$$\text{def } L_o = \text{def } L_{o-} + \text{def } L_{o+} - 2n. \qquad (2.114)$$

The results (ii) and (iii) together yield the deficiency indices of an operator L_0 with two singular end points.

5. Weyl's alternative, singular points of limit point type and limit circle type for second order differential operators

Because of their wide physical applications second order differential operators need to be considered in detail. The result (ii) that in the case of a second order operator with one singular end point, L_0 has deficiency indices (1,1) or (2,2) is known as Weyl's alternative. When the deficiency indices are (1,1) the singular end point is said to be of the limit point type, and when the deficiency indices are (2,2) the singular end point is said to be of the limit circle type . In defining self-adjoint extensions of L_0 no boundary conditions are needed at a singular end point of limit point type because the requirement of square integrability takes the place of a boundary condition. At a singular point of limit circle type a boundary condition is needed and has to be supplied from physical considerations.

In the case of an L_0 defined on (a,b) with both end points singular, the singular end point a is classified as limit point type or limit circle type by considering the operator L_{0-}, and the end point b is analogously classified by considering L_{0+}. In fact, the radial equations occurring in quantum mechanics usually have singular end points $r = 0$ and $r = \infty$ with $r = \infty$ of the limit-point type and $r = 0$ sometimes of the limit point type, sometimes of the limit circle type.

We begin by proving Weyl's alternative for an L_0 defined on (a,b) with 'a' a regular end point and 'b' a singular end point, i.e., by proving that L_0 cannot have deficiency index (0,0). We show that there exists at least one $f(x) \in D_L$ such that

$$Lf = \lambda f, \; \mathrm{Im}\lambda > 0 \tag{2.115}$$

We know that $\ell f = \lambda f$ has solutions $f_1(x)$, $f_2(x)$ subject to the boundary conditions

$$\begin{aligned} &f_1(a) = 1, \; (p_0 f_1')(a) = 0, \\ &f_2(a) = 0, \; (p_0 f_2')(a) = 1, \end{aligned} \tag{2.116}$$

with Wronskian W(a) = 1. Since

$$\frac{dW}{dx} = f_1(p_o f_2')' - f_2(p_o f_1')'$$

$$= f_1(p_1-\lambda)f_2 - f_2(p_1-\lambda)f_1 = 0, \tag{2.117}$$

W(x) = 1 for all x. Apart from an overall multiplicative constant, the general solution of $\ell f = \lambda f$ is

$$f(x) = f_1(x) + \mu f_2(x), \tag{2.118}$$

where μ = constant. Multiplying the eqn. $\ell f = \lambda f$ by f^* and integrating from a to β, where $\beta < b$, an integration by parts gives

$$-f^* p_o f' \Big|_a^\beta + \int_a^\beta dx(p_o|f'|^2 + (p_1-\lambda)|f|^2) = 0. \tag{2.119}$$

Using the boundary conditions at a on f_1 and f_2, and taking imaginary parts in the above eqn, we obtain

$$F(\mu,\beta) = -\text{Im}\mu + \text{Im}\lambda \int_a^\beta dx|f|^2, \tag{2.120}$$

where

$$F(\mu,\beta) \equiv -\text{Im}(f^* p_o f')\Big|_\beta$$

$$= A|\mu|^2 + B\,\text{Re}\,\mu + C\,\text{Im}\,\mu + D. \tag{2.121}$$

Here,

$$A = -\text{Im}(f_2^* p_o f_2')\Big|_\beta = \text{Im}\lambda \int_a^\beta dx|f_2|^2, \tag{2.122}$$

the last equality following from (2.119) with $f = f_2$. Further,

$$B = -\text{Im}(f_2^* p_o f_1' + f_1^* p_o f_2')\Big|_\beta,$$

$$C = -\text{Re}(f_1^* p_o f_2' - f_2^* p_o f_1')\Big|_\beta,$$

$$D = -\text{Im}(f_1^* p_o f_1')\Big|_\beta \tag{2.123}$$

For a given β, the locus of $F(\mu,\beta) = 0$ is a circle in the complex μ-plane with centre $(-B/(2A), -C/(2A))$ and radius $(B^2 + C^2 - 4AD)^{1/2}/(2A)$. A little calculation gives

$$B^2 + C^2 - 4AD = |f_2 p_o f_1' - f_1 p_o f_2'|^2\Big|_\beta = W^2(\beta) = 1 \tag{2.124}$$

Thus the circle has radius $1/(2A)$ which decreases monotonically as β increases towards b. Inside the circle $F(\mu,\beta) < 0$, and outside it $F(\mu,\beta) > 0$. Further (2.120) shows that for a given μ, $F(\mu,\beta)$ increases with β. Hence the disc $F(\mu,\beta) \leqslant 0$ contracts as β increases. If the circle $F(\mu,\beta) = 0$ contracts to a point μ_b when $\beta \to b$, i.e., $A \to \infty$ for $\beta \to b$, then the end point b is said to be of the limit point type; if the circle tends to a limit circle when $\beta \to b$, i.e., $A \to$ finite value for $\beta \to b$, then the end point b is said to be of the limit circle type:

$$\lim_{\beta \to b} \int_a^\beta dx|f_2|^2 < \infty; \text{ 'b' limit circle type} \tag{2.125}$$

$$\int_a^\beta dx|f_2|^2 \to \infty \text{ for } \beta \to \infty; \text{ 'b' limit point type.} \tag{2.126}$$

In the limit point case there is a unique μ_b with $F(\mu_b,b) \leqslant 0$, i.e.,

$$\int_a^b dx|f|^2 \leqslant (\mathrm{Im}\,\mu_b)/(\mathrm{Im}\lambda) < \infty, \tag{2.127}$$

i.e., $f = f_1 + \mu_b f_2 \in D_L$. Since $f_2 \notin L^2$, for $\mu \neq \mu_b$, $f_1 + \mu f_2 \notin L^2$. Thus L_o has deficiency index (1,1) in the limit point case. In the limit circle case all values of μ in the limiting disc have $F(\mu,b) \leqslant 0$, and hence for $f = f_1 + \mu f_2$,

$$\int_a^b dx|f|^2 \leqslant \mathrm{Im}\mu/\mathrm{Im}\lambda. \tag{2.127}$$

Since such μ is nonunique, L_o has deficiency index (2,2) in the limit circle case. Eqns. (2.125), (2.126), viz. square integrability of

f_2 may be used as a criterion to determine whether an end point is of the limit point or limit circle type.

6. Boundary conditions for self-adjointness

Since L_o has deficiency indices (m,m), Von Neumann's theorem shows that all self-adjoint extensions of it L_U are characterized by m x m unitary matrices U with the domain of L_U being given by (2.83). In order to specify the domain in terms of boundary conditions, note that $L_o \subset L_U \subset L = L_o^\dagger$. Hence (2.100) gives

$$[f,g]\Big|_a^b = 0 \text{ for } f \in D_{L_o},\ g \in D_L, \tag{2.128}$$

Further $D(L_U^\dagger)$ (ofcourse $L_U^\dagger = L_U$) consists of those $h \in D_L$ for which

$$[f_u,h]\Big|_a^b = 0, \text{ for all } f_U \in D(L_U). \tag{2.129}$$

Now substitute (2.83)(for deficiency indices (m,m))

$$f_U = f + \sum_{j=1}^{m} C_j(\phi_{+j} - \sum_{k=1}^{m} U_{jk}\phi_{-k}), \tag{2.130}$$

where $L\,\phi_{\pm j} = \pm i\,\phi_{\pm j}$, the $\phi_{\pm j}$ are orthonormal and C_j are constants.

Due to (2.128) $[f,h]\Big|_a^b = 0$, and the condition (2.129) for all $\{C_j\}$ means that $D(L_U)$ consists of those $h \in D_L$ which obey the boundary conditions

$$[\phi_{+j} - \sum_{k=1}^{m} U_{jk}\,\phi_{-k},h]\Big|_a^b = 0,\ j = 1,..,m\ . \tag{2.131}$$

Since the $p_i(x)$ are real, we may choose $\phi_{-k} = \phi_{+k}^*$ x const., where the const. is of modulus unity.

These boundary conditions may simplify in special cases. E.g., for a second order differential operator on (a,b) with 'a' a regular end point, b a singular end point of the limit point type, there is a one parameter family of self-adjoint extensions L_α characterized by the boundary condition

$$h(a)\cos\alpha + p(a)h'(a)\sin\alpha = 0, \qquad (2.132)$$

where α is any real parameter. Explicit results for many differential operators of quantum mechanics are given in R8,A3.

II.4. Dirac Operators.

Dirac Hamiltonians have wide physical applications. We give explicitly the boundary conditions for self-adjointness of some Dirac operators (R7).

1. 2x2 Dirac operator. Consider the formal Dirac differential operator

$$H_{diff} = -i\sigma_2 \frac{d}{dx} + \phi(x), \quad x \in (a,b), \qquad (2.133)$$

where $\phi(x)$ is a bounded hermitian 2x2 matrix function and σ_2 a Pauli matrix. In the Jackiw-Rebbi (J2) discussion of fractional fermion number in 1+1 dimension, $\phi(x) = g\sigma_1\phi_1(x)$, and in the Goldstone-Wilczek discussion (G5) $\phi(x) = g(\sigma_1\phi_1(x) + \sigma_3\phi_3(x))$. We find in the case (b-a) finite that if $\Psi_1(x)$, $\Psi_2(x)$ belong to the maximal domain D(H) which can be associated with (2.133), then

$$(\Psi_2, H_{diff}\Psi_1) - (H_{diff}\Psi_2, \Psi_1) = -i\Psi_2^\dagger(x)\sigma_2\Psi_1(x)\Big|_a^b . \qquad (2.134)$$

Here the Hilbert space H is defined to have the scalar product

$$(\chi, \Psi) = \int_a^b dx\, \chi^\dagger(x)\Psi(x). \qquad (2.135)$$

It is convenient to write, for i=1,2,

$$\Psi_i(x) = \Psi_{i+}(x)\chi_+ + \Psi_{i-}(x)\chi_-, \qquad (2.136)$$

where

$$\chi_\pm \equiv \frac{1}{\sqrt{2}}\begin{pmatrix} 1 \\ \pm i \end{pmatrix}, \quad \sigma_2\chi_\pm = \pm\chi_\pm. \qquad (2.137)$$

Then,

$$(\Psi_2, H_{diff}\Psi_1) - (H_{diff}\Psi_2, \Psi_1) = -i(A_2^\dagger A_1 - B_2^\dagger B_1), \tag{2.138}$$

where

$$A_i = \begin{pmatrix} \Psi_{i+}(b) \\ \Psi_{i-}(a) \end{pmatrix}, \quad B_i = \begin{pmatrix} \Psi_{i-}(b) \\ \Psi_{i+}(a) \end{pmatrix} \tag{2.139}$$

For self-adjointness we seek a boundary condition on A_1, B_1 such that the vanishing of the right-hand side of (2.137) implies an identical boundary condition on A_2, B_2. We find that

$$A_1 = U\,B_1 \Rightarrow A_2 = (U^\dagger)^{-1}B_2 \tag{2.140}$$

Hence the general self-adjoint operator H_U associated with the formal operator H_{diff} has the domain characterized by a unitary matrix U,

$$D(H_U) = \{\Psi(x)|\ \Psi \text{ absolutely continuous}, \\ \Psi_c \text{ and } H_{diff}\Psi \in H,\ A_\Psi = UB_\Psi\}, \tag{2.141}$$

where

$$A_\Psi = \begin{pmatrix} \Psi_+(b) \\ \Psi_-(a) \end{pmatrix}, \quad B_\Psi = \begin{pmatrix} \Psi_-(b) \\ \Psi_+(a) \end{pmatrix}. \tag{2.142}$$

When one of the boundary points, say $b \to \infty$, the unitary matrix U becomes a single complex number of modulus unity and the boundary condition for self-adjointness becomes $\Psi_-(a) = U\Psi_+(a)$.

2. <u>NxN Dirac operator.</u> Consider on (a,b) the formal NxN Dirac differential operator

$$H_{diff} = \alpha p + \beta(x)\ ,\ p = -i\,\frac{d}{dx}, \tag{2.143}$$

where α is a constant NxN hermitian matrix and $\beta(x)$ an NxN hermitian matrix function of x which is bounded on (a,b). With the scalar product defined as in (2.135) and (b-a) finite, let H be the maximal operator associated with H_{diff}. For $\Psi_1, \Psi_2 \in D(H)$,

$$(\Psi_2, H_{diff}\Psi_1) - (H_{diff}\Psi_2, \Psi_1) = -i\Psi_2^\dagger(x)\alpha\Psi_1(x)\Big|_a^b . \quad (2.144)$$

Let χ_i, $i = 1,\ldots N$, denote orthonormal eigenvectors of α with eigenvalues λ_i. Each χ_i is an N component spinor eigenfunction of α, i.e.,

$$\alpha\chi_i = \lambda_i\chi_i, \text{ (no sum)}, i=1,\ldots N. \quad (2.145)$$

Let $\lambda_i \geq 0$ for $i=1,\ldots M$, and $\lambda_i < 0$ for $i=M+1,\ldots N$. Denoting

$$\Psi(x) = \sum_{i=1}^{N} U_i(x)\chi_i, \quad (2.146)$$

and proceeding as in the 2x2 case we find that the general self-adjoint operator H_U associated with the formal Dirac Operator has a domain $D(H_U)$ consisting of absolutely continuous $\Psi(x)$ such that Ψ, $H_{diff}\Psi \in H$, and Ψ obeys the boundary condition

$$\begin{pmatrix} \sqrt{\lambda_1}\; U_1(b) \\ \sqrt{\lambda_M}\; U_M(b) \\ \sqrt{|\lambda_{M+1}|}\; U_{M+1}(a) \\ \sqrt{|\lambda_N|}\; U_N(a) \end{pmatrix} = U \begin{pmatrix} \sqrt{|\lambda_{M+1}|}\; U_{M+1}(b) \\ \sqrt{|\lambda_N|}\; U_N(b) \\ \sqrt{\lambda_1}\; U_1(a) \\ \sqrt{\lambda_M}\; U_M(a) \end{pmatrix}, \quad (2.147)$$

where U is any NxN unitary matrix.

III. Fractional Fermion Number.

Jackiw and Rebbi (J2) discovered that when a Dirac fermion is coupled to a soliton, the soliton has half-integral fermion number. Further, Witten (W1), Grossman (G4) and Yamagishi (Y1) have established that for fermion-monopole interactions in the presence of CP-violation the monopole becomes a dyon of fractional charge. The occurrence of fractional vacuum charge is now confirmed in several field theoretical models, and condensed matter systems (G5,J3,S4,S5,R9,T2,H1,B1,I1,G6). The dependence on boundary conditions in finite volume is explored in

(K2,R6,R7,S3,R10,G7).

III.1. Fermion monopole system in infinite volume.

Witten (W1) showed that in the presence of a CP-violating angle Θ_o the monopole acquires a charge $-e\Theta_o/(2\pi)$ where e is the fermion charge. In an independent line of research (K1,C3,G3) it was discovered that the fermion-Dirac monopole Hamiltonian in the lowest angular momentum state is self-adjoint only when a boundary condition at the origin (monopole position) involving the CP-vidating parameter θ_o is imposed. Two major consequences are the "inevitable failure of helicity conservation" (G3) (related to the Rubakov-Callan effect in the nonabelian case (R3,C4)) and the confirmation by Grossman (G4) and Yamagishi (Y1) of the Witten effect with a precise connection to the r=0 boundary condition. The monopole becomes a helicity leaking dyon of fractional charge, the fraction being irrational in general but half-integer or integer when CP is conserved.

We use the Wu-Yang (W5) vector potentials $\vec{A} = \vec{A}_a(\vec{r})$ for $\vec{r} \varepsilon R_a (\theta \neq \pi)$, and $\vec{A}_b(\vec{r})$ for $\vec{r} \varepsilon R_b$ $(\theta \neq o)$ where

$$\vec{A}_a(\vec{r}) = \vec{A}_b(-\vec{r}) = g\, \hat{z} \times \hat{r}/(r(1+\hat{r}.\hat{z})), \tag{3.1}$$

with g = monopole strength. The Dirac wave section $\Psi(=\Psi_a$ in R_a, Ψ_b in $R_b)$ obeys

$$H\Psi(\vec{r},t) = i\partial\Psi(\vec{r},t)/\partial t, \quad H = \vec{\alpha}.\vec{\pi}+\beta M, \tag{3.2}$$

where

$$\vec{\pi} = -i\vec{\nabla} - e\vec{A}, \quad \vec{\alpha} = \begin{pmatrix} 0 & \vec{\sigma} \\ \vec{\sigma} & 0 \end{pmatrix}, \quad \beta = \begin{pmatrix} 1 & 0 \\ 0 & -1 \end{pmatrix}, \tag{3.3}$$

the $\vec{\sigma}$ are Pauli matrices and e,M denote the charge and mass of the fermion. The angular momentum operator commuting with this Hamiltonian is

$$\vec{J} = \vec{r} \times \vec{\pi} - q\hat{r} + \frac{1}{2}\vec{\sigma}, \quad q \equiv eg, \tag{3.4}$$

$\vec{J}^2$ has eigenvalues $j(j+1)$, with

$$j = |q| - \frac{1}{2}, \ |q| + \frac{1}{2}, \ \ldots . \tag{3.5}$$

Further, the operator K introduced by P. Rossi (similar to Dirac's K operator for the Coulomb problem)

$$K = \beta(\vec{\Sigma}.(\vec{r}\times\vec{\pi}) + 1), \ [K,\vec{J}] = [K,H] = 0, \tag{3.6}$$

where $\vec{\Sigma} = \mathrm{diag}(\vec{\sigma},\vec{\sigma})$, allows us to look for a simultaneous eigenfunction of $H,\vec{J}^2,J_z,K$ in the form

$$\Psi = \frac{1}{r}\begin{pmatrix} \phi_1(\theta,\phi)\,F(r) \\ \phi_2(\theta,\phi)\,G(r) \end{pmatrix}. \tag{3.7}$$

The eqn. $(E-H)\Psi = 0$ then yields the radial eqn.

$$(E-H_r)\left(\frac{\chi(r)}{r}\right) = 0, \quad \chi(r) = \begin{pmatrix} F(r) \\ G(r) \end{pmatrix}, \tag{3.8}$$

with

$$H_r = \begin{pmatrix} M & \eta\left(p_r - i\,\frac{\lambda_K}{r}\right) \\ \eta\left(p_r + i\,\frac{\lambda_K}{r}\right) & -M \end{pmatrix}, \ p_r = -\frac{i}{r}\frac{d}{dr}\,r, \tag{3.9}$$

$\eta = -1$ for $j > |q| - \frac{1}{2}$ and $\eta = \mathrm{sgn}\,q$ for $j = |q| - \frac{1}{2}$. Here λ_K denotes the eigenvalue of K,

$$\lambda_K = \pm[(j + \tfrac{1}{2})^2 - q^2]^{1/2}. \tag{3.10}$$

Before defining the boundary conditions which make H self-adjoint, we note, following Goldhaber (G3), the following "formal" symmetry properties

$$\vec{\Sigma}.\vec{\pi}H\vec{\Sigma}.\vec{\pi} = H, \ (CP)H(CP)^{-1} = -H, \tag{3.11}$$

$$(CT)H(CT)^{-1} = -H, \ (PT)H(PT)^{-1} = H.$$

Here $\vec{\Sigma}.\vec{\pi}$ is related to the helicity. The discrete transformations

C,P,T are defined by

$$\psi_{a,b}(\vec{x},t) \xrightarrow{P} \psi^{P}_{a,b}(\vec{x},t) = \eta_P \beta \psi_{b,a}(-\vec{x},t), \tag{3.12}$$

$$\psi(\vec{x},t) \xrightarrow{C} \psi^{C}(\vec{x},t) = \eta_C \beta \alpha_2 \psi^{\star}(\vec{x},t), \tag{3.13}$$

$$\psi(\vec{x},t) \xrightarrow{T} \psi^{T}(\vec{x},t) = \eta_T \Sigma_2 \psi^{\star}(\vec{x},-t), \tag{3.14}$$

$$P\vec{A}(\vec{x})P^{-1} = C\vec{A}(\vec{x})C^{-1} = T\vec{A}(\vec{x})T^{-1} = -\vec{A}(\vec{x}), \tag{3.15}$$

where $|\eta_P| = |\eta_C| = |\eta_T| = 1$. The subscripts a,b in Eq.(3.12) refer to regions R_a, R_b. The corresponding subscripts in Eqs.(3.13)-(3.15) are omitted, since the same subscript occurs throughout each equation. It follows that the wave sections,

$$\psi^{PT}_{a,b}(\vec{x},t) = \eta_P \eta_T \beta \Sigma_2 \psi^{\star}_{b,a}(-\vec{x},-t), \tag{3.16}$$

$$\psi^{CT}(\vec{x},t) = - \eta_C \eta_T^{\star} \beta \alpha_2 \Sigma_2 \psi(\vec{x},-t), \tag{3.17}$$

and

$$\psi^{CP}_{a,b}(\vec{x},t) = - \eta_C \eta_P^{\star} \alpha_2 \psi^{\star}_{b,a}(-\vec{x},t) \tag{3.18}$$

obey the same Dirac Eqn. as $\psi(\vec{x},t)$. In the lowest angular momentum state $j = |q| - \frac{1}{2}$, and with m = eigenvalue of J_z,

$$\psi(\vec{x},t) = \frac{1}{r} \begin{pmatrix} F(r) & \eta_{j,m}(\theta,\phi) \\ G(r) & \eta_{j,m}(\theta,\phi) \end{pmatrix} e^{-iEt}$$

$$\equiv \frac{1}{r} \chi(r) \otimes \eta_{jm}(\theta,\phi) e^{-iEt}, \tag{3.19}$$

where

$$H_o \chi(r) = E\chi(r), \; H_o = -i \; \mathrm{sgn}(q) \sigma_1 \frac{d}{dr} + M\sigma_3. \tag{3.20}$$

Using the expressions (K1) for the 2 component angular sections η_{jm} in terms of the monopole harmonics $Y_{q\ell m}$, and the formulae

$$(Y^{*}_{q\ell m}(-\hat{r}))_b = (-1)^{\ell+m}(Y_{q,\ell,-m}(\hat{r}))_a, \tag{3.21}$$

$$\sigma_2(\eta^{*}_{j,m}(-\hat{r}))_b = i(-1)^{j+m+1}(\eta_{j,-m}(\hat{r}))_a, \tag{3.22}$$

we have the PT, CT and CP transforms of Eq.(3.19)

$$\psi^{PT}(\vec{x},t) = i(-1)^{j+m+1}\eta_T\eta_P\,\frac{1}{r}\,\chi^{PT}(r)\otimes\eta_{j,-m}(\hat{r})e^{-iEt}, \tag{3.23}$$

$$\psi^{CT}(\vec{x},t) = -i\eta_C\eta^{*}_T\,\frac{1}{r}\,\chi^{CT}(r)\otimes\eta_{j,m}(\hat{r})e^{+iEt}, \tag{3.24}$$

$$\psi^{CP}(\vec{x},t) = i(-1)^{j+m}\eta_C\eta^{*}_P\,\frac{1}{r}\,\chi^{CP}(r)\otimes\eta_{j,-m}(\hat{r})e^{+iEt}, \tag{3.25}$$

where

$$(\chi^{PT}(r),\ \chi^{CT}(r),\ \chi^{CP}(r)) \equiv (\sigma_3\chi^{*}(r),\ \sigma_2\chi(r),\ \sigma_1\chi^{*}(r)). \tag{3.26}$$

A consequence of $\psi^{PT,CT,CP}(\vec{x},t)$ obeying the Dirac eqn. is that

$$H_o(\chi^{PT},\ \chi^{CT},\ \chi^{CP}) = E(\chi^{PT},\ -\chi^{CT},\ -\chi^{CP}). \tag{3.27}$$

Similarly from the formal commutation of $\vec{\Sigma}.\vec{\pi}$ with H, and the relation in the lowest j state

$$\vec{\Sigma}.\vec{\pi}\psi(\vec{x},t) = -\frac{i}{r}\,\mathrm{sgn}(q)\,\frac{d\chi(r)}{dr}\otimes\eta_{jm}(\theta,\phi), \tag{3.28}$$

we have formal helicity conservation,

$$[h_o,H_o] = 0,\quad h_o \equiv -i\,\frac{d}{dr}\,\mathrm{sgn}(q) \tag{3.29}$$

To go from formal symmetries to actual symmetries we discuss now the boundary conditions. We easily deduce the result of Callias and Goldhaber that whereas h_o cannot be made self-adjoint with any boundary condition, self-adjointness of H_o requires the boundary condition

$$F(0) + G(0) = (F(0) - G(0))\exp(i(\theta_o - \frac{\pi}{2})). \tag{3.30}$$

This does not allow even hermiticity (symmetry) of h_o. Despite the formal eq.(3.29), helicity is not conserved; more accurately helicity is not an observable. A regularization of H_o which makes helicity observable has quite different consequences from (3.30) (B2). The condition (3.30) allows PT invariance, but CP and CT invariance are violated except for $\theta = \eta\pi$. By the procedure we explain in greater detail for finite volume, the Witten-Grossman-Yamagishi result for (infinite volume) vacuum charge is shown to be:

$$N_o = q\theta_o/\pi. \tag{3.31}$$

III.2. <u>Fermion-Monopole System In a Bag.</u> Consider now the same system except that the fermion is confined to a spherical bag of radius R (R10). <u>Self-adjointness of H_o</u> now requires the boundary condition

$$\begin{pmatrix} F(R)+G(R)\mathrm{sgn}q \\ F(0)-G(0)\mathrm{sgn}q \end{pmatrix} = U \begin{pmatrix} F(R)-G(R)\mathrm{sgn}q \\ F(0)+G(0)\mathrm{sgn}q \end{pmatrix} \tag{3.32}$$

where U is the 2x2 unitary matrix

$$U \equiv e^{i\alpha} \begin{pmatrix} \cos\lambda e^{i(\beta+\frac{\pi}{2})} & \sin\lambda e^{i\gamma} \\ \sin\lambda e^{-i\gamma} & -\cos\lambda e^{-i(\beta+\frac{\pi}{2})} \end{pmatrix} \tag{3.33}$$

with $\alpha,\beta,\gamma,\lambda$ being arbitrary real parameters. These boundary conditions define a four parameter family of self-adjoint Hamiltonians. <u>Self-adjointness of helicity.</u> Similarly, the boundary conditions which make h_o self-adjoint are found to be:

$$\begin{pmatrix} F(R)+G(R)\mathrm{sgn}q \\ F(R)-G(R)\mathrm{sgn}q \end{pmatrix} = V \begin{pmatrix} F(0)+G(0)\mathrm{sgn}q \\ F(0)-G(0)\mathrm{sgn}q \end{pmatrix}, \tag{3.34}$$

where V is an arbitrary 2x2 unitary matrix.

Simultaneous self-adjointness of hamiltonian and helicity. For this purpose the parameters $\alpha,\beta,\gamma,\lambda$ must be such that Eqs.(3.32) and (3.34) are equivalent (i.e., they imply each other). This happens if and only if $\sin\lambda = \pm 1$. Hence the common domain of self-adjointness of the hamiltonian and helicity is specified by the simple boundary condition

$$\begin{pmatrix} F(R) \\ G(R)\,\mathrm{sgn}q \end{pmatrix} = \exp(i\gamma_1 + i\alpha\sigma_1)\begin{pmatrix} F(0) \\ G(0)\,\mathrm{sgn}q \end{pmatrix}, \tag{3.35}$$

where

$$\exp(i\gamma_1) \equiv \exp(i\gamma)\sin\lambda, \quad \sin\lambda = \pm 1. \tag{3.36}$$

Both Eqs.(3.32) and (3.34) are equivalent to (3.35) with,

$$U = e^{i\alpha}\begin{pmatrix} 0 & e^{i\gamma_1} \\ e^{-i\gamma_1} & 0 \end{pmatrix}, \quad V = e^{i\gamma_1}\begin{pmatrix} e^{i\alpha} & 0 \\ 0 & e^{-i\alpha} \end{pmatrix}. \tag{3.37}$$

Discrete Symmetries. For CP invariance (3.11) to hold if $\chi\varepsilon D(H_o)$, then $\chi^{cp}\varepsilon D(H_o)$, and vice versa. Similar conditions hold for CT and PT invariance. Thus conditions for the discrete symmetries to hold are given by

$$\text{PT: } U=U^T; \quad \text{CT: } U=-\sigma_3 U^{\dagger}\sigma_3; \quad \text{CP: } U=-\sigma_3 U^{*}\sigma_3. \tag{3.38}$$

The infinite space results follow from (3.38) by setting $|U_{22}| = 1$, and other $U_{ij} = 0$. For finite R, Eqs.(3.38) yield:

$$\begin{aligned} &\text{PT: } \sin\lambda\sin\gamma = 0 \\ &\text{CT: either } \sin\alpha = \cos\lambda\sin\beta = 0 \text{ or } \cos\alpha = \cos\beta = \sin\lambda = 0 \\ &\text{CP: either } \sin\alpha = \cos\lambda\sin\beta = \sin\lambda\sin\gamma = 0, \\ &\qquad \text{or } \cos\alpha = \cos\lambda\cos\beta = \sin\lambda\cos\gamma = 0. \end{aligned} \tag{3.39}$$

For CT or CP invariance we need $\sin\alpha\cos\alpha = 0$. If $\sin\alpha = 0$, CP invariance implies CT invariance. If $\cos\alpha = 0$, CT invariance implies

CP invariance. The most symmetric case, with helicity self-adjoint and conserved, and all the discrete symmetries PT, CT, CP valid corresponds to:

$$h_o,\text{PT},\text{CT},\text{CP}: \sin\alpha = \cos\lambda = \sin\gamma = 0, \text{ i.e. } U = \pm\sigma_1, \tag{3.40}$$

i.e., periodic or antiperiodic boundary conditions.

Vacuum Charge Eigenvalues. The Dirac field operator $\Psi(\vec{x},t)$ in the Heisenberg representation obeys the same differential eqn. as the c-no. wave function $\Psi(\vec{x},t)$ and has the charge conjugation property

$$\Psi_\alpha(\vec{x},t) \xrightarrow{C} \Psi^C_\alpha(\vec{x},t) = \eta_C(\beta\alpha_2)_{\alpha\beta}\Psi^\dagger_\beta(\vec{x},t). \tag{3.41}$$

The total charge operator N (odd under C) and its eigenvalue N_o in a vacuum state (with a convenient regularization (G7,N2) are given by

$$N = \frac{1}{2}\int d^3x[\Psi^\dagger_\alpha(\vec{x},t), \Psi_\alpha(\vec{x},t)], \tag{3.42}$$

$$N_o = \lim_{\varepsilon\to 0+}\left(-\frac{1}{2}\sum_{\substack{n\\ E_n\neq 0}} (\text{sgn}E_n)e^{-\varepsilon|E_n|}\right) + N_{o,o}, \tag{3.43}$$

where $N_{o,o}$ is the contribution of zero energy levels to vacuum charge. When there are ν zero energy C-no. wave functions the vacuum state is 2^ν fold degenerate and the corresponding $N_{o,o}$ vary from $-\nu/2$ to $\nu/2$ in steps of unity. In such a case we shall calculate the N_o corresponding to the choice of the vacuum with $N_{o,o} = -\nu/2$; the other vacua will have charges differing from this by integers.

We shall only calculate the contribution to vacuum charge N_o of the lowest angular momentum levels. The results will however give the full vacuum charge whenever the boundary conditions for the higher j levels respect CP or CT leading to $E \to -E$ symmetry of these levels and hence their zero contribution to N_o.

For the most general boundary condition given by (3.32), (3.33), we find that the energy levels are given by

$$\cos(kR)\cos\alpha + \frac{\sin(kR)}{k}[E\sin\alpha + M\cos\lambda\cos\beta] = \cos\zeta, \tag{3.44}$$

where

$$k^2 \equiv E^2-M^2, \cos\zeta \equiv \sin\lambda\cos\gamma. \tag{3.45}$$

Vacuum charge for M = 0. In this case,

$$E(M=0) = \frac{\alpha+\zeta+2n\pi}{L} \text{ and } \frac{\alpha-\zeta+2n\pi}{L}, \quad n = 0, \pm 1, \pm 2, \ldots . \tag{3.46}$$

Each of these levels is $2j+1 = 2|eg| = 2|q|$ fold degenerate. A simple calculation yields

$$N_o(M=0) = 2|q|\left[\frac{\alpha}{\pi} - 1 - \mathrm{Int}(\frac{\alpha+\zeta}{2\pi}) - \mathrm{Int}(\frac{\alpha-\zeta}{2\pi}) \right], \tag{3.47}$$

with $\mathrm{Int}x \equiv$ largest integer $\leq x$.

Vacuum Charge For M > 0. In this case, for given U, $E = E(M=0) + O(1/n)$, for $n \to \infty$. We can then show that $dN_o(M)/dM = 0$ except at those values $M = M_i$ where one or more eigenvalues E pass through zero leading to an integer jump in $N_o(M) \equiv N_o$. We thus find

$$N_o(M) = N_o(M=0) + I, \tag{3.48}$$

where I is an integer which can be calculated explicitly (R10). Helicity conservation allows arbitrary α and hence arbitrary fractional value for vacuum (monopole) charge. So does PT conservation. For either CP or CT to be conserved $\sin\alpha\cos\alpha = 0$, and hence the vacuum charge is $\frac{1}{2}$ x integer, where the integer can be even or odd. This is due to the fact that the non-zero energy levels then occur in pairs due to $E \to -E$ symmetry and do not contribute to the vacuum charge. If helicity, CP, CT and PT are conserved then $\sin\alpha = 0$, and the vacuum charge must be integral. (Table 1)

TABLE 1

Dependence of fractional part of vacuum charge Eq.(3.47) on symmetries of the lowest angular momentum fermion-monopole bag hamiltonian assuming CP (or CT) invariance of the higher angular momentum hamiltonians.

Conserved quantities	Helicity	CP	CT	PT	Helicity, CP,CT,PT
Fractional part of vacuum charge	arbitrary	$0, \frac{1}{2}$	$0, \frac{1}{2}$	arbitrary	0

This extension of the Callias-Goldhaber and Witten-Grossman-Yamagishi results demonstrates that the change from infinite to finite space volume has interesting consequences for the vacuum charge eigenvalues and symmetry properties. E.g., if we consider $U = \text{diag}(\exp(i\eta(\theta_0' - \frac{\pi}{2})), \exp(-i\eta(\theta_0 - \frac{\pi}{2})))$ which Yamagishi used in discussing the $R\to\infty$ limit, we find that N_0 has a fractional part $2q(\theta_0'-\theta_0+\pi)/(2\pi)$, with a nontrivial dependence on the boundary condition parameter θ_0' at $r = R$, no matter how large R may be! Further, more symmetry properties are possible to satisfy for finite R than for $R = \infty$, each symmetry restricting boundary conditions and vacuum charge eigenvalues. In particular it is possible to have a helicity conserving dyon of fractional charge. CP violation forces fractional charge but does not force helicity violation. Further, <u>the simultaneous conservation of helicity, CP, CT and PT forces the monopole charge to be integral.</u>

Since the vacuum charge eigenvalues are independent of R but depend nontrivially on boundary conditions at r=R (as well as r=0) they have nonunique $R \to \infty$ limit. A similar boundary condition dependence of the vacuum charge will be obtained for the 1+1 dimensional fermion soliton system.

III.3. Soliton induced fractional fermion number in infinite volume.

Consider the original Jackiw-Rebbi (J2) example in 1+1 dimension with Hamiltonian

$$\hat{H} = \int_{-\infty}^{\infty} dx \left(\frac{1}{2}\pi^2 + \frac{1}{2}\left(\frac{\partial\Phi}{\partial x}\right)^2 + \Psi^\dagger(-i\sigma_2 \frac{\partial}{\partial x} + g\sigma_1\Phi)\Psi + V(\Phi) \right), \tag{3.49}$$

where Φ is a neutral scalar field, $\pi = \dot{\Phi}$ its canonical conjugate, Ψ a fermion field, g a constant and

$$V(\Phi) = \frac{\lambda^2}{2\mu^2}(\mu^2-\Phi^2)^2. \tag{3.50}$$

The classical minima of V being at $\Phi = \pm\mu$, spontaneous breaking of the reflection symmetry $V(\Phi) = V(-\Phi)$ occurs. For a time independent configuration $\pi = 0$ and the contribution of the scalar field to the classical energy is

$$E_{scalar} = \int_{-\infty}^{\infty} dx \left(\frac{1}{2}\left(\frac{\partial\Phi}{\partial x}\right)^2 + V(\Phi) \right). \tag{3.51}$$

Classical configurations with finite E_{scalar} ($\Phi^2-\mu^2$ and $\partial\Phi/\partial x \to 0$ for $x \to \pm\infty$) extremising E_{scalar} are:

$$\Phi = \pm\mu, \; E_{scalar} = 0, \quad \text{(Vacuum Sector)} \tag{3.52}$$

$$\Phi = \pm\mu \tanh(x\lambda), \; E_{scalar} = \frac{4}{3}\mu^2\lambda, \quad \text{(Soliton Sector)} \tag{3.53}$$

$\Phi = \pm\mu$ giving zero classical energy, called the vacuum solutions, $\Phi = \pm\mu \tanh(x\lambda)$ the soliton solution, and $\Phi = -\mu \tanh(x\lambda)$ the anti-soliton solution are pictured below.

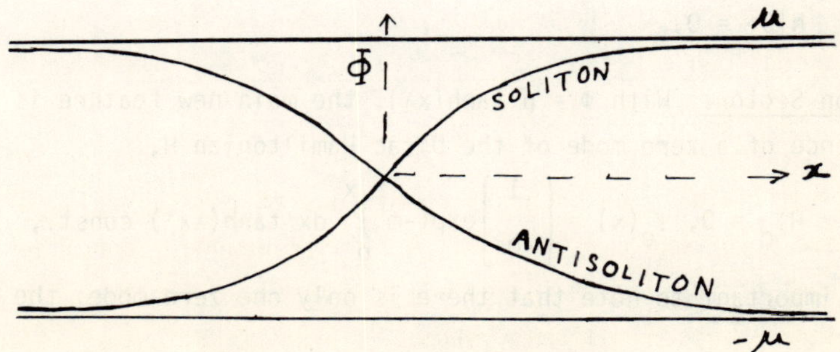

Quantizing according to the semiclassical method with these classical solutions as background fields we obtain the vacuum and soliton sectors of the theory. The Dirac field $\Psi(x,t)$ is constructed to obey canonical equal time anticommutation relations and solve the equation

$$i \frac{\partial\Psi(x,t)}{\partial t} = H\Psi(x,t),\ H = -i\sigma_2 \frac{d}{dx} + g\sigma_1\Phi, \tag{3.54}$$

where Φ is the appropriate classical solution.

Vacuum Sector. Consider $\Phi = \mu$, and set $g\mu = m$. Then the Dirac field Ψ has the standard free field form

$$\Psi(x,t) = \frac{1}{\sqrt{2\pi}} \int_{-\infty}^{\infty} dk \sqrt{\frac{m}{E_i}} \left(U_k(x)e^{-iE_k t} bk + V_k(x)e^{iE_k t} d_k^\dagger \right), \tag{3.55}$$

with $E_k = (k^2+m^2)^{1/2}$, $V_k = \sigma_3 U_k^*$, and the operators b_k, d_k obey the anticommutation rules

$$\{b_k,b_{k'}\} = \{d_k,d_{k'}\} = \delta(k-k'),\ \text{rest } 0. \tag{3.56}$$

Under charge conjugation $\Psi_\alpha \to \Psi_\alpha^c = (\sigma_3)_{\alpha\beta}\Psi_\beta^\dagger$. The fermion number operator odd under charge conjugation is then given by

$$N = \int_{-\infty}^{\infty} dx \frac{1}{2} \left(\Psi_\alpha^\dagger, \Psi_\alpha \right). \tag{3.57}$$

Substituting (3.55) we obtain

$$N = \int_{-\infty}^{\infty} dk (b_k^\dagger b_k - d_k^\dagger d_k). \tag{3.58}$$

The vacuum state, annihilated by b_k, d_k has zero fermion number

$$N|0> = 0. \tag{3.59}$$

Soliton Sector. With $\Phi = \mu \tanh(x\lambda)$, the main new feature is the existence of a zero mode of the Dirac Hamiltonian H,

$$H\Psi_0 = 0,\ \Psi_0(x) = \begin{pmatrix} 1 \\ 0 \end{pmatrix} \exp(-m \int_0^x dx' \tanh(\lambda x')\ \text{const.}, \tag{3.60}$$

It is important to note that there is only one zero mode: the second linearly independtdent solution of $H\Psi = 0$

$$\binom{0}{1}\exp(m \int_0^x dx'\tanh(\lambda x')) \tag{3.61}$$

is not normalizable, and has exponential growth at infinity. The solution Ψ_o is self-charge conjugate. $\sigma_3\Psi_o^* = \Psi_o$. Expanding in the complete set of eigenmodes of H we obtain,

$$\Psi = a\Psi_o(x) + \frac{1}{\sqrt{2\pi}} \int_{-\infty}^{\infty} dk \sqrt{\frac{m}{E_k}} \left(e^{-iE_k t} U_k(x)\, b_k + e^{iE_k t} U_k(x) d_k^\dagger\right), \tag{3.62}$$

with $E_k = (m^2+k^2)^{1/2}$, b_k, d_k obeying the same anticommutation rules as before, and further

$$\{a,a^\dagger\} = 1, \; a^2 = 0, \tag{3.63}$$

with 'a' anticommuting with b_k, $b_k^\dagger$, d_k, $d_k^\dagger$. The ground states in the soliton sector consist of a degenerate soliton doublet $|S\rangle$ and $a^\dagger|S\rangle$, where

$$a|S\rangle = b_k|S\rangle = d_k|S\rangle = 0 \tag{3.64}$$

The fermion number operator is again (3.57) which now simplifies to

$$N = a^\dagger a + \int_{-\infty}^{\infty} dk\,(b_k^\dagger b_k - d_k^\dagger d_k) - \frac{1}{2}. \tag{3.65}$$

The c-number term -1/2 in N cannot be removed by redefinition of N because we require N to be odd under charge conjugation. Hence,

$$N|S\rangle = -\frac{1}{2}|S\rangle, \; Na^\dagger|S\rangle = \frac{1}{2}a^\dagger|S\rangle. \tag{3.66}$$

The fractional fermion number of the soliton states is a direct consequence of the existence of an odd number of zero modes of H. It is interesting to note that this number is invariant under perturbations of g (x) which keep sign ($\Phi(\pm\infty)$) invariant. The Dirac H is of the form occurring in the index theorem C5,B3 :

$$H = \begin{pmatrix} 0 & L \\ L^{\dagger} & 0 \end{pmatrix}, \; L = -\frac{d}{dx} + g\Phi(x). \tag{3.67}$$

The zero modes may generally be of two classes, those with $\sigma_3 = \pm 1$, (say n_+ in number)

$$\Psi = \begin{pmatrix} \Psi_+ \\ 0 \end{pmatrix}, \; L^+\Psi_+ = 0, \tag{3.68}$$

and those with $\sigma_3 = -1$, (say n_- in number)

$$\Psi = \begin{pmatrix} 0 \\ \Psi_- \end{pmatrix}, \; L\Psi_- = 0. \tag{3.69}$$

The index of L is defined as

$$\text{index}(L) = n_- - n_+. \tag{3.70}$$

The Callias-Bott-Seeley [C5,B3] index theorem states that

$$\text{index}(L) = -\frac{1}{2}\left(\frac{\Phi(+\infty)}{|\Phi(+\infty)|} - \frac{\Phi(-\infty)}{|\Phi(-\infty)|} \right). \tag{3.71}$$

The fermion number of the soliton state is them

$$N|S\rangle = \text{index}(L)|S\rangle, \tag{3.72}$$

and the index theorem (3.71) enables calculation of this fermion number even for $\Phi(x)$ for which n_- and n_+ are difficult to calculate.

The Jackiw-Rebbi example is C-invariant and leads to half-integral fermion number. It is instructive to note also that there exist C-noninvariant Hamiltonians which lead to fermion numbers which are arbitrary fractions. E.g. in the Goldstone-Wilczek example (G5)

$$H = -i\sigma_2 \frac{d}{dx} + g(\sigma_1\Phi_1(x) + \sigma_3\Phi_3(x)), \tag{3.73}$$

the ground-state fermion number is, for $g \neq 0$,

$$\frac{\theta(\infty)-\theta(-\infty)}{2\pi}, \tag{3.74}$$

where $\theta(x) = \tan^{-1}(\Phi_3(x)/\Phi_1(x))$.

III.4. Boundary condition induced fractional fermion number in finite volume.

In order to elucidate the fractional fermion number phenomenon Kivelson and Schrieffer (K2) and Rajaraman and Bell (R6) discussed the problem of a Dirac fermion interacting with an idealised soliton (i.e. a step function) confined to one dimensional box of finite length L. They imposed periodic boundary conditons. Their conclusion was that the total vacuum charge is integral. Only partial charge operators in a cleverly defined limit can have fractional eigenvalues. Roy and Singh (R7) confirmed their result about the total charge being integral for periodic boundary conditions, but pointed out that the essential aspects of the fractional charge phenomena require for their elucidation a consideration of more general boundary conditions. The boundary condition dependence of vacuum charge is quite natural. The topological effects for the <u>whole</u> infinite space problem reflect the far away situation. For <u>finite</u> space the analogous role is played by the boundary conditions. We will explicitly show how the finite one-dimensional box problem with appropriate boundary conditions model different infinite one-dimensional space situations.

We first discuss all possible self-adjoint Goldstone-Wilczek Hamiltonians describing a Dirac fermion in a bounded background field confined to a one dimensional finite box i.e. $x \varepsilon [L_1, L_2]$ and having the formal Hamiltonian (3.73). From the results of Sec.II.4 we deduce that

I. The Goldstone-Wilczek Hamiltonian (3.73) for $x \varepsilon [L_1, L_2]$ is self-adjoint over the domain $D(H_U)$ consisting of the spinors χ which satisfy (i) $\| \chi \| < \infty$, (ii) $\| H \| < \infty$ and (iii)

$$\begin{pmatrix} \langle e_+|\chi(L_2)\rangle \\ \langle e_-|\chi(L_1)\rangle \end{pmatrix} = U \begin{pmatrix} \langle e_-|\chi(L_2)\rangle \\ \langle e_+|\chi(L_1)\rangle \end{pmatrix} \tag{3.75}$$

where

$$\sqrt{2}\ \langle e\pm| \equiv (1, \mp i), \tag{3.76}$$

and U is a 2x2 unitary matrix i.e. $U^+U = UU^+ = 1$.

We thus get a four parameter family of self adjoint Hamiltonians for the case of a fermion moving in bounded potentials in a finite box. The choice $U = \sigma_1$ corresponds to periodic boundary conditions. For charge-conjugation invariance one needs not only that $\Phi_3 = 0$ but also that $\psi^C(x)$ obeys the same boundary conditions as $\psi(x)$ i.e. we must also have $U = U^*$. We shall denote the eigenfunctions of H corresponding to an eigenvalue E_i by χ_i.

We shall now calculate the vacuum total charge N_o for all the allowed boundary conditions given above. As is well-known the expression for N_o requires a regularisation. We shall use the regularisation, given by (N2,G7,P1)

$$N_o = \lim_{\beta\to 0+}\left(-\frac{1}{2} \sum_{\substack{n \\ E_n\neq 0}} (\mathrm{Sgn}E_n)\, e^{-\beta|E_n|} \right) + N_{o,0} \tag{3.77}$$

Alternative regularisations in terms of the -invariant of Atiyah, Patodi and Singer (A4) or the point splitting regularisation of Schwinger lead to identical results. Here $N_{o,0}$ is the contribution of the zero energy states to the vacuum total charge. If there are C-number eigenfunctions of zero energy then the vacuum state has a degeneracy equal to 2^ν and $N_{o,0}$ for these states varies from $\frac{\nu}{2}$ to $-\frac{\nu}{2}$ in steps of unity. In view of this degeneracy we shall from now on deal with $N_{o,0} = -\nu/2$ state. Other vacuum states differ by integer in total charge from this state.

For ease in calculation we shall model soliton fields as idealised step functions following earlier practice. We take with

$L_2>0>L_1$,

$$\Phi \equiv \Phi_1 + i\Phi_3 = \Phi_+ \text{ step } (x) + \Phi_- \text{ step } (-x), \tag{3.78}$$

where $\Phi_\pm$ are complex constants. The soliton is "located" at x = 0. It is easy to show that the eigenfunctions of H must be continuous at x = 0. The eigenvalues of H corresponding to the self-adjoint extension labelled by the unitary matrix U parametrised by

$$U = e^{i\alpha}\begin{pmatrix} \text{Cos}\lambda e^{i\beta} & \text{Sin}\lambda e^{i\gamma} \\ \text{Sin}\lambda e^{-i\gamma} & -\text{Cos}\lambda e^{-i\beta} \end{pmatrix} \tag{3.79}$$

with $\alpha,\beta,\gamma,\lambda$ = real,

are given by G(E) = 0 where

$$\begin{aligned} G(E) \equiv & -\text{Sin}\lambda\text{Cos}\gamma + G_{CC}\text{Cos}(k_+L_2)\text{Cos}(k_-L_1) \\ & + G_{SS}\text{Sin}(k_+L_2)\text{Sin}(k_-|L_1|) + G_{SC}\text{Sin}(k_+L_2)\text{Cos}(k_-L_1) \\ & + G_{CS}\text{Cos}(k_+L_2)\text{Sin}(k_-|L_1|), \end{aligned} \tag{3.80}$$

with

$$\begin{aligned} G_{CC} = \text{Cos}\alpha,\ k_+k_-G_{SS} = & \ g^2\text{Re}[e^{i\alpha}(\Phi_-\Phi_+^*)] - E^2\text{Cos}\alpha \\ & + g\ E\ \text{Cos}\lambda\ \text{Im}(e^{i\beta}(\Phi_-^*-\Phi_+^*)) \end{aligned} \tag{3.81}$$

$$k_+G_{SC} = E\ \text{Sin}\alpha + g\ \text{Cos}\lambda\ \text{Re}(\Phi_+^* e^{i\beta})$$

$$k_-G_{CS} = E\ \text{Sin}\alpha + g\ \text{Cos}\lambda\ \text{Re}(\Phi_-^* e^{i\beta})$$

and

$$E^2 = k_\pm^2 + |\Phi_\pm|^2 . \tag{3.82}$$

We must now evaluate N_0 using these eigenvalues. We first attempt this for the simple case in which $g\Phi_\pm = 0$ i.e. no soliton is present. The eigenvalue equation simplifies to

$$0 = - \mathrm{Sin}\lambda\mathrm{Cos}\gamma + \mathrm{Cos}(EL-\alpha), \tag{3.83}$$

where $L = L_2 - L_1$ is the total length of the box. The energy eigenvalues are now given by,

$$E(g=0) = (2n\pi + \alpha \pm \zeta)/L \equiv E_{n\pm}, n=0,\pm1,\pm2, \ldots, \tag{3.84}$$

where,

$$\mathrm{Cos}\ \zeta \equiv \mathrm{Cos}\gamma\ \mathrm{Sin}\lambda. \tag{3.85}$$

It is straightforward to calculate N_0 now and we obtain

$$N_0(g=0) = \frac{\alpha}{\pi} - 1 - \left[\frac{\alpha+\zeta}{2\pi}\right]_{INT} - \left[\frac{\alpha-\zeta}{2\pi}\right]_{INT} \tag{3.86}$$

where $[a]_{INT} \equiv$ largest integer $\leq a$. We thus see that the fractional part of $N_0(g=0)$ is equal to the fractional part of $\frac{\alpha}{\pi}$ and $N_0(g=0)$ is thus not integral in general. For periodic boundary conditions $U = \sigma_1$ and it follows that $N_0(g=0) = -1$. Note that the fractional part of $N_0(g=0)$ is thus totally fixed by det U.

We next show that the inclusion of soliton background field (i.e. $g\Phi_\pm \neq 0$) does not change the fractional part of $N_0 = N_0(g)$, i.e. $N_0 - N_0(0)$ = integer, or

$$\text{fractional part of } (N_0 - \frac{\alpha}{\pi}) = 0. \tag{3.87}$$

We first note that large eigenvalues of the Goldstone-Wilczek Hamiltonian are given by

$$E = E_{n\pm} + \frac{h_\pm}{n} + O\left(\frac{1}{n^2}\right) \text{ as } |n| \to \infty. \tag{3.88}$$

where $h_\pm$ are bounded functions of n. Now we have

$$N_0(g) = -\frac{1}{2}\sum_{\substack{i=1 \\ E_i\neq 0}}^{M} \mathrm{Sgn}E_i - \frac{1}{2}\lim_{\beta\to 0+}\sum_{M+1}^{\infty} \mathrm{Sgn}E_i e^{-\beta|E_i|} + N_{0,0}(g), \tag{3.89}$$

where we choose M to be some large integer in some ordering such that for i > M we can use asymptotic estimates for eigenvalues.

$$\frac{dN_o(g)}{dg} = + \frac{1}{2} \lim_{\beta\to 0+} \beta \sum_{i=M+1}^{\infty} e^{-\beta|E_i|} \frac{dE_i}{dg} , \tag{3.90}$$

except at those g values where one (or more) eigenvalue vanishes. Using asymptotic estimates $|E_i| = \alpha\ |i|)$ and $|\frac{dE_i}{dg}| = 0(\frac{1}{|i|})$ we obtain a value zero for the right hand side of (3.90). We thus see that $N_o(g) - N_o(g=0)$ is a piecewise constant function of g which always takes integral values. The discontinuities of this function occur for g-values at which one (or more) eigenvalue of H crosses zero.

For the Jackiw-Rebbi Hamiltonian (with $\Phi_3 = 0$, $\Phi_+ + \Phi_- = 0$ and $L_2 + L_1 = 0$) we find that

$$N_o(g) = N_o(g = 0) \tag{3.91}$$

since the condition for the existence of a zero energy level, i.e. $\det(1 - U\sigma_1) = 0$, is independent of g. Hence no levels will pass through E = 0 as g is varied. For <u>periodic</u> boundary conditions (i.e. $U = \sigma_1$) we have two zero energy eigenfunctions of the Dirac operator given by ($m = g|\Phi_\pm|$)

$$\chi_{0,1} = \sqrt{\frac{m}{1-e^{-mL}}} \begin{pmatrix} e^{-m|x|} \\ 0 \end{pmatrix}, \chi_{0,2} = \sqrt{\frac{m}{e^{mL}-1}} \begin{pmatrix} 0 \\ e^{m|x|} \end{pmatrix} . \tag{3.92}$$

We thus have integral eigenvalues for the charge operator. Note that $\chi_{0,2}$ disappears in the $L \to \infty$ limit.

We now specialise to those boundary conditions which preserve C-invariance and allow E = 0 solutions. These are given by $U = \pm\mathrm{Cos}\theta_o + i\sigma_2\ \mathrm{Sin}\theta_o$ apart from the periodic boundary condition $U = \sigma_1$. These have det U = 1 (except for $U = \sigma_1$) and hence half-odd integral total vacuum charge. There is only one zero energy solution:

$$\chi_o \sim (1+\mathrm{Sin}\theta_o \pm \mathrm{Cos}\theta_o)\ \chi_{0,1} + i(1+\mathrm{Sin}\theta_o \mp \mathrm{Cos}\theta_o) e^{-mL/2}\ \chi_{0,2} \tag{3.93}$$

Only for U = 1 (i.e. $\mathrm{Cos}\theta_o = \pm 1$) is the admixture of ψ_{02}, which is not normalisable for L = ∞, absent. It is thus clear that if we have to model the physics of the infinite space solitonic Jackiw-Rebbi case by finite L system with appropriate boundary conditions then we should choose U = 1. The various other boundary conditions we have can reproduce infinite space Goldstone-Wilczek results. We regard it as very natural and likely that the sets of possible boundary conditions allowed by self adjointness consideration would appropriately model the physics of different nontrivial topologies for the background field. Notice that the fractional fermion number α/π is independent of (i) L, the box length, and (ii) Φ, the bounded background field, but depends non-trivially on the boundary conditions, thus making the infinite volume limit non-unique.

IV. Theta-vacua for charged Klein-Gorden field.

Nontrivial vacuum structure in gauge theories (J1,C1,R1,R2) arises from the requirement that under a large (i.e., homotopically nontrivial) gauge transformation the vacuum wave functional is not invariant but multiplied by a phase factor related to the vacuum angle.

Here we perform (R5) a new quantization of an elementary system: the charged Klein-Gordon field. The simple idea providing a point of departure from the standard route is the following. Allow the state vector to change by a phase factor under 2π rotations in charge (iso-spin) space, but insist on self-adjointness of the field theory Hamiltonian. A family of Hamiltonians each characterized by θ-parameters is obtained. It exhibits interesting field theoretical phenomena drastically different from a conventionally quantised free field theory. For general θ-values charge conjugation (C-) invariance is violated and number operators have fractional eigenvalues. The phenomenon is quite distinct from fractional Fermion number and CP-noninvariance induced by solitons (J2) or boundary conditions at spatial boundaries (R7). The present Bosonic phenomenon is connected only with charge space boundary conditions.

Consider the classical Hamiltonian

$$H = \int d\vec{x}[\pi^\dagger\pi + \nabla\phi^\dagger.\nabla\phi + \mu^2\phi^\dagger\phi], \tag{4.1}$$

with Klein Gordon field $\phi = \phi(\vec{x},t)$, and its' canonically conjugate momentum $\pi = \pi(\vec{x},t)$, describing charged mesons of mass μ. To quantize it we require the usual equal time commutation rules (ETCR)

$$[\phi(\vec{x},t),\pi(\vec{x}',t)] = i\delta(\vec{x}-\vec{x}'),[\phi(\vec{x},t),\pi^\dagger(\vec{x}',t)] = 0, \tag{4.2}$$

etc. The equations of motion lead to $\pi(\vec{x},t) = \partial\phi^\dagger/\partial t$. In addition the field $\phi(\vec{x},t)$ satisfies Klein-Gordon equation. For simplicity we assume the field to be confined to a box of volume $V = L^3$ and to obey periodic boundary conditions. Then the ETCR are solved by writing

$$\phi(\vec{x},t) = \sum_{\vec{k}}(V\, 2\omega_k)^{-1/2}(a_k e^{-ik.x}+b_k^\dagger e^{ik.x}), \tag{4.3}$$

where,

$$k.x \equiv \omega_k t - \vec{k}.\vec{x},\ \omega_k = \sqrt{k^2+\mu^2},\ \vec{k} = 2\pi\vec{n}/L, \tag{4.4}$$

with $\vec{n} = (n_1,n_2,n_3)$, $n_i = 0,\pm1,\pm2, \ldots$, and

$$[a_k,a_{k'}^\dagger] = [b_k,b_{k'}^\dagger] = \delta_{\vec{k}\vec{k}'},\ \text{rest zero.} \tag{4.5}$$

The charge operator Q, odd under C, is then given by, $(C\phi C^{-1} = \phi^\dagger)$,

$$Q = \frac{i}{2}\int d\vec{x}(\{\pi,\phi\} - \{\phi^\dagger,\pi^\dagger\}) = \sum_{\vec{k}}(b_k^\dagger b_k - a_k^\dagger a_k), \tag{4.6}$$

and the Hamiltonian becomes,

$$H = \sum_{\vec{k}} \omega_k(a_k^\dagger a_k + b_k^\dagger b_k + 1). \tag{4.7}$$

We introduce new coordinates $q_{\alpha k}$ and momenta $p_{\alpha,k}$ $(\alpha = 1,2)$ defined by

$$a_k = [q_{1k}+iq_{2k} + i(p_{1k}+ip_{2k})]/2$$

$$b_k = [i(q_{1k}-iq_{2k})-(p_{1k}-ip_{2k})]/2. \tag{4.8}$$

They satisfy

$$[q_{\alpha,k}, p_{\beta k'}] = i\, \delta_{\alpha\beta}\delta_{kk''}, \tag{4.9}$$

other commutators = 0.

The Hamiltonian H and the charge operators Q are now given by

$$H = \frac{1}{2}\sum_{\vec{k}} \omega_k H_k \text{ with } H_k \equiv \sum_{\alpha=1}^{2} (q^2_{\alpha k}+p^2_{\alpha k}) \tag{4.10}$$

$$Q = \sum_{\vec{k}} Q_k \qquad \text{with } Q_k \equiv q_{1k}p_{2k}-q_{2k}p_{1k} \tag{4.11}$$

4.1 Charge Space Boundary Conditions.

The departure from the usual quantisation of the K.G. field consists of using the following representation for the momentum operators $p_{,k}$,

$$p_{\alpha,k} = -i\frac{\partial}{\partial q_{\alpha k}} + \frac{\partial}{\partial q_{\alpha k}}\sum_{\vec{k}'}(\alpha_{k'}, \chi_{k'}) \tag{4.12}$$

where χ_k is the phase of $q_{1k} + iq_{2k}$ i.e.

$$q_{1k} + iq_{2k} \equiv \rho_k \exp(i\chi_k),\ \rho_k \text{ and } \chi_k \text{ real}, \tag{4.13}$$

and α_k being arbitrary real parameters. The usual scheme of quantisation can be recovered by putting all $\alpha_k = 0$ and restricting the wave functions to be single valued. The ρ_k and χ_k are the cylindrical coordinates in charge space for the k-th mode.

In terms of the variables ρ_k and χ_k we have

$$H_k = \rho_k^2 - [\frac{\partial^2}{\partial\rho_k^2} + \frac{1}{\rho_k}\frac{\partial}{\partial\rho_k}] + \frac{Q_k^2}{\rho_k^2} \tag{4.14}$$

$$Q_k = -i(\frac{\partial}{\partial\chi_k} + i\alpha_k). \tag{4.15}$$

Note that H_k differs from the Hamiltonian for a circular oscillator only by replacement $\partial/\partial\chi_k \to \partial/\partial\chi_k + i\alpha_k$. The parameters α_k may be therefore interpreted as Aharonov-Bohm potentials in charge space.

To specify a domain for H that makes it self-adjoint we choose the simple alternative of making each H_k self-adjoint. Since (4.13) is not well defined for $\rho_k = 0$, the starting domain must exclude functions which are non-zero at $\rho_k = 0$. For the azimuthal dependence of the state vector Ψ we impose

$$\Psi(\ldots, \chi_k=2\ , \ldots) = \exp(i2\pi\beta_k)\ \Psi(\ldots, \chi_k=0, \ldots),$$

$$\partial\Psi/\partial\chi_k(\ldots, \chi_k=2\pi, \ldots) = \exp(i2\rho\beta_k)\partial\Psi/\partial\chi_k(\ldots, \chi_k=0, \ldots). \tag{4.16}$$

Here the dots in the parentheses denote all the ρ_k's and all χ_k, with $\vec{k}' \neq k$, and have the same values on both sides of (4.16). The β_k are real parameters. Nonintegral β_k imply non-singlevaluedness in charge space. It will appear immediately that only the sums

$$\theta_k \equiv \alpha_k + \beta_k,\ \theta_k\ \varepsilon\ [0,1), \tag{4.17}$$

rather than α_k and β_k separately, influence physical results. Hence the results may be interpreted to arise purely from Aharonov-Bohm potentials in charge space, on purely from non-siglevaluedness in charge space. There is no loss of generality in restricting to $\theta_k\ \varepsilon\ [0,1)$ as the values θ_k and θ_k + integer give the same physical results.

IV.2. Fractional eigenvalues of number operators, Spectrum and Eigenfunctions.

We look for energy eigenvectors of the form,

$$\Psi \equiv \prod_{\vec{k}} \psi_k, H_k\psi_k = \varepsilon_k\psi_k, \tag{4.18}$$

$$H\Psi = E\Psi,\ E = \sum_{\vec{k}} \frac{1}{2}\omega_k\varepsilon_k, \tag{4.19}$$

and

$$\psi_k(\rho_k,\chi_k) = v(\rho_k)\exp(i(m_k+\beta_k)\chi_k), \tag{4.20}$$

$$m_k = 0, \pm 1, \pm 2, \ldots . \tag{4.21}$$

The resulting radial Hamiltonian is H_k with the replacement $Q_k \to M_k$, where

$$M_k \equiv m_k + \theta_k . \tag{4.22}$$

It has two singular end points ($\rho_k = 0$ and $\rho_k = \infty$). It is self-adjoint without any boundary conditions for $|M_k| \geq 1$. For $|M_k| < 1$ we obtain a one parameter family of self-adjoint radial hamiltonians corresponding to the boundary condition

$$[\text{Cos}\, \gamma_k \, \rho_k^{|M_k|} (\rho_k \, d/d\rho_k - |M_k|) + \text{Sin}\, \gamma_k \, \rho_k^{-|M_k|} (\rho_k \, d/d\rho_k + |M_k|]\; v(\rho_k) \xrightarrow{\rho_k \to 0} 0, \tag{4.23}$$

where

$$\gamma_k \, \varepsilon [0, \pi), \quad |M_k| < 1. \tag{4.24}$$

We give explicit results here for the choice $\gamma_k = 0$ which corresponds to energy eigenfunctions being bounded at $\rho_k = 0$.

$$v(\rho_k) = \text{Const.}\; \rho_k^{|M_k|} \exp(-\tfrac{1}{2} \rho_k^2) \times {}_1F_1(\tfrac{1}{2} |M_k| + \tfrac{1}{2} - \tfrac{\varepsilon_k}{4}, |M_k| + 1, \rho_k^2). \tag{4.25}$$

Here ${}_1F_1$ denotes the confluent hypergeometric function. The condition of square integrability (measure $\rho_k \, d\rho_k$) at $\rho_k = \infty$ yields the energy eigenvalues for the kth mode,

$$\varepsilon_k = e(|M_k| + 2(n_r)_k + 1), \; (n_r)_k = 0,1,2,\ldots . \tag{4.26}$$

Hence the total energy and charge eigenvalues are

$$E = \sum_{\vec{k}} \omega_k (|m_k| + \theta_k| + 2(n_r)_k + 1), \tag{4.27}$$

$$Q = \sum_{\vec{k}}(m_k + \theta_k), \tag{4.28}$$

which correspond to fractional eigenvalues of number operators:

$$a_k^\dagger a_k = |m_k + \theta_k| \operatorname{step}(-m_k - \theta_k) + (n_r)_k, \tag{4.29}$$

$$b_k^\dagger b_k = |m_k + \theta_k| \operatorname{step}(m_k + \theta_k) + (n_r)_k. \tag{4.30}$$

The state vectors ψ_k characterized by the integers $(n_r)_k$ and $(m_r)_k$, yield a complete orthonormal set for each mode $\vec{k}$, with the following normalization:

$$\psi_{n_r,m} = d_{n_r,m}\rho^{|M|}\exp(-\tfrac{1}{2}\rho^2)\,{}_1F_1(-n_r, |M| + 1, \rho^2)$$
$$\times \exp(i(m+\beta)\chi)$$
$$\begin{cases} c_{n_r,m}(a^\dagger)^{n_r}(b^\dagger)^{n_r+m}\rho^\theta\exp(-\tfrac{1}{2}\rho^2)\exp(i\beta\chi), & m \geq 0 \\ c_{n_r,m}(b^\dagger)^{n_r}(a^\dagger)^{n_r+|m|-1}\rho^{1-\theta}\exp(-\tfrac{1}{2}\rho^2) & \\ \times \exp(i(\beta-1)\chi), & m \leq -1 \end{cases} \tag{4.31}$$

where

$$|c_{n_r,m}| = [\pi(n_r)!\Gamma(n_r+1+|M|)]^{-1/2},$$

$$d_{n_r,m} = c_{n_r,m}(|M|+1)(|M|+2)\ldots(|M|+n_r)i^{n_r} \times \begin{cases} (-i)^m, & m \geq 0 \\ 1, & m \leq -1 \end{cases} \tag{4.32}$$

We have suppressed the mode subscript k in Eqs.(4.31), (4.32).

Discussion

(i) Each energy level E in the spectrum of H is specified by giving $(n_r)_k$ and m_k for all $\vec{k}$. The contribution to E of a given mode $\vec{k}$ is $\omega_k\varepsilon_k/2$. The levels with $Q_k > 0$ and $Q_k < 0$ form two separate ladders each with separation ω_k, the lowest energy positive and negative charge levels being respectively $\psi_{0,0}$ and $\psi_{0,-1}$. The energy separation between $\psi_{0,-1}$ and $\psi_{0,0}$ is $\omega_k(1-2\theta_k)$. If $\theta_k < \frac{1}{2}$ for all k, the ground state is nondegenerate and corresponds to $(n_r)_k = m_k = 0$ for all k. As ususal, the ground state energy is infinite and must be subtracted out. The ground state charge is also infinite unless $\Sigma_k\theta_k$ is finite.

(ii) How is it that the energy separation between $\psi_{0,-1}$ and $\psi_{0,0}$ is a fractional multiple of ω_k and the number operators have fractional eigenvalues despite the usual commutation rules obeyed by them? This is possible due to the fact that although $a\psi_{0,0} = b\psi_{0,-1} = 0$, $b\psi_{0,0}$ and $a\psi_{0,-1}$ are nonzero and outside the domain of the hamiltonian H_k due to unacceptable $\rho_k = 0$ behaviour:

$$b\psi_{0,0} = i\,\frac{\theta}{\rho}\exp(-i\chi)\psi_{0,0}; \quad a\psi_{0,-1} = \frac{1-\theta}{\rho}\exp(i\chi)\psi_{0,-1}. \tag{4.33}$$

(iii) The spectrum is charged conjugation symmetric not only for the usual case, all $\theta_k = 0$, but also when all $\theta_k = 1/2$. The case all $\theta_k = \frac{1}{2}$ is very interesting. The levels $\psi_{0,0}$ and $\psi_{0,-1}$ are then degenerate for each k. Hence the ground state becomes 2^∞-fold degenerate, the various ground states being distinguished by their different charges.

V. Skyrmions, Hall Effect and Anomalies.

Boundary condition related phenomena extend far beyond the topics we have discussed. It may be useful to record here a few references on topics we have not touched upon. First, an account of index theorems useful in many such investigations is given in [H2, E1, C5, B3]. The questions of baryon number in chiral bag models are discussed in [C6], in the Skyrme model in [S6], [B5,W6], and in hybrid of chiral bag and Skyrme models in [G7]. The observed quantum Hall effect [K3,T3] has stimulated much theoretical work [L2,L1,H3,H4]. Related work is on anomalies in odd space time dimensions [R11,A5,D1,J4,S7,F1,W7,I3,N2,B6]. Anomalies in even space-time dimensions are discussed in [F2,A6,Z2,B6]. Boundary condition effects in soliton mass corrections in supersymmetric theories are discussed in [S8,K4,I2,C7]. Fractional Witten-indices are discussed in [N4,Y2,B7].

ACKNOWLEDGEMENT

I am grateful to Virendra Singh in collaboration with whom all my ideas in the subject were developed. I thank warmly R. Ramachandran for the invitation to give these lectures and for the excellent hospitality at I.I.T. Kanpur.

REFERENCES

A1. Y. Aharonov, A. Casher and S. Yankielowicz, Phys. Rev. D30, 386 (15) (1984).

A2. D. Arovos, J.R. Schrieffer and F. Wilczek, Phys. Rev. Lett. 53, 722 (1984).

A3. N.I. Akhiezer and I.M. Glazman, "Theory of Linear Operators in Hilbert Space", Vol.2, (N.Y. 1963).

A4. M.F. Atiyah, V.K. Patodi and I.M. Singer, Bull. Lond. Math. Soc. 5, 229 (1973).

A5. L. Alvarez-Gaume and E. Witten, Nucl. Phys. B234, 269 (1983).

A6. L. Alvarez-Gaume and P. Ginsparg, Nucl. Phys. B243, 449 (1984).

B1. W.A. Bardeen, S. Elitzur, Y, Frishman, E. Rabinovici, Nucl. Phys. B218, 445 (1983).

B2. A.P. Balachandran, S.M. Roy and V. Singh, Phys. Rev. D28, 2669 (1983).

B3. R. Bott and R. Seeley, Comm. Math. Phys. 62, 235 (1978).

B4. R. Blankenbecler and D. Boyanovsky, "Fractional charge and spectral asymmetry in 1-D: a closer look", SLAC-PUB-3430 (1984), and Phys. Rev. D (to appear).

B5. A.P. Balachandran, V.P. Nair, S.G. Rajeev and A. Stern, Phys. Rev. Lett. 49, 1124 (1982).

B6. D. Boyanovsky and R. Blankenbecler, SLAC-PUB-3567 (1985).

B7. D. Boyanovsky and R. Blankenbecler, Phys. Rev. D30, 1821 (1984).

C1. C.G. Callan Jr., R.F. Dashen and D.J. Gross, Phys. Lett. 63B, 334 (1976).

C2. K.M. Case, Phys. Rev. 80, 797 (1950).

C3. C.J. Callias, Phys. Rev. D16, 3068 (1977).

C4. C.G. Callan Jr., Phys. Rev. D25, 2141 (1982);
D26, 2058 (1982); Nucl. Phys. B212, 391 (1983);
C.G. Callan and S.R. Das, Phys. Rev. Lett. 51 1155 (1983).

C5. C.J. Callias, Comm. Math. Phys. 62, 213 (1978).

C6. A. Chodos and C.B. Thorn, Phys. Rev. D12, 2733 (1975).

C7. A. Chatterjee and P. Majumdar, Phys. Rev. D30, 844 (1984), and SINP/TNP/85-5 preprint.

D1. S. Deser, R. Jackiw and S. Templeton, Ann. Phys. (N.Y.) 140, 372 (1982).

E1. T. Eguchi, R. Gilkey and A. Hanson, Phys. Rep. 66, 213 (1980).

F1. M. Friedman, J. Sokoloff, A. Widom and Y. Srivastava, Phys. Rev. Lett. 52, 1587 (1984).

F2. K. Fujikawa, Phys. Rev. Lett. 42, 1195 (1979); Phys. Rev. D21, 2848 (1980); Phys. Rev. Lett. 44, 1733 (1980).

G1. J. Glimm and A. Jaffe, "Quantum Physics: a functional integral point of view", N.Y. Springer Verlag (1981).

G2. C.C. Gerry and V.A. Singh, Phys. Rev. D20, 2550 (1979).

G3. A.S. Goldhaber, Phys. Rev. D16, 1815 (1977).

G4. B. Grossman, Phys. Rev. Lett. 50, 464 (1983).

G5. J. Goldstone and F. Wilczek, Phys. Rev. Lett. 47, 986 (1981).

G6. I.B. Goldberg et al, J. Chem. Phys. 70, 1132 (1979).

G7. J. Goldstone and R.L. Jaffe, Phys. Rev. Lett. 51, 1518 (1983).

H1. A. Heeger, Comments Solid State Phys. 10, 53 (1981).

H2. M. Hirayama, Prog. Theor. Phys. 70, 1444 (1983); M. Hirayama and T. Torii, Prog. Theor. Phys. 68, 1354 (1982).

H3. F.D. Haldane, Phys. Rev. Lett. 51, 605 (1983).

H4. B.I. Halperin, Phys. Rev. B25, 2185 (1981).

I1. S.Ikehata et al, Phys. Rev. Lett. 45, 1123 (1980).

I2. C. Imbimbo and S. Mukhi, Nucl. Phys. B247, 471 (1984).

I3. K. Ishikawa, Phys. Rev. Lett. D53, 1615 (1984).

J1. R. Jackiw and C. Rebbi, Phys. Rev. Lett. 37, 172 (1976).

J2. R. Jackiw and C. Rebbi, Phys. Rev. D13, 3398 (1976).

J3. R. Jackiw and J.R. Schrieffer, Nucl. Phys. B190, 253 (1981).

J4. R. Jackiw, Phys. Rev. D29, 2375 (1984).

K1. Y. Kazama, C.N. Yang and A.S. Goldhaber, Phys. Rev. D15, 2287 (1977).

K2. S. Kivelson and J.R. Schrieffer, Phys. Rev. B25, 6447 (1982).

K3. K.V. Klitzung, G. Dorda and M. Pepper, Phys. Rev. Lett. 45, 494 (1980).

K4. R.K. Kaul and R. Rajaraman, Phys. Lett. 131B, 357 (1983).

L1. R.B. Laughlin, Phys. Rev. B23, 5632 (1981) and Phys. Rev. Lett. 50, 1395 (1983).

L2. H. Levine et al, Nucl.Phys. B240 [FS12] 30, 49, 71 (1984).

N1. M.A. Naimark, "Linear differential operators, part II", G.G. Harap & Co. Ltd., London (1968).

N2. A.J. Niemi and G.W. Semenoff, Phys. Rev. Lett. 51, 2077 (1983) and Phys. Rev. D30, 809 (1984); A.J. Niemi, M.B. Paranjape and G.W. Semenoff, Phys. Rev. Lett. 53, 515 (1984).

N3. A.J. Niemi and G.W. Semenoff, Phys. Rev. Lett. 54, 873 (1985).

N4. A.J. Niemi and L.C.R. Wijewardhana, Phys. Lett. 138B, 389 (1984).

P1. M.B. Paranjape and G.W. Semenoff, Phys. Lett. 132B, 369 (1983).

R1. H.J. Rothe and J.A. Swieca, Nucl. Phys. B138, 26 (1978).

R2. H.J. Rothe and J.A. Swieca, Nucl. Phys. B149, 237 (1979).

R3. V.A. Rubakov, Pis'ma Zh. Eksp. Teor. Fiz. 33, 658 (1981) [JETP Lett. 33, 644 (1981)]; Nucl. Phys. B203, 311 (1982).

R4. S.M. Roy and V. Singh, Phys. Rev. Lett. 51, 2069 (1983).

R5. S.M. Roy and V. Singh, 'Theta vacua for charged Klein-Gordon field: a new quantization', TIFR preprint TH/85-2.

R6. R. Rajaraman and J.S. Bell, Phys. Lett. B116, 151 (1982).

R7. S.M. Roy and V. Singh, Phys. Lett. 143B, 179 (1984); Pramana 23, 333 (1984).

R8. R.D. Richtmyer, "Principles of Advanced Mathematical Physics", Vol.1, Springer Verlag, N.Y. (1978).

R9. M.J. Rice, Phys. Lett. A71, 152 (1979).

R10. S.M. Roy and V. Singh, "Fractionally charged non-leaking dyons & fermions in a bag", Pramana (to appear).

R11. A.N. Redlich, Phys. Rev. Lett. 52, 18 (1984) and Phys. Rev. D29, 2366 (1984).

S1. L.S. Schulman, Phys. Rev. 176, 1558 (1968).

S2. L.S. Schulman, "Techniques and applications of path integration", N.Y., John Wiley (1981).

S3. V. Singh, report at XXII Intern. Conf. on High Energy Physics, Leipzig (1984).

S4. W.P.Su, J.R. Schrieffer and A.J. Heeger, Phys. Rev. Lett. 42, 1698 (1979); Phys. Rev. B22, 2099 (1980).

S5. W.P. Su & J.R. Schrieffer, Phys. Rev. Lett. 46, 681 (1981).

S6. T.H.R. Skyrme, Proc. Roy. Soc. A260, 127 (1961).

S7. Y. Srivastava and A. Widom, Nuov. Com. 37, 77 (1983).

S8. J.F. Schonfeld, Nucl. Phys. B161, 125 (1979).

T1. R. Tao and Y.-S. Wu, 'Fractional statistics and fractional quantized Hall effect', Seattle preprint 1984.

T2. H. Takayama, Y.R. Lin-Liu, and K. Maki, Phys. Rev. B21, 2388 (1980).

T3. D. Tsui, H. Stormer and A. Gossard, Phys. Rev. Lett. 48, 1559 (1982).

W1. E. Witten, Phys. Lett. 86B, 283 (1979).

W2. F. Wilczek & A. Zee, Phys. Rev. Lett. 51, 2250 (1983).

W3. Y.S. Wu, Phys. Rev. Lett. 52, 2103 (1984), 53, 111 (1984) and (Erratum) 53, 1028 (1984).

W4. A.S. Wightman, in Proc. 5th Ann. East. Th. Phys. Conf., Ed. D. Feldman (N.Y. 1967) p.120, and in 1964 "Cargese lectures in theoret. phys; high energy electromagnetic interactions and field theory".

W5. T.T. Wu and C.N. Yang, Phys. Rev. D12, 3845 (1975).

W6. E. Witten, Nucl. Phys. B223, 433 (1983).

W7. A. Widom and T.D. Clark, Phys. Lett. 90A, 280 (1983).

Y1. H. Yamagishi, Phys. Rev. D27, 2383 (1983); Phys. Rev. Lett. 50, 458 (1983).

Y2. H. Yamagishi, Phys. Rev. D29, 2975 (1984).

Z1. Ya. B. Zel'dovich and V.S. Popov, Usp. Fiz. Nauk 105, 403 (1971) [Sov. Phys. Usp. 14, 673 (1972)].

Z2. B. Zumino, Les Houches lectures Aug. 1983, and B. Zumino, Y.S. Wu and A. Zee, Nucl. Phys. B239, 477 (1984).

DYNAMICS NEAR SECOND ORDER PHASE TRANSITIONS

Richard A. Ferrell

Department of Physics, University of Maryland
College Park, Maryland 2074, USA
&
Jayanta K. Bhattacharjee

Department of Physics
Indian Institute of Technology, Kanpur 208016, India

1. STATICS RECALLED

The second order phase transition[1)] is characterized by an order parameter field which acquires a nonzero expectation below the critical temperature T_c i.e., in the ordered phase. As T_c is approached the fluctuations in the order parameter become correlated over large distances. This is expressed in terms of a correlation length ξ which becomes larger with approaching T_c. The inverse of the correlation length $\kappa = \xi^{-1}$ plays the role of the mass in the field theory of phase transitions. At $T = T_c$, the correlation length becomes infinity and one has a massless theory. The correlation function is the propagator of the field theory. The Fourier transformed correlation function $g(k,\kappa)$ is a function of the momentum k and the mass κ and to a very good approximation is given by the Ornstein-Zernike form

$$g(k,\kappa) = (k^2 + \kappa^2)^{-1} \tag{1.1}$$

The correlation length has a power law behavior as $T \to T_c$ i.e., $\kappa \to 0$ and this is expressed by

$$\xi \sim (T - T_c)^{-\nu} \tag{1.2}$$

The thermodynamic susceptibility χ is the long wavelength limit of the correlation function and from Eq. (1.1),

$$\chi = g(k, \kappa)\,|_{k=0} = \kappa^{-2} \propto (T-T_c)^{-\gamma} , \tag{1.3}$$

where $\gamma = 2\eta$. A more accurate treatment which takes deviations from the form of Eq. (1.1) into account would yield $\gamma = (2-\eta)\nu$, where η is the anomalous dimension exponent. The exponent η is generally very small (0(0.04)) and difficult to measure accurately in an experiment. The indices γ, η, ν, etc. are the critical exponents. They depend only on the dimensionality D of space and the dimensionality n of the order parameter. They are independent of the details of the interactions at the atomic level and this feature is called universality. Universality classes are determined by the value of n and D. Let us take n=1. This implies that the order parameter has just one component - it is a scalar. Examples are binary liquids, Ising model, liquid-gas systems. In a binary liquid, the relevant fluctuations that become very strong at the critical point, the so called consolute point in the binary liquid, are concentration fluctuations. At the consolute point the liquids tend to separate and produce co-existing phases. One phase is rich in one of the components and the other phase is rich in the other component. In the Ising model the order parameter is the magnetisation, with the magnetisation allowed only one direction (spin 'up' or spin 'down') which makes it a scalar. In the liquid-gas system the order parameter is the density which is clearly a scalar. The point to note is that in spite of the very different natures of these systems, they all exhibit identical critical behaviour for a given D since they all have scalar order parameters, n=1. This is universality.

The prime example of a n=2 transition is the lambda transition in liquid Helium. The order parameter is the condensate wavefunction which is complex and hence characterized by two components. The isotropic Heisenberg ferromagnet forms the best example of a n=3 system.

We now show that there is an intimate connection between the statistical mechanical problem of phase transitions in a D-dimensional Euclidean space and a D-1 dimensional quantum theory in Minkowski space. To do so, we note that the partition function in the statistical mechanical problem can be written as,

$$Z = \sum_{\{\phi\}} \exp(-\int d^D x\, F(\phi)) \;, \tag{1.4}$$

where the sum is over all the configurations of the field ϕ and $F(\phi)$ is the free energy functional that defines the probability. Writing 'it' for one of the coordinates x we can write Eq. (1.4) as

$$Z = \sum_{\text{paths}} \exp(i \int dt \int d^{D-1}x\, L) \tag{1.5}$$

where L is the Lagrangian density. The sum over paths leads to the unitary time evolution operator and hence

$$Z \propto \text{Tr} \exp(-i\, t\, H) = \text{Tr} \exp(-lH) = \exp(-lE_g), \tag{1.6}$$

where l is the length of the system in the direction which had been changed to 'it' and E_g is the ground state energy of the system. The last step in Eq. (1.6), follows from the fact that in the thermodynamic limit when $l \to \infty$, only the lowest eigenvalue will survive. The thermodynamic free energy is the logarithm of the partition function and is thus equal to the ground state energy of the D-1 dimensional quantum system. Taking two derivatives of the energy of the quantum system yields the propagator, taking two derivatives of the free energy yields the correlation function - the propagator of the field theory.

The first exactly solvable model which gave non-trivial exponents was the two-dimensional Ising model[2-4]. Spins which can only point 'up' or 'down' are placed on the sites of two-dimensional square lattice and they interact with their nearest neighbours. Denoting the horizontal and vertical coordinates by i and j respectively, the energy of the Ising lattice can be written as

$$E = - T J \sum_{ij} \sigma_{ij} \sigma_{i+1j} - T J' \sum_{ij} \sigma_{ij+1} \sigma_{ij}, \tag{1.7}$$

where T is the temperature and TJ and TJ´ are the interaction strengths in the two directions.

The technique of dealing with the problem of calculating the partition function is the method of transfer matrix. It is basically a way of working through the system under study proceeding in a certain direction. The transfer matrix method has been greatly simplified by Schultz et al.[3] who noted that the dichotomy in the Ising spin variable (up or down) could be transformed to a ´yes-or-no´ question referring to the presence or absence of certain fictitious particles. In order to avoid the undesired possibility of multiple occupation of a site by these particles, they are assumed to be ferminos and obey the Pauli exclusion principle. These fictitious particles are governed by a very simple interaction - it is one corresponding to the presence of a field which spontaneously emits and absorbs pairs of particles possessing zero net total momentum. This is the kind of effect that is encountered in superconductivity in the BCS theory and can be handled by methods developed for that. Instead of that, the problem can be further simplified[4] by the help of a transformation which makes it identical to the problem of non-interacting spinless fermions in one dimension. What one is now required to do is find the ground state energy of the fermion problem and that provides the free energy of the statistical mechanical problem we started with.

The ground state energy E_g of the fermion problem is found by filling the negative-energy Dirac sea and thus

$$E_g = - \sum_q (\kappa^2 + q^2)^{1/2} = \frac{L}{\pi} \int_o^{q_c} dq(\kappa^2 + q^2)^{1/2} \tag{1.8}$$

where the mass κ of the Dirac particles is found to be proportional to $T-T_c$, and where T_c is the transition temperature. This yields $\nu = 1$. To obtain the specific heat we need two derivatives of E_g, the free energy of the Ising system and this yieds as $\kappa \to 0$,

$$C \propto \ln (q_c/\kappa), \tag{1.9}$$

i.e., the specific heat is logarithmically divergent. This simplified version does not allow one to compute γ or η. Using the full force of the Onsagar technique one finds $\gamma = 1.75$ and $\eta = 0.25$ which satisfies the scaling law $\gamma = (2 - \eta)\nu$.

The exact solution to the three-dimensional Ising model has not been found yet. For dimensions between 2 and 4, the ε expansion, in conjunction with instanton techniques to yield high order terms and Borel summation procedures, give fairly accurate answers for the critical exponents. For $D \simeq 4$ and $n \simeq 4$ it is found that the specific heat is logarithmic once more. Calculations also show a logarithmic specific heat for $D=3$ and $n \simeq 2$. For the superfluid transition $n=2$ and Ahlers[5)] measurement of the specific heat near the transition temperature T_λ does indeed show a very nearly logarithmic specific heat.

The specific heat we have talked about so far is the thermodynamic specific heat which is obtained as the $k \to 0$ limit of the k and κ-dependent correlation function of the square of the order parameter. The momentum-dependent function can be conveniently described by the spectral function. The physical interpretation of the spectral function is the pair production rate and hence it has a threshold. The pair mass is 2κ and hence the relevant variable u which has the dimensions of momentum squared has to be greater than $4\kappa^2$. The decay rate is obtained from the 'Golden rule' of perturbation theory and is equal to the product of the density of states and the square of the matrix element. For the two-dimensional Ising model just considered the quantum mechanical problem[6)] is in one dimension and hence the density of states is dp/dE, with $E^2 = p^2 + m^2$ for these relativistic particles. The matrix element is proportional to p/E and hence the decay rate F is given by

$$F \propto (p/E)^2 \cdot dp/dE = P/E = v \tag{1.10}$$

At D=4, when one has a free boson theory, the same answer as in Eq. (1.10) is obtained for the rate. The phase space in the quantum problem now corresponds to D-1 = 3 and the matrix element is proportional to E^{-1} and thus we get back Eq. (1.10). The scaling functions for the specific heat is thus identical in the two cases where the thermodynamic specific heat is logarithmic. One is tempted to believe that for the lambda transition where the specific heat is nearly logarithmic the above rate would give very nearly the correct answer. The scaling function for the specific heat is further discussed elsewhere in this volume. Light scattering near the lambda point probes the specific heat scaling function. We will not discuss here the order-parameter correlation function[8)].

2. INTRODUCING DYNAMICS

We now turn to the question of how long does it take an order-parameter fluctuation of wavelength λ (wave-number k) to return to equilibrium once it is disturbed from it. As the fluctuations become more and more long ranged with approaching T_c, the longer it takes to return to equilibrium. The relaxation rate γ goes to zero as T_c is approached (i.e., $\kappa \to 0$), in a characteristic manner. Alternatively, we can set $\kappa = 0$ ($T=T_c$) and study the relaxation rate as a function of the wavenumber k. As the wavenumber of the fluctuation decreases, i.e., the wavelength increases, the relaxation time becomes larger and in the limit $k \to 0$, the rate $\gamma \to 0$ in a manner which is described by

$$\gamma(k) \propto k^{z_o}, \tag{2.1}$$

where z_o is the dynamic scaling exponent[9)]. The simplest model for dynamics consists of writing down Langevin equations for the order-parameter. A linear equation of motion for the kth Fourier component $\phi(k)$ of the order parameter has the form

$$\dot{\phi}(k) = -\Gamma \chi^{-1} \phi(k) + N, \tag{2.2}$$

where the noise N comes from averaging over the short range fluctuations. It is a Gaussian white noise which ensures that the equilibrium distribution is left unaltered by the above equation of motion. The suscetibility χ is very well approximated by the Ornstein-Zernike form $\chi = (k^2 + \kappa^2)^{-1}$. The Onsager coefficient Γ has to be written as Γk^2 if the order parameter follows a conservation law as is the case with the liquid-gas system, the binry liquid or the ferromagnet. The mean field or Van-Hove theory assumes that is a constant independent of k and κ. Setting $T=T_c$ or $\kappa = 0$, we immediately find that

$$z_o = \begin{cases} 2 \text{ for non-conserved order parameter} \\ 4 \text{ for conserved order parameter} \end{cases} \tag{2.3}$$

None of these exponents are borne out by experiments.

The error lies in the assumption of a non-critical Γ. Nonlinear terms that have been ignored in the equation of motion (Eq. (2.2)) invest Γ with a k-and κ-dependence. The first nonlinearity that comes to mind is the one coming from the quartic term in the free energy - the effect of this nonlinearity is miniscule for the order parameter correlations. Its importance lies in the discussion of the frequency dependent specific heat. The important nonlinearities are the reversible terms in the equation of motion. We shall discuss the binary liquid and the single-component fluid as examples of systems with conserved order parameters and the lambda transition of liquid He^4 as the example of a system with non-conserved order parameter. It is in these two systems that certain issues remain open, requiring further investigation. Some of the recent advances, controversies and open issues will be described in Sec. III for the binary liquid and single-component fluid and in Sec. IV for the lambda transition of ^4He. The phenomenon of ultrasonic attenuation and dispersion has recently been well understood for all these systems. This will be the topic of discussion in Sec. V.

In the following sections the equation of motion will be used with the reversible terms alone[10]. This is perfectly permissible for those problems in dynamics in which the contributions of the dissipative nonlinear terms are negligible. The dissipative linear term and the noise can also be dropped because the reversible terms give rise to an effective rate which dominates over the non-critical rate of Eqs. (2.1) and (2.3), unless one is very from the critical region. The reversible terms are always constructed in a manner[10] that the equilibrium free energy is left invariant in time and that allows for consistent definition of time-dependent equilibrium correlation functions. If the order parameter fluctuation of wavelength k has the value $\phi(k, t_1)$ at time t_1 and the value $\phi(k, t_2)$ at time t_2, because of the time development of the system, the equilibrium time-dependent correlation function is

$$G(k, \kappa, t_{12}) = \langle \phi(k, t_1) \; \phi(-k, t_2) \rangle , \tag{2.4}$$

and the study of this and related correlations will be the topic of the next two sections.

3. BINARY LIQUID AND THE LIQUID-GAS SYSTEM

We first consider the binary liquids. The order parameter is the difference in concentration from a critical value. The concentration satisfies a conservation law and is thus coupled to the velocity functions. The Eulerian derivative $d/dt = \partial/\partial t + (\vec{v} \cdot \vec{\nabla})$ couples the concentration and velocity fluctuations. We ignore for the present purpose the noncritical diffusion $D\nabla^2 c$ and the noise (the two terms on the right hand side of Eq.(2.2)) as explained towards the end of the previous section and consider the equation of motion in the form

$$\frac{\partial c}{\partial t} = -(\vec{v} \cdot \vec{\nabla})c. \tag{3.1}$$

The Fourier component $c(k)$ satisfies

$$\dot{c}(k) = -i \sum_{p} \vec{k} \cdot \vec{v}(\underset{\sim}{p}) \, c(\underset{\sim}{k} - \underset{\sim}{p}) \tag{3.2}$$

The time-dependent correlation function of the order parameter has the form

$$g_c(k,t_{21}) = \langle c(\underset{\sim}{k},t_2)c(-\underset{\sim}{k},t_1)\rangle$$

$$= k_B T(k^2+\kappa^2)^{-1} \exp(-Dk^2\, t_{21}) , \qquad (3.3)$$

where D is the effective diffusion constant arising from the nonlinear term in Eq. (3.1) or (3.2). The velocity field is transverse because of the incompressibility condition $\vec{\nabla}\cdot\vec{v} = 0$. The correlations of the transverse velocity decay according to the law

$$g_v(k,t_{21}) = \langle v_i(k,t_2)v_j(-k,t_1)\rangle$$

$$= \frac{1}{m}\, k_B T \quad \delta_{ij} \exp(-\eta k^2 t_{21}), \qquad (3.4)$$

where η is the shear viscosity. To close the system the equation of motion for the transverse velocity has to bedetermined. This is conveniently done by requiring that the free energy functional F = $\int d^D x\,(1/2\ \kappa^2\ \phi^2 + 1/2\,(\nabla\ \phi)^2 + 1/2\ v^2)$ be time-independent.

Having obtained the equations of motion, the time-dependent correlation function of Eq. (3.3) is obtained as a perturbation series. The negative of the self-energy thus obtained corresponds to the relation rate Dk^2. Standard techniques[11) lead at the one-loop level to the expression

$$D(k,\kappa) = \chi^{-1}\int \frac{d^3p}{(2\pi)^3}\int_0^\infty dt_{21}\, g_c\,(\underset{\sim}{k}-\underset{\sim}{p},\ t_{21})g\,(\underset{\sim}{p},\ t_{21})\sin^2\theta$$

$$= k_B T\,(k^2+\kappa^2)\int \frac{d^3p}{(2\pi)^3}\ \frac{\sin^2\theta}{[(k-p)^2+\kappa^2]}\ \frac{1}{[D(p-k)^2+\eta p^2]}$$

$$\simeq k_B T\,(k^2+\kappa^2)\int \frac{d^3p}{(2\pi)^3}\cdot\frac{\sin^2\theta}{(k-p)^2+\kappa^2}\cdot\frac{1}{\eta p^2} \qquad (3.5)$$

In Eq. (3.5), θ is the angle between k and p and the last step follows since ηp^2 is much greater (orders of magnitude) than $D(p-k)^2$. Because of certain geometric factors the higher loop terms in the perturbation theory are at least one order of magnitude smaller and hence Eq. (3.5) provides answers which compare favourably with accurate experiments.

We first note that if we assume the viscosity to be non-critical, then Eq. (3.5) yields[11,12)]

$$D(k, \kappa) \propto \begin{cases} k & \text{for } k \gg \kappa \\ \kappa & \text{for } \kappa \gg k \end{cases} \tag{3.6}$$

Evaluation of the integral leads to the relaxation rate

$$\gamma(k, \kappa) = k^2 D(k, \kappa) = k^2 \frac{k_B T \kappa}{6\pi\eta} F(x) \quad , \tag{3.7}$$

where

$$F(x) = (3/4x^2)\{1 + x^2 + (x^3 - 1/x)\tan^{-1}x\}. \tag{3.8}$$

It is seen from Eqs. (3.7) and (3.8) that $\gamma(k,0) \propto k^3$, which implies that $z_o = 3$ for the binary liquid as opposed to the $z_o = 4$ expected on the basis of the mean field theory.

The assumption of a non-critical viscosity needs to be investigated next. Using the equations of motion one constructs the self-energy for the time-dependent velocity autocorrelation function and using $\gamma(k) \sim k^3$, one finds η has a logarithmic divergence[11)] as $T \to T_c$ or as $k \to 0$ at $T = T_c$. A logarithmic divergence at leading order generally implies a small critical exponent. Assuming that $\eta(k) \sim k^{-z_\eta}$, we find that

$$z_o = 3 + z_\eta \tag{3.9}$$

and z_η is to be determined by a self-consistency requirement on the order-parameter and velocity self-energies[13,14]. A two loop ε expansion[15] leads to $z_\eta = 0.065$, while a two-loop calculation directly in three dimensions leads to $z_\eta \simeq 0.051$. The three dimensional calculation is expected to have an accuracy of few percent.

Turning to the experimental scene, we see that all experiments show a distinct rise in viscosity with approaching T_c. Typical data[16] is shown in Fig. 1. Since the binary liquid and the liquid gas system are in the same universality class, an identical behavior is expected of the viscosity in the liquid gas system. The measurements in the liquid-gas systems[17,18] generally favour a viscosity exponent z_η near 0.05, while the measurements in the

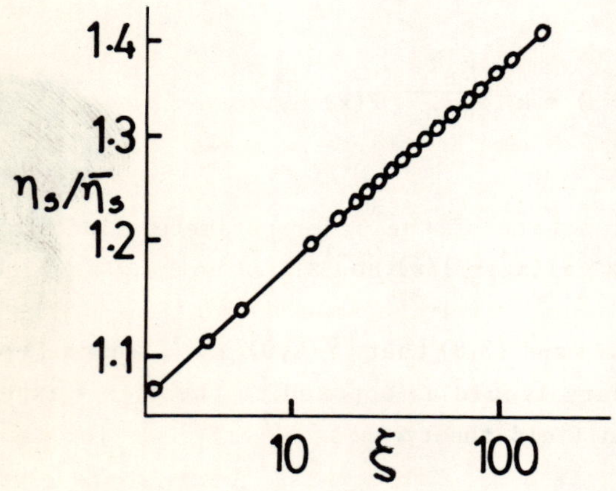

Fig. 1 : Divergence of shear viscosity.

binary liquids[16] favour an exponent close 0.06. An indirect measurement of z_η comes from the light scattering experiments. The frequency resolved light scattering at different wave vectors and temperatures close to T_c provide a measurement of the relaxation rate $\gamma(k, \kappa)$. Correct extrapolation[14] to T_c of the highly accurate light scattering work of Burstyn and Sengers[19] gives $z_\eta \simeq 0.054$.

The issue remains open. More accurate hydrodynamic measurements and light scattering experiments on different binary mixtures are required for a proper confrontation between theory and experiment.

The fact that $\gamma(k) \sim k^{3+z_\eta}$ requires small corrections to the scaling form of Eq. (3.7). The systematic experimental deviations from Eq. (3.7) because of the small exponent z were noted by Burstyn et.al.[20] A theoretical scaling function which is in good agreement with the data of Burstyn et.al.[19] was constructed[21], giving

$$\gamma(k, \kappa) = k^2 \cdot \frac{k_B T\kappa}{6\pi\eta} \cdot R(1 + b^2x^2)^{z/2} \cdot F(x) , \tag{3.10}$$

where F(x) is defined in Eq. (3.8), R = 1.03 and b = 0.55.

A further manifestation of the critical viscosity is seen in the shape of the fluctuation spectrum. On the basis of Eq. (3.3), one expects a Lorentzian line with $D(k, \kappa)k^2$ giving the line width. Turning to Eq. (3.5) we see that $D(k, \kappa)$ is the zero-frequency limit of a self-energy which in general would be frequency dependent. The shear viscosity η has a divergence only if the frequency is much less than the relaxation rate of the order-parameter fluctuations. As we approach the critical point and consider the long wavelength fluctuations, it is possible that the frequency of the probe will exceed the relaxation rate. The viscosity now ceases to diverge at long wavelengths and acquires a limiting value determined by the external frequency. The larger the frequency, the earlier will be the rise of the viscosity cut off, which will give a smaller viscosity. Turning now to Eq. (3.5) we see that a frequency-dependent diffusion would be caused by a frequency-dependent viscosity, and hence at higher frequencies we expect a higher diffusion coefficient. The result of a frequency-dependent diffusion will be deviations from the Lorentzian shape. In real time this implies non-exponential decay of the fluctuations. We have just seen that the decay rate will be higher at shorter time and hence the semi-log plot of the correlation function against time will curve up. The effect is of the order of a few percent[22] and has been verified[19] experimentally for the binary liquid 3-methylpentane + nitroethane.

As noted earlier, the liquid-gas transition falls in the same universality class as the binary liquid. The role of the order parameter is played by the entropy and the analogue of the diffusion constant D for the binary liquid is the heat diffusion λ/C_p, where λ is is thermal conductivity and C_p is the specific heat at constant pressure. For $\dot{\kappa} >> k$, it follows from this universality and Eq. (3.6) that

$$\lambda \propto C_p \; \kappa \propto \xi^2 \; \xi^{-1} = \xi \; . \tag{3.11}$$

Thus the thermal conductivity diverges as T_c is approached. The experimental demonstration of this fact by Sengers[23)] marked the beginning of research in dynamic critical phenomena.

A recent development[24)] has been the experimental study of transport coefficients near the liquid-vapour transitions of mixtures. It was found by Cohen et al.[24)] that the divergence of the thermal conductivity is exactly the same as that of the pure fluid. A theoretical understanding is lacking.

4. THE LAMBDA TRANSITION

The lambda transition in ^{4}He has drawn considerable attention in the last few years. This followed the work of DeDominicis and Peliti[25)], which pointed out the possibility of violation of dynamic scaling in this system where the principle of dynamic scaling was first established[26)]. The phase of the complex order parameter changes by coupling to the entropy fluctuations. This is the content of the Josephson equation. The closure of the equations of motion is ensured by writing down the two-fluid equation for the rate of change of entropy. For the low frequency properties of interest here, the incompressibility condition is always assumed. Below the lambda temperature T_λ , the linearized equations of motion allow the propagation of second sound waves. These are the Goldstone modes of the system. The second sound velocity is seen to be

$$c_2 \propto \left(\frac{\rho_s}{C_p}\right)^{\frac{1}{2}} \propto \kappa^{D/2 - 1} \tag{4.1}$$

where the second step follows by ignoring the comparatively weak critical variation in C_p and using the Josephson relation $\rho_s \sim \kappa^{D-2}$. This relation is valid in the hydrodynamic region where $\kappa >> k$. As one approaches the critical point and crosses the line $\kappa = k$ to values $\kappa << k$. As the correlation length exceeds the wavelength, no further changes occur in the responses of the system and hence for $\kappa \to 0$ at a finite k, c_2 reaches the limiting value

$$c_2 \propto k^{D/2 - 1} \text{ for } k >> \kappa . \tag{4.2}$$

The lambda point dispersion relation now reads

$$\omega = c_2 k \propto k^{D/2} \tag{4.3}$$

The dynamic scaling exponent is $z_o = D/2$ instead of $z_o = 2$ as expected from the Van Hove theory.

The second sound will in general be damped, giving a complex frequency of propagation $\omega = c_2 k + i D_2 k^2$, where D_2 is the damping coefficient. Clearly $D_2 k^2$ must have the same dimensions as $c_2 k$ and hence at $T = T_\lambda$, $D_2 k^2 \sim k^{D/2}$ or $D_2 \sim k^{D/2-2} = k^{-\varepsilon/2}$ where $\varepsilon = 4-D$. If we now proceed to the hydrodynamic region below T_λ, the dynamic scaling hypothesis asserts that for $\kappa >> k$

$$D_2 \propto \kappa^{-\varepsilon/2} \propto t^{-\frac{\varepsilon\nu}{2}} = t^{-1/3} \text{ , for } D = 3 \tag{4.4}$$

The propagating mode below T_λ becomes the diffusive Rayleigh peak above T_λ. D_2 now represents the thermal diffusion λ/C_p where λ is the thermal conductivity and C_p the specific heat at constant pressure. For $\kappa >> k$ and $T > T$ we then have

$$\lambda \propto C_p \kappa^{-\varepsilon/2} \propto C_p t^{-1/3} \text{ for } D = 3 \tag{4.5}$$

The predictions of Eqs. (4.5) and (4.4) were verified experimentally by Ahlers[27)] and Tyson[28)] soon after their publication in 1967.

Subsequent more refined measurements by Ahlers[29] of the thermal conductivity showed an exponent about 20% higher than the 1/3 of Eq. (4.5). A decade after the dynamic scaling predictions, DeDominicis and Peliti[25] made an important discovery. An approach equivalent to theirs is to consider a model with n-component order parameter $\{\psi_i\}(i=1,2, \ldots n)$. The equation of motion for ψ is taken to be a rotation generated by an antisymmetric tensor S_{ij} with $n(n-1)/2$ components. The equation of motion for S_{ij} is determined by the constraint that the free energy density $\frac{1}{2}\kappa^2\sum_i \psi_1^2+\frac{1}{2}\sum_i(\nabla\psi_i)^2+\frac{1}{2}\sum_{ij} S_{ij}^2$ is independent of time. This model can be analysed in D-dimensions. The lambda transition of He^4 corresponds to n=2 and D=3. Dynamic scaling gives for the momentum-dependent relaxation rates

$$\gamma_s = a_s\, k^{D/2}$$

and

$$\gamma_\psi = a_\psi\, k^{D/2} \tag{4.6}$$

where a_s and a_ψ are constants. Writing down the expressions for the two relaxation rates one finds that a smaller γ_ψ implies a larger γ_s and a larger γ_s leads to a smaller γ_ψ. This is a destabilizing feedback and hence for some values of n and it will be possible to have $w = a_\psi / a_s = 0$. This curve which is the locus of points where w = 0 in the n- ε plane, divides the plane into a scaling and non-scaling region. For a given n if $n > n_c$, the value on the curve, one has dynamic scaling, for $n < n_c$ the scaling solution is unstable. DeDominicis and Peliti evaluated this boundary to two loop order and found that

$$n_c = 3/2 + 0.41\,\varepsilon \tag{4.7}$$

Thus at $\varepsilon = 1$, $n_c = 1.91$. Helium, with n = 2, is very close to the critical value. A three loop calculation[30] which gives the quadratic term in Eq. (4.8) has recently been computed. The coefficient of ε^2 is very small and the critical value for $\varepsilon = 1$ is $n_c \simeq 1.9$. This keeps He^4 safely on the scaling side, but being this close to the instability causes a number of observable effects, which we discuss below.

It has been known for quite some time that the asymptopic scaling laws such as those expressed in Eqs. (1.2), (1.3), (2.1) etc. are not the complete story. These are to be seen for extremely small values of t or k, as the case may be. If one is not very close to the critical point, it is necessary to include correction terms of the form

$$A(k) = A_{scaling}(k)\left[1 + \sum_{i,m} a_{im} k^{m\omega_i}\right], \tag{4.8}$$

where A is any physical quantity showing critical behaviour and $A_{scaling}(k)$ is the asymptotic scaling form. The exponents ω_i are called the correction to scaling exponents and are universal quantities. The amplitudes a_i are however non-universal. The exponents $\omega_i > 0$, so that for small k, the correction terms vanish. The number of ω_i in dynamics is equal to the number of dynamical variables and m in Eq. (4.8) is an integer running from 1 to infinity. In the case of ^{4}He with two dynamical variables, the order parameter and the entropy S, there are two transients denoted by ω_w and ω_f. On the boundary given in Eq. (4.7), $\omega_w = 0$, if $n > n_c$, $\omega_w > 0$, while for $n < n_c$, $\omega_w < 0$ signifying the instability of the scaling solution. Now if $n > n_c$ but very close to it, we will have a situation where ω_w is positive, but very small. This implies that correction terms of the form k^{ω_w} will be significant even when k is small and the onset of the scaling region will be postponed[31]. This explains qualitatively why the accurate measurements of the thermal conductivity show an exponent different from the 1/3 found in Eq. (4.5).

To proceed quantitatively we note that for $\kappa \simeq 0$

$$\lambda(\kappa) = C_p^{\frac{1}{2}} \kappa^{-\frac{1}{2}} (1 + a_1 \kappa^{\omega_w} + O(\kappa^{2\omega_w})) \tag{4.9}$$

The magnitude of a_1 is non-universal but its sign can be easily settled. Since $w \ll 1$ for $t \simeq 0$, the order parameter relaxation rate is very small compared to the entropy rate. In the noncritical region $t \simeq 0.1$, the two rates are very nearly the same since both would correspond to a diffusion constant of $O(h/m)$. It can be shown from the theory that if the order parameter relaxation rate is written as

$$\gamma_{\psi}(\kappa) \propto \kappa^{3/2} \left(1 + b_1 \kappa^{\omega_w} + O(\kappa^{2\omega_w})\right) , \tag{4.10}$$

then a_1/b_1 is negative. The situation consistent with the above description is $b_1 > 0$ and $a_1 < 0$. With the sign of a_1 being negative the schematic drawing of $\lambda(\kappa)$ in the approximation of Eq. (4.9) is as shown by the appropriate curve in Fig. (2). Since $a_s > a_{\psi}$, this curve is very much above the lower one.

We now drop down to the other end i.e., the noncritical one. One can set up a systematic expansion[32] in powers of κ^{-1} to express the thermal conductivity as

$$\lambda = \lambda_B + \sum_{n=1}^{\infty} \frac{\lambda_n}{\kappa^n} \tag{4.11}$$

The coefficient λ_n is obtained from a n-loop calculation and λ_B is the noncritical background. This form implies that far from the asymptopic critical region the thermal conductivity vs κ^{-1} (i.e., $t^{-2/3}$) plot should be linear. This is indeed true in the range $10^{-1} < t < 10^{-3}$, after which the correlation length κ^{-1} is large enough to make the higher order terms important. The κ^{-1} addition to the background is called the precritical term and the curve for the thermal diffusivity with the precritical rise included is shown as the curve in Fig. 2.

What one now needs is an interpolation curve connecting the behaviour of Eq. (4.10) valid for very small to the behaviour in the precritical region. There are several technical prescriptions[33,34] for doing this - here we point out the obvious fact which requires no calculation. The curve interpolating between the scaling part and precritical rise part of Fig. 2 must needs have an inflection point.

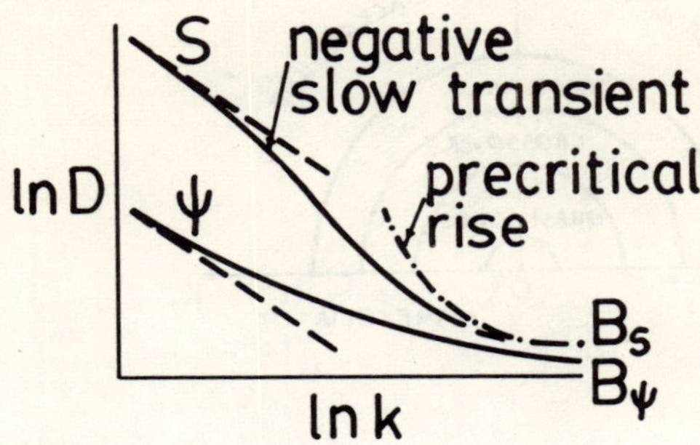

Fig. 2 : The enitropy(s) and order parameter (ψ) kinetic coefficients $D_{s,\ \psi}$ vs k. Dashed lines are the asymptotic behaviours and dot dashed line is the precritical rise.

Simply by the nature of the inflection point, the index $d \ln \lambda / d \ln \kappa$ will be a constant around $\kappa = \kappa_o$ where the inflection point occurs. That means around $\kappa = \kappa_o$, we will appear to have a scaling behaviour with the exponent $(d \ln \lambda / d \ln \kappa)_{\kappa=\kappa_o}$. This region is termed the quasi-scaling region[35]. The experiments (Fig. 3) we have been talking about happen to be performed in this quasi-scaling regime and hence they seem to suggest a scaling with an exponent larger than that obtained from the dynamic scaling predictions. The asymptopic scaling region seems to be confined to inaccessible temperatures. The situation is summarised in Fig. 3.

Turning to the frequency-resolved light scattering we see that the finite wavenumber associated with the scattering ensures that the experiment will never probe the critical region. Consequently the spectra of Tarvin, Vidal and Greytak do not show the broadening of the Rayleigh line that one would normally expect to see as one approached T. The observations of Tarvin et.al[7] have been satisfactorily explained quantitatively[36] as well.

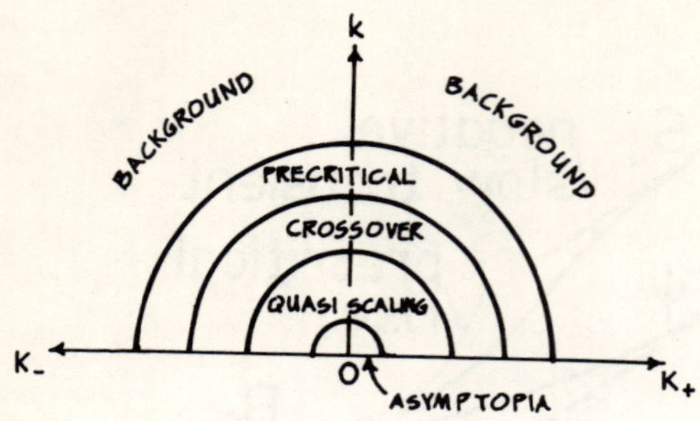

Fig. 3 : Different regions of the k-κ plane

Finally we address the question of second sound damping[37,38]. The real challenge here is in understanding the damping relatively farther away from the lambda point in the temperature range $10^{-3} < t < 10^{-1}$. Several effects like viscous damping[39], contribution from the order parameter relaxation[40] etc. have to be taken into account before agreement with measurements can be obtained. None of these effects are contained in model E or F of Hohenberg and Halperin[9] - yet calculations[34,41] based on those models claim good agreement with experiments. This remains a puzzle.

5. CRITICAL ULTRASONICS

A. Binary Liquids

In this case the sound velocity does not show any appreciable critical dependence, but the attenuation shows a marked increase as the critical consolute point is approached. The increase of the attenuation above the classical Stokes-Kirchoff formula is due to the slow energy exchange between the short-ranged fluctuations which define the local temperature, and the long-ranged fluctuations characteristic of the critical region. The resultig relaxation phenomenon involving a continuous set of relaxation rates $\gamma(k, \kappa)$ gives rise to a complex frequency-dependent specific heat[42-44] and the excess sound absorption. Using the adiabatic condition[45] of sound propagation, the compressibility can be related to the specific heat. The propagation veclocity, which is proportional square root of the compressibility, is thus related to the specific heat and becomes complex in the critical region. The imaginary part of the velocity gives the critical attenuation; the real part yields the dispersion.

We first note that the critical consolute point T_c changes with pressure and we get a "lambda" line of critical points $T_c(p)$. The temperature variable will be taken as $\Delta T = T - T_c(p)$. To determine the sound velocity[46], we need $\partial V/\partial P$ under entropy clamping. The volume V is found from the Gibbs free energy as

$$V = \left.\frac{\partial G}{\partial P}\right)_T = \left.\frac{\partial G}{\partial P}\right)_{\Delta T} + \left.\frac{\partial G}{\partial \Delta T}\right)_P \left.\frac{\partial \Delta T}{\partial P}\right)_T$$

$$= \left.\frac{\partial G}{\partial P}\right)_{\Delta T} + T_c' S \tag{5.1}$$

where $T_c = d\, T_c'/d\, P$. Integrating away from the λ -line at constant pressure yields

$$G(P, \Delta T) = G_c(P) - \int_0^{\Delta T} S(P, \Delta T')d(\Delta T') \tag{5.2}$$

The required derivative in Eq. (5.1) is parallel to the lambda line. We can consequently assume $\left(\frac{\partial S}{\partial P}\right)_{\Delta T} \simeq \frac{\partial S_c}{\partial P} = S_c'$ along the lambda line, and obtain from Eq. (5.2)

$$\left.\frac{\partial G}{\partial P}\right)_T = G_c' - S_c' \Delta T \tag{5.3}$$

Coupled with Eq. (5.1), this yields

$$V = G_c' - S_c' \Delta T + T_c' \Delta S \tag{5.4}$$

Differentiation yields

$$\frac{\partial V}{\partial P}\Big)_S = G_c'' + T_c'' S_c + T_c'' \Delta S - S_c'' \Delta T - S'_c \frac{\partial \Delta T}{\partial P}\Big)_{S'} \quad (5.5)$$

Dropping the first order terms ΔS and ΔT and noting that

$$\frac{\partial(\Delta T)}{\partial P}\Big)_S = -\left(\frac{\partial S}{\partial P}\right)_T / \left(\frac{\partial S}{\partial T}\right)_P \cong -\frac{T_c S'_c}{C_P} = -g \frac{V_c}{C_P} ,$$

where

$$g = \frac{T_c S'_c}{V_c} , \quad (5.6)$$

we can write the isentropic compressibility $\beta_S = -\frac{1}{V} \frac{\partial V}{\partial P}\Big)_S$ as

$$\beta_S = \beta_S^c - \frac{V_c}{T_c} \cdot \frac{g^2}{C_P} \quad (5.7)$$

This quantity has been measured directly by Tanaka et.al.[47)] and excellent agreement with the above result found. The second term is much smaller than the first and since the critical behavior is contaied entirely in the second term via the specific heat C_P the critical effect is small. Since $u^2 = \beta_s^{-1}$,

$$u^{-2} = u_c^{-2} - \frac{V_c}{T_c} \cdot \frac{g^2}{C_P} , \quad (5.8)$$

where u_c is the critical point velocity. The smallness of the second term allows us to write

$$u = u_c + \frac{g^2 u_c^3}{2T_c C_P} = u_c + \Delta u \quad (5.9)$$

This connection between the sound velocity and the specific heat forms the corner stone of the theory of critical ultrasonics in binary liquids and in the lambda transition of liquid helium.

Turning to Eq. (5.9) we note that a complete picture of dispersion and attenuation can be had if C_p is known as a function of frequency and temperature (measured in terms of κ). The theory of C_p is presented in a separate article in this volume. Here we point out the salient features. In the thermodynamic limit, $\omega = 0$, the specific heat can be written as

$$C_p(\kappa) = A\,\kappa^{-\alpha/\nu} + B. \tag{5.10}$$

The characteristic fluctuations in the system have a wavenumber and temperature dependent relaxation rate $\gamma(k, \kappa)$ as discussed in Sec. 3. At a given temperature the characteristic rate is the rate evaluated at $k = \kappa$. Denoting this by $\gamma(\kappa)$, we see that $\gamma(\kappa) \sim \kappa^{z_o}$. As we approach the critical point $\gamma(\kappa)$ becomes smaller and at some point equals the external frequency ω. Further lowering of κ cannot change the responses any further since it cannot respond to the fast oscillations. The response is now independent of κ and determinend by ω alone. The thermodynamic specific heat can be written in terms of γ as

$$C_p = \tilde{A}\,\gamma^{-\alpha/z_o\nu} + B, \tag{5.11}$$

and since it reaches a limiting value when $\gamma \simeq \omega$, the value for $\omega >> \gamma$ is

$$C_p(\omega) = \tilde{A}\left(\frac{-i\omega}{a}\right)^{-\alpha/z_o\nu} + B, \tag{5.12}$$

where ´a´ is a number of O(1) and γ has been replaced by $-i\omega$ because the system is dissipative (i.e., the concentration fluctuations are relaxational).

To obtain the critical point attenuation[48] we write the critical part Δu of the sound velocity as

$$\Delta u(T_c, \omega) = \frac{g^2 u_c^3}{2 T_c}\Big[\tilde{A}\Big(\frac{\omega}{a}\Big)^{-\frac{\alpha}{z_o \nu}} \cos \frac{\pi\alpha}{2 z_o \nu} + B + i\, \tilde{A}\Big(\frac{\omega}{a}\Big)^{-\frac{\alpha}{z_o \nu}} \sin \frac{\pi\alpha}{2 z_o \nu}\Big]$$

$$= u_1 + i\, u_2 \,, \tag{5.13}$$

where

$$u_2 = - \frac{g^2\, u_c^3\, \tilde{A}(\omega/a)^{-\alpha/z_o \nu} \sin \frac{\pi\alpha}{2 z_o \nu}}{2T_c \Big[\, \tilde{A}(\omega/a)^{-\alpha/z_o \nu} \cos \frac{\pi\alpha}{2 z_o \nu} + B\,\Big]^2} \tag{5.14}$$

since imaginary part is small and allows linearization. Further $\alpha / 2 z_o \nu << 1$ since α is small and we obtain the attenuation per wavelength $\alpha\lambda$ at the critical point as

$$(\alpha\lambda)_c = - \frac{u_2}{u_c} = \frac{g^2\, u_c^2\, \alpha_o\, \tilde{A}}{4\, z_o \nu} \; \frac{(\omega/a)^{-\alpha_o/z_o \nu}}{u\; u_1^2} \,, \tag{5.15}$$

where $u_1 = A(\omega /a)^{-\alpha_o/z_o} + B$ and the specific heat exponent has been changed to α_o to avoid confusion. When u_1 does not have an appreciable frequency dependence, which is generally true in the binary liquids, because of the relatively large value of B, the attenuation per wavelength decreases with increasing frequency as $\omega^{-\alpha_o/z_o\nu}$. The index $\alpha_o/z_o\nu \simeq 0.058$. The ω-dependence consequently is weak.

At temperatures away from T_c, we need the full frequency and temperature dependent specific heat

$$C(\omega\,, \kappa) = \tilde{A}\, \gamma^{-\alpha_o/z_o \nu}\, L(\Omega) + B \,, \tag{5.16}$$

where $L(\Omega)$ is a function of the scaling variable $\Omega = \omega/\gamma(\kappa)$. The limiting form of $L(\Omega)$ for $\Omega >> 1$ is $L(\Omega) = (z/a)^{-\alpha_o/z_o\nu}$, with $z = -i\,\Omega$ to conform to the form of Eq. (5.12). The function $L(\Omega)$ is in general complex and can be written as $L(\Omega) = L_1(\Omega) +$

$iL_2(\Omega)$. The attenuation per wavelength normalised to $(\alpha\lambda)_c$ is

$$\frac{(\alpha\lambda)}{(\alpha\lambda)_c} = F(\Omega) \quad , \tag{5.17}$$

where $F(\Omega) = 1$ if $\Omega \to \infty$. Since λ does not change appreciably with temperature

$$\frac{\alpha}{\alpha_c} = F(\Omega) \quad , \tag{5.18}$$

and the normalised attenuation data at various frequencies and temperatures are expected to collapse on a single universal curve[48)] when plotted against the reduced frequency (Fig. 4).

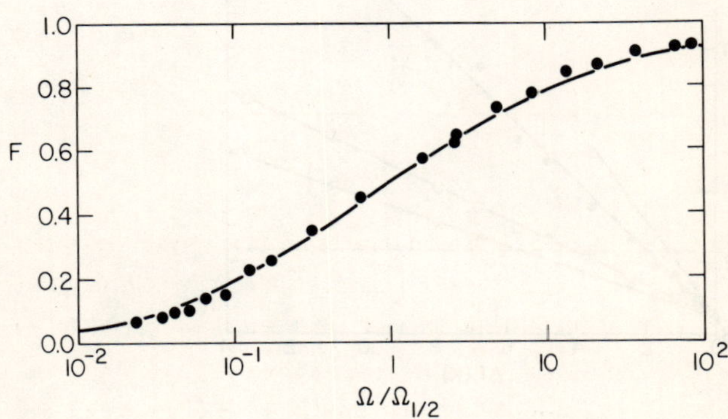

Fig. 4 : Universal attenuation function.

It is straightforward to establish that $F(\Omega) \sim \Omega$ for $\Omega << 1$. For $\Omega >> 1$, i.e., at temperatures very close to T_c, the first departure of the function $F(\Omega)$ from 1 is given by the form

$$F(\Omega) = 1 - \frac{C_1}{\Omega^{1/2}} \quad , \tag{5.19}$$

where C_1 is a constant. Now $z_o = 3$ and $\nu \simeq 2/3$ for the binary liquids and hence at a given frequency $\Omega^{1/2} \simeq \Delta T$. Thus from the low frequency form of $F(\Omega)$, we see that

$$\left(\frac{\alpha_c}{\alpha}\right)^{\frac{1}{2}} \propto \Delta T \text{ if } \Delta T \text{ is large} \tag{5.20}$$

and Eq. (5.19) shows

$$\sqrt{\frac{\alpha_c}{\alpha}} = 1 + \text{const}\,(\Delta T) \text{ if } \Delta T \text{ is small} \tag{5.21}$$

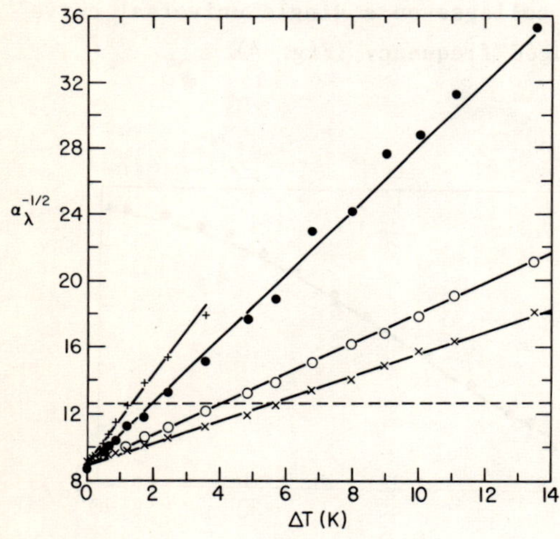

Fig. 5 : $\alpha_\lambda^{-\frac{1}{2}}$ vs ΔT for 3 MP-NE.

We thus expect the plot of $(\alpha_c/\alpha)^{1/2}$ vs. ΔT should be linear for a given frequency over a large range of temperature. This is convincingly demonstrated for the binary mixture 3-methyl pentane-nitroethane (3 MP-NE) in Fig. 5. Combining the $\Omega \ll 1$ and $\Omega \gg 1$ limits we propose the form[46)]

$$F(\Omega) = \left(1 + \frac{C_1}{\Omega^{\frac{1}{2}}}\right)^{-2} \tag{5.22}$$

for the universal scaling function. This is a one-parameter family of functions. A universal number should be the half-attenuation point $\Omega_{1/2}$, for which $F(\Omega_{1/2}) = 1/2$. A two-loop calculation yields $\Omega_{1/2} \simeq 2$ and hence the proposed universal function becomes

$$F(\Omega) = (1 + \frac{0.6}{\Omega^{1/2}})^{-2} \tag{5.23}$$

Note that for the data to be plotted as a function of Ω, the relaxation rate must be known from light scattering experiments. The accurate knowledge of $\gamma(\kappa)$ for 3 MPNE via light scattering, makes it the favoured candidate[49,50] for testing theories. For the other binary mixtures[51,52], the scaling laws are verified but for a comparison with the theory with no adjustable parameters, the relaxation rate must be accurately known.

B. Liquid Helium Near Lambda Transition

The sound velocity near T_λ follows a relation identical to Eq. (5.9). The differences lie in the form of the specific heat and the relaxation rate. The thermodynamic specific heat is very nearly logarithmic near the lambda transition and can be written in the thermodynamic limit as

$$C_P(\kappa) = A \ln \frac{t_o}{t} \;\tilde{=}\; A \ln \frac{\gamma_o}{\gamma}, \tag{5.24}$$

where $\gamma(\kappa)$ is the characteristic relaxation rate and γ_o is a constant. When we pass on to the high frequency limit ($T = T_\lambda$), we have γ replaced as before by $-i\omega/a$ and this yields

$$C_P(\omega) = \tilde{A} \ln \frac{\omega_o}{-i\omega} = \tilde{A}\left[\ln \frac{\omega_o}{\omega} + \frac{i\pi}{2}\right] \tag{5.25}$$

The attenuation per wavelength at the λ-point is now

$$(\alpha\lambda)_c = \frac{\pi}{4} \frac{g^2 u_c^2}{T_c \tilde{\tilde{A}}} \frac{1}{(\ln \omega_o/\omega)^2 + \pi^2/4} \tag{5.26}$$

Thus the attenuation per wavelength increases with increasing frequency. However there is no definite scaling law. If we define the effective exponent d ln($\alpha\lambda$)c/d ln ω , then we find that it increases[53)] with increasing frequency - this is the reason why experimentalists working at low frequencies reported an exponent close to 0.15, while those working at high frequencies reported an exponent near 0.3. The truth is that there is no single exponent for the entire frequency range. The course of the attenuation function is given by Eq. (5.26) and as seen from Fig. 6, it accounts for all the data[54-57)] over several decades of frequencies.

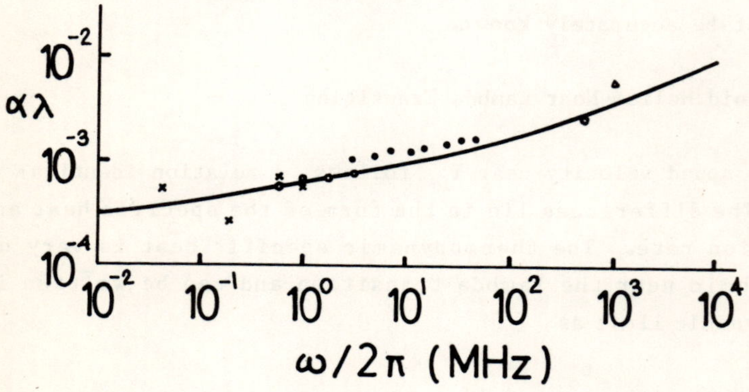

Fig. 6 : λ point attenuation vs frequency.

The extension to temperatures away from T_λ requires the complete computation of C(ω, κ). This is carried out in a separate article in this volume. The results explain[53,58)] all the existing ultrasonics data near the superfluid transition both above and below T_λ .

C. Liquid-Gas System

The density fluctuations are the order parameter fluctuations in this case and the sound velocity is very strongly affected near the critical point. The sound velocity u has the form

$$u^2 = C_o/C_v \tag{5.27}$$

where C_v is the specific heat at constant volume and C_o is a constant. For $T \simeq T_c$, C_v diverges in the same manner as C_p in the binary liquid and hence the zero frequency velocity vanishes as the critical point is reached. The complex sound velocity is determined by the frequency dependent C_v. Because the single-component fluid and the binary liquid are in the same universality class, the behavior of C_p for binary liquid and C_v for the single-component one are identical in the critical region. The thermodynamic C_v can be written as

$$C_v = \tilde{A}\gamma^{-\alpha_o/z_o\nu} - \tilde{B} \ , \tag{5.28}$$

with $\tilde{B} < 0$ for the liquid gas systems. $\tilde{B}$ is a non-universal system-dependent number just as B in Eq. (5.11) was for the binary liquid. The frequency dependence at $T = T_c$ is of the form

$$C_v(\omega) = \tilde{A}\left(-\frac{i\omega}{a}\right)^{-\alpha_o/z_o\nu} - \tilde{B} \tag{5.29}$$

leading[59] to the attenuation per wavelength

$$(\alpha\lambda)_c = \frac{\pi^2\alpha_o}{4} \cdot \frac{1}{1-(\tilde{B}/\tilde{A})(\omega/a)^{\alpha_o/z_o\nu}} \tag{5.30}$$

If $\tilde{B}$ were to be zero, the attenuation per wavelength at $T = T_c$ would be a universal number $\pi^2\alpha_o/4$. The fact that the divergence of the specific heat is weak, makes $\tilde{B}$ sizeable in most cases, hence deviations from this number are seen according to Eq. (5.30).

That is not all. The single-component fluids are further plagued by the fact that the line width exhibits crossover behavior[60] in the frequency range of interest. The background contribution to the linewidth is considerable in all known cases, and in the background region the exponent is $z_o = 4$. In terms of our frequency variable it implies that as we go from low frequencies to high frequencies the z_o changes from 3 to 4. The crossover frequency depending upon the system can be somewhere near 1 or 10 MHz - which is in the middle of the experimental range. The complications produced by this effect have recently been studied[61].

REFERENCES

1. For reviews see Wilson, K.G. and Kogut, J., Phys. Rep. 12C, 75 (1974); Amit, D.J., "Field theory of critical phenomena", NcGraw-Hill (1977), Fisher, M.E. in "Critical Phenomena", Ed. Hahne, F.J.W., Springer-Verlag (1983).
2. Onsager, L., Phys. Rev. 65, 117 (1977).
3. Schultz, T.D., Mattis, D.C., and Lief, E.H., Rev. Mod. Phys. 36, 856 (1964).
4 Ferrell, R.A., Jour. of Stat. Phys. 8, 265 (1973).
5. Ahlers, G., Phys. Rev. A8, 530 (1973).
6. Ferrell, R.A., Jour. of Stat. Phys. 25, 361 (1981).
7. Tarvin, J., Vidal, F., and Greytak, T.J., Phys. Rev. B15, 4193 (1977).
8. Chang, R.F., Burstyn, H.C., and Sengers, J.V., Phys. Rev. A19, 866 (1979).
9. Discussion on variety of different systems can be had in Hohenberg, P.C., and Halperin, B.I., Rev. Mod. Phys. 49, 435 (1977).
10. Ferrell, R.A., and Bhattacharjee, J.K., J. of Low Temp. Phys. 36, 166 (1979).
11. Kawasaki, K., Ann. Phys.(N.Y.), 61, 1 (1970).
12. Ferrell, R.A., Phys. Rev. Lett. 24, 1169 (1970).
13. Ohta, T., and Kawasaki, K., Prog. Theor. Phys. 55, 1384 (1976).
14. Bhattacharjee, J.K., and Ferrell, R.A., Phys. Rev. A28, 2363 (1983).
15. Siggia, E.D., Hohenberg, P.C., and Halperin, B.I., Phys. Rev. B13, 2110 (1976).

16 Tsai, B.C., and McIntyre, D., J. Chem. Phys. 60, 937 (1974).

17. Zozulya, V.N., and Blagoi, Yu P., Sov. Phys. - JETP, 39, 99 (1974).

18. Rivkin, S.L., Levin, A.Ya., and Izrailevsky, L.B., Tplofiz. Svoista Veshchestv Mater. 10, 232 (1976).

19. Burstyn, H.C., and Sengers, J.V., Phys. Rev. A27, 1071 (1983).

20. Burstyn, H.C., Sengers, J.V., Bhattaccharjee, J.K., and Ferrell, R.A., Phys. Rev. A28, 1567 (1983).

21. See also Paladin, G. and Peliti, L., J. Phys. (Paris), Lett. 43, L-15 (1982).

22. Bhattacharjee, J.K. and Ferrell, R.A., Phys. Rev. A23, 1511 (1981).

23. Sengers, J.V., Thesis (Assen, 1962), see also Michels, A. and Sengers, J.V., Physica 28, 1238 (1962).

24. Cohen, L.H., Dingus, M.L. and Meyer, H., Phys. Rev. Lett. 50, 1058 (1983).

25. De Dominicis, C. and Peliti, L., Phys. Rev. B18, 353 (1978).

26. Ferrell, R.A., Menyhard, N., Schmidt, H., Schwabl, F. and Szepfalusy, P., Ann. Phys. (N.Y.), 47, 565 (1968), also Halperin, B.I. and Hohenberg, P.C., Phys. Rev. 177, 952 (1968).

27. Ahlers, G., Phys. Rev. Lett. 21, 1159 (1968).

28. Tyson, J.A., Phys. Rev. Lett. 21, 1235 (1968).

29. Ahlers, G.,"Proceedings of the 12th International Conference on low temperature Physics" Ed. Kanda, E., 1971.

30. Korth, M.S., Thesis, Univ. of Maryland, 1984.

31. Ferrell, R.A. and Bhattacharjee, J.K., Phys. Rev. Lett. 42, 1638 (1979).

32. Ferrell, R.A. and Bhattacharjee, J.K., "Proceedings of Conference on Critical Dynamics, Geneva 1979", Ed. Enz, C.P. Springer (1979).

33. Dohm, V. and Folk, R., Phys. Rev. Lett. 46 349 (1981).

34. Ahlers, G., Hohenberg, P.C. and Korublit, A., Phys. Rev. B25, 3136 (1982).

35. Ferrell, R.A. and Bhattachaarjee, J.K., Phys. Rev. B28, 121 (1983)

36. Ferrell, R.A. and Bhattacharjee, J.K., Phys. Rev. B20, 3690 (1979).

37. Crooks, M.J. and Robinson, B.J., Physica (Utrecht), 107B, 339 (1981).
38. Mehrotra, R. and Ahlers, G., Phys. Rev. Lett. 51, 2116 (1983).
39. Ferrell, R.A. and Bhattacharjee, J.K., Phys. Rev. Lett. 51, 487 (1983).
40. Ferrell, R.A. and Bhattacharjee, J.K., Phys. Rev. B (in press).
41. Dohm, V. and Folk, R.A., Zeit. fur Phys. B45, 129 (1981).
42. Herzfeld, K.E. and Rice, F.O., Phys. Rev. 31, 691 (1928).
43. Fixman, M., J. Chem. Phys. 36, 1961 (1962).
44. Mistura, L., "Proceedings of the Varenna Summer School on Critical Phenomena", Ed. Green, M.S., (Academic, N.Y. 1971).
45. Laplace, P.S., Ann. Chim. Phys. Tome iii 238 (1816).
46. Ferrell, R.A. and Bhattacharjee, J.K., Phys. Rev. A (in press).
47. Tanaka, H., Wada, Y. and Nakajima, H., J. Chem. Phys. 68 223 (1982); 75, 37 (1983).
48. Bhattacharjee, J.K. and Ferrell, R.A., Phys. Rev. A24, 1643 (1981).
49. Garland, C. and Sanchez., J. Chem. Phys. 79, 3090 (1983).
50. Harada, Y., Suzuki, Y. and Ishida, Y., J. Phys. Soc. Japan, 48, 703 (1980).
51. Fenner, D.B., Phys. Rev. A23, 1931 (1981).
52. Tanaka, H., Wada, Y. and Nakajima, H., J. Chem. Phys. 68, 223 (1982)
53. Ferrell, R.A. and Bhattacharjee, J.K., Phys. Rev. Lett. 44, 403 (1980).
54. Williams, R.D. and Rudnick, I., Phys. Rev. Lett. 25, 276 (1970).
55. Lambert, B., Legros, P. and Salin, D., J. Phys. (Paris) Lett. 41, 507 (1980).
56. Tozaki, K. and Ikushima, A., J. Low Temp. Phys. 32, 379 (1978).
57. Roe, D.B., Meyer, H. and Ikushima, A., J. Low Temp. Phys. 32, 67 (1978).
58. Ferrell, R.A. and Bhattacharjee, J.K., Phys. Rev. B23, 2434 (1981).
59. Bhattacharjee, J.K. and Ferrell, R.A., Phys. Lett. 88A, 77 (1982).
60. Swinney, H.L. and Henry, D.L., Phys. Rev. A8, 2486 (1973).
61. Bhattacharjee, J.K. and Ferrell, R.A., to be published.

THEORY OF WAVENUMBER AND FREQUENCY DEPENDENT SPECIFIC HEAT NEAR SECOND ORDER PHASE TRANSITIONS

Richard A. Ferrell

Department of Physics, University of Maryland
College Park, Maryland 20742, USA
&

Jayanta K. Bhattacharjee

Department of Physics
Indian Instituteof Technology, Kanpur 208016, India

1. INTRODUCTION

Second order phase transitions are characterised by the existence of an order parameter, whose expectation value vanishes for temperatures T greater than the critical temperature T_c and assumes non-zero values for $T < T_c$. As T approaches T_c, large scale fluctuations occur in the order parameter and it is the long-ranged nature of these fluctuations which makes the theory of such transitions difficult. The order parameter is generally described in terms of a n-component field variable $\phi_i(x)(i = 1,2, \ldots n)$. For the magnetic transition from a paramagnetic to a ferromagnetic state it is the magnetisation vector (n=3), for the superfluid transition in ^4He it is the complex wave function of the condensate (n=2), while for the liquid-gas transition or near the consolute point of a binary liquid mixture it is a scalar quantity (n=1) being the difference in density between the liquid and gas phase or the difference in concentrations of the components of the binary mixture. The correlation between $\phi_i(x_1)$ and $\phi_i(x_2)$ is the quantity of interest which can be characterised by a correlation length ξ. This length diverges (becomes infinite) as T approaches T_c. The behavior for $T \simeq T_c$ can be characterised as

$$\xi = \kappa^{-1} \propto (T-T_c)^{-\nu} \tag{1.1}$$

The divergencce of the correlation length leads to divergences in thermodynamic quantities that can be expressed as integrals over correlation functions. The susceptibility, χ which is the integral of the two point function has the strongest divergence and is characterized as

$$\chi \propto (T - T_c)^{-\gamma} \tag{1.2}$$

The specific heat which is the integral of the correlation of the square of the order parameter has a weaker divergence and is characterized as

$$C \propto (T - T_c)^{-\alpha} \tag{1.3}$$

The long ranged nature of the two point function $G(x_{12}) = \langle \phi_i(x_1) \phi_i(x_2) \rangle$ at $T = T_c$, is expressed by

$$G(x_{12}) \sim \frac{e^{-x_{12}/\xi}}{x_{12}^{D-2\xi\eta}} \tag{1.4}$$

where D is the dimensionality of space, and η is the anomalous dimension exponent. For $T < T_c$, the order parameter acquires an expectation value which grows as T is lowered. This is expressed by the relation

$$m = \langle \phi \rangle \propto | T - T_c |^{\beta} \tag{1.5}$$

The indices α , β , γ , ξ and ν are not all independent, they are related by so called scaling laws

$$\gamma = (2 - \eta)\nu \tag{1.6}$$

$$\alpha + 2\beta + \gamma = 2 \tag{1.7}$$

The exponent η is extremely small (~ 0.04) and difficult to obtain experimentally.

Experiments near T_c can measure the thermodynamic quantities χ & C but, more importantly, light scattering and neutron scattering techniques can actually probe the correlation function at different length scales and at different temperatures. This implies probing the function $G(x_{12}, \xi)$, or, more precisely, the Fourier transform thereof defined as

$$g(k, \kappa) = \int d^D x_{12}\, G(x_{12}, \xi) \exp(ik.x_{12}) \tag{1.8}$$

Clearly for $\xi = \infty$ or $\kappa = 0$,

$$g(k) \sim k^{-2+\eta}, \tag{1.9}$$

while for k = 0,

$$g(\kappa) \sim \kappa^{-2+\eta} \tag{1.10}$$

The static scaling hypothesis now asserts that $g(k, \kappa)$ is a homogeneous function of k and κ, so that introducing the ratio $x = k/\kappa$, we have

$$g(k, \kappa) = k^{-2+\eta} f(x) \tag{1.11}$$

with

$$f(x) = \begin{cases} \text{const.} & \text{for } x \gg 1 \\ x^{2-\eta} & \text{for } x \ll 1 \end{cases} \tag{1.12}$$

The function f(x) is called the susceptibility scaling function and is much more difficult to compute than the exponents introduced above.

Light scattering near the lambda transition probes a different scaling function. The order parameter is the quantum mechanical wavefunction ψ, while light scatters from the density fluctuations which are fluctuations in $|\psi|^2$. Thus the correlation function involved in the scattering of light near the lambda transition is the correlation of the square of the order parameter $G_2(x_{12}, \xi) = \langle \phi^2(x)\phi^2(x)\rangle$ or more precisely the Fourier transform $G_2(k, \xi)$ defined as

$$G_2(k, \kappa) = \frac{1}{(2\pi)^D} \int d^D x_{12} \exp(ik.x_{12}) G_2(x_{12}, \xi) \qquad (1.13)$$

The subscript denotes the connected part of the equilibrium average. The local (k = 0) limit is the thermodynamic specific heat $C(\kappa)$. Thus,

$$G_2 \; (\kappa) \;\propto\; \kappa^{-\alpha/\nu} \qquad (1.14)$$

The static scaling hypothesis now asserts that

$$G_2(k, \kappa) = \kappa^{-\alpha/\nu} C(k/\kappa) \qquad (1.15)$$

where

$$C(x) = \begin{cases} \text{const} & \text{for } x \ll 1 \\ x^{-\alpha/\nu} & \text{for } x \gg 1 \,. \end{cases} \qquad (1.16)$$

The function C(x) is the specific heat scaling function and $G_2(k, \kappa)$ will henceforth be denoted by $C(k, \kappa)$ - the wavenumber dependent specific heat to within a multiplicative constant.

The theory[1)] of second order phase transitions proceeds by writing down the phenomenoligical free energy functional, which can be used to define the correlation functions and expectation values. The standard form is

$$F(\phi) = \int d^D x \left[\tfrac{1}{2} \tau \sum_i \phi_i^2 + \tfrac{1}{2} \sum_i (\nabla\phi_i)^2 + \frac{u}{4} (\sum_j \phi_j^2)^2 \right], \qquad (1.17)$$

where $\tau \propto (T - T'_c)$, T'_c being the mean field transition temperature and u a coupling constant. The equilibrium average of any functional $A(\phi)$ is now defined quantitatively as the functional integral

$$\langle A(\phi)\rangle = 1/Z \int D\,[\phi]\,\exp\{-F(\phi)\}\,A(\phi) \qquad (1.18)$$

where $Z = \int D\,[\phi]\,\exp(-F(\phi))$ is the partition function. The presence of the quartic term in $F(\phi)$ makes exact evaluation of such correlations impossible in general. Perturbation in u is not feasible in general because for D = 3 (real world) u is not small. It was shown by Wilson and Fisher[2], however, that for evaluation of asymptotic scaling behaviors, u is very small near D=4 and vanishes at D=4, when the Gaussian model (u=0) becomes exact. Hence it is possible to set up a perturbation theory near D=4, using the value of u as

$$u = \frac{2\pi^2 \varepsilon \Lambda^{\varepsilon}}{n+8} \neq O(\varepsilon^2)\ , \qquad (1.19)$$

where ε = 4-D and a factor of $C_D/(2\pi)^D$ has been absorbed in Λ. Using this well known ε expansion, exponents and scaling functions may be evaluated. The leading order result for various exponents are

$$\gamma = 1 + \varepsilon\,\tfrac{1}{2}\,\frac{n+2}{n+8} \qquad (1.20)$$

$$\nu = \tfrac{1}{2} + \tfrac{1}{4}\,\varepsilon\,\frac{n+2}{n+8} \qquad (1.21)$$

$$\eta = O(\ ^2) \qquad (1.22)$$

$$\alpha = \frac{4-n}{2(n+8)}\,\varepsilon \qquad (1.23)$$

$$\beta = \tfrac{1}{2} - \frac{3\varepsilon}{n+8} \qquad (1.24)$$

Since η is $O(\varepsilon^2)$, calculation of the susceptibility scaling function involves evaluation of two loop integrals which is technically difficult. This calculation is not very rewarding either because the experimental data are not yet precise enough to detect significant differences from the simple Ornstein-Zernike form $g(x) = (1+ x^2)^{-1}$ – a result obtained by dropping the quartic term. Thus the effect of u on the shape of g(x) is too small to be readily accessible[3-6] experimentally.

The specific heat exponent is small, being close to zero for $n \simeq 4$ for $D \simeq 4$, so that the specific heat there is nearly logarithmic. If we evaluate α for D=3, n=2, by taking higher terms in Eq. (1.22), $\alpha \simeq 0$ once more so that for the lambda transition in ^{4}He the specific heat is very nearly logarithmic. It is customary to write the specific heat as

$$C\ (\kappa) = \frac{\tilde{A}\nu}{\alpha}\left[\ (\frac{\Lambda}{\kappa})^{-\alpha/\nu} - 1\ \right] + \tilde{B}\ , \tag{1.25}$$

so that for $\alpha/\nu \rightarrow 0$

$$C\ (\kappa) = \tilde{A}\ \ln(\frac{\Lambda}{\kappa}) + \tilde{B} \tag{1.26}$$

The wavenumber dependent specific heat can in general be written as

$$C(k, \kappa) = A\ (\frac{\Lambda}{\kappa})^{-\alpha/\nu} C(x) + B\ , \tag{1.27}$$

where $A = \tilde{A}\nu/\alpha$. There is no simple form for the scaling function C(x) as there is for g(x). Consequently, it is a more interesting exercise to calculate C(x)[7-9]. The fact that C(x) can actually be measured by light scattering near the superfluid transition makes such a calculation of the scaling function useful. In Section 2, we will deal with the computation of C(x).

Turning now to the dynamics[10], as the fluctuations become long ranged, they also take a very long time to relax to thermal equilibrium, once disturbed from it. The characteristic relaxation rate γ of the fluctuations depends on the temperature (measured in terms of κ) and the wavenumber k. In the long wavelength ($k \rightarrow 0$) limit, the relaxation rate at $T = T_c$ is a function of k and goes to zero as $k \to 0$. It does so in a characteristic manner which is expressed by

$$\gamma(k) \propto k^{z_o} \tag{1.28}$$

The dynamic scaling hypothesis[11,12] asserts that the temperature and momentum dependent relaxation rate $\gamma(k, \kappa)$ is a homogeneous function of k and κ and can be expressed as

$$\gamma(k,\kappa) = k^{z_o} f(x) \tag{1.29}$$

where,

$$f(x) = \begin{cases} \text{const} & \text{for } x \gg 1 \\ x^{-z_o+2} \text{ or } x^{-z_o} & \text{for } x \ll 1, \end{cases}$$

for conserved and nonconserved order parameters respectively. (1.30)

The exponent z_o and the function f(x) are well established for all of the well known systems like single component classical fluid and binary liquid[13,14], ^{4}He near the superfluid transition[15-17], ferromagnet[18-19] antiferromagnet[20] etc.

A recent development in the theory of dynamic critical phenomena is a better understanding of ultrasonic attenuation in the classical fluids[21-24] as well as the superfluid[25,26] as well as the superfluid. The theory for sound attenuation and dispersion is based on the frequency dependent specific heat - an approach advocated by Fixman[27], Mistura[28] and others. At temperatures well above T_c, the relaxation rate is large, in particular it is much larger than the frequency ω of the sound wave and one obtains the usual thermodynamic specific heat as given in Eq. (24). As the temperature is brought closer to T_c, the relaxation rate decreases and becomes smaller than the frequency of the sound wave. Further change in temperature cannot cause any change in the response of the system which, in the limit $\omega \gg \gamma$ is determined by ω alone. Thus we have a frequency and temperature dependent specific heat $C(\omega, \kappa)$ for which the dynamic scaling hypothesis asserts that

$$C(\omega, \kappa) = A\, \kappa^{-\alpha/\nu} L(\Omega) + B \,, \tag{1.31}$$

where

$\Omega = \omega/\gamma(\kappa)$ and

$$L = \begin{cases} \text{const} & \text{for } \Omega \ll 1 \\ L\, \Omega^{-\alpha/z_o \nu} & \text{for } \Omega \gg 1 \end{cases} \tag{1.32}$$

In Section 3 of this article we show how the frequency dependent scaling function can be computed for different systems.

2. STATIC SCALING FUNCTION

A. $T > T_c$

In the absence of interaction the single-loop graph of Fig. 1(a) yields a wave number dependent specific heat of the form

$$C(k, \kappa) \propto \frac{\Lambda^{\varepsilon}}{C_D} \int_{p<\Lambda} \frac{d^D p}{(p^2 + \kappa^2)(p'^2 + \kappa^2)}$$

$$= \frac{1}{\varepsilon} \{ I_D(k,\kappa) - 1 \} \tag{2.1}$$

where

$$I_D(k, \kappa) = \frac{\varepsilon \Lambda^{\varepsilon}}{C_D} \int \frac{d^D p}{(p^2 + \kappa^2)(p'^2 + \kappa^2)} \tag{2.2}$$

is not restricted by the cutoff. $C_D/D = \pi^{D/2}/(D/2)!$ is the volume of the unit sphere in D dimensions. The convolution integral is restricted by the momentum conservation condition **p + p′ = k.** The dimensional constant Λ^{ε} has been introduced so as to render the resulting expression dimensionless. For the purpose of taking the interaction into account, it is useful to introduce the function

$$K_D(k,\kappa) = 1/2 \ln I_D(k,\kappa) \ , \qquad (2.3)$$

which puts the specific heat into the exponentiated form

$$C(k,\kappa) \propto \frac{1}{\varepsilon} \left(e^{\varepsilon K_D} - 1\right) \qquad (2.4a)$$

To two-term accuracy, this can be expanded to give

$$C(k,\kappa) \propto K_D + \frac{\varepsilon}{2} K_D{}^2 \qquad (2.4b)$$

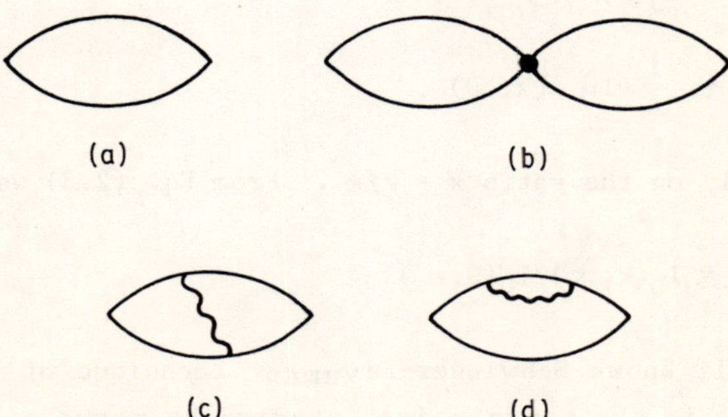

Fig. 1 : One and two loop graphs for specific heat ($T > T_c$)

The lowest order effect of interaction of the fluctuations, as illustrated by Fig. 1(b), is to add to Eq. (2.4) a negative term proportioanl to $K_D{}^2$, yielding

$$C(k,\kappa) \propto K_D + \frac{\alpha}{2\nu} K_D^2 , \tag{2.5a}$$

with $\alpha/\nu = \varepsilon(4-n)/(n+8)$ the critical exponent of Eqs. (1.23) and (1.21). It is natural, in general, to again put $C(k,\kappa)$ into exponentiated form by writing Eq. (2.5a) and

$$C(k,\kappa) \propto \frac{\nu}{\alpha}\left(e^{\frac{\alpha}{\nu}K_D} - 1\right) \tag{2.5b}$$

For the special case of liquid He^4, however, we take the limit $\alpha \to 0$, which reduces Eq. (2.5b) to

$$C(k,\kappa) \propto K_D(k,\kappa) , \tag{2.5c}$$

a result which can be obtained directly from Eq. (2.5a) by dropping the second term.

Comparing $K_D(k,\kappa)$ to its thermodynamic limit, $K_D(0,\kappa)$, yields the scaling function

$$\begin{aligned} L(x, D) &= K_D(k,\kappa) - K(0,\kappa) \\ &= \frac{1}{\varepsilon} \ln J(x, D) , \end{aligned} \tag{2.6}$$

depending only on the ratio $x = k/\kappa$. From Eq. (2.3) we have

$$J(x, D) = I_D(k,\kappa)/I_D(0,\kappa) \tag{2.7}$$

Using the well known Schwinger-Feynman technique of combining the denominators in the convolution integral by means of an auxiliary variable, we find explicitly

$$J(x, D) = \int_o^1 \frac{dy}{[\,1+x^2y(1-y)\,]^{\varepsilon/2}} \tag{2.8}$$

Expanding Eq. (2.8) in powers of ε as

$$J(x, D) = 1 + \varepsilon J_1(x) + \varepsilon^2 J_2(x) + \ldots \tag{2.9}$$

and substituting into Eq. (2.6) gives

$$L(x, D) = L_o(x) + \varepsilon L_1(x) + \ldots \tag{2.10}$$

where the lowest order term is the Bray function

$$L_o(x) = J_1(x) = B(x) = 1 -(1 + 4/x^2)^{1/2} \ln\left[\frac{x}{2}+\left(1+\frac{x^2}{4}\right)^{\frac{1}{2}}\right] \tag{2.11a}$$

The first-order correctin has the form

$$L_1(x) = J_2(x) - 1/2\, J_1{}^2(x). \tag{2.11b}$$

An explicit expression for $J_2(x)$ will not be attempted here, as it involves spence functions.

It is, however, feasible to examine closely the passage to the λ point, $x \to \infty$, for which Eqs. (2.8) and (2.6) take on the asymptotic forms

$$J(x, D) \sim \left(\frac{a}{x}\right)^{\varepsilon} \tag{2.12a}$$

and $$L(x, D) \sim -\ln x + \ln a \tag{2.12b}$$

where

$$\ln a = (1/\varepsilon) \ln M(D) \tag{2.13}$$

and $$M(D) = \int_0^1 y^{-\varepsilon/2}(1 - y)^{-\varepsilon/2}\, dy$$

$$= \beta\left(1 - \frac{\varepsilon}{2},\ 1 - \frac{\varepsilon}{2}\right) = \frac{\Gamma^2\left(1 - \frac{\varepsilon}{2}\right)}{\Gamma(2 - \varepsilon)} \tag{2.14}$$

The ε expansion of Eq. (2.13) brings in the second derivative of the gamma function at unit argument, from which we obtain

$$\ln a = 1 + \varepsilon/2\,(1 - \frac{\pi^2}{12}) + \ldots \tag{2.15}$$

Substituting the two-term expansion

$$a \simeq a_o + a_1\,\varepsilon \;= a_o\,(1 + \frac{a_1}{a_o}\,\varepsilon\,) \tag{2.16a}$$

into Eq. (2.15) yields

$$a_o = e = 2.718 \tag{2.16b}$$

$$\text{and}\quad a_1/a_o = 1/2\,(1 - \pi^2/12) \simeq 0.1. \tag{2.16c}$$

Thus, the effect of the interaction, to this order of accuracy, evaluated for $\varepsilon = 1$ ($D = 3$) is to increase a, relative to its a_o value, by approximately 10%.

A different approach involves calculating the single-loop integral at D=3 rather than in the ε expansion. Then Eq. (2.6) and (2.7) lead to the approximant

$$L(x) = -\ln(x/2) + \ln(\tan^{-1} x/2) \tag{2.17}$$

In this approximation

$$a = \pi = 3.14\;, \tag{2.18}$$

or approximately a 15% increase relative to a_o. It may be noted that a 3-loop calculation by Nicoll[9)] yields a number close to π for ´a´.

As explained at the end of the next subsection, knowledge of ´a´ allows one to determine the amplitude ξ_o of the correlation length from the light scattering data. Carrying out the light scattering at different pressures one would thus know the pressure dependence of ξ_o. The amplitude C_o of the specific heat is known experimentally as a function of pressure. Thus, the knowledge of ξ_o would provide a direct experimental test of the two-scale-factor universality which asserts that

$$R = \frac{A}{k_B} \xi_o^3 \tag{2.19}$$

is a universal quantity. Using the value of A_o and ξ_o below T_λ, obtained from purely thermodynamic measurements, Singaas and Ahlers[29] have recently demonstrated the validity of two-scale factor using over a wide range of pressures.

B. $T < T_c$

In this broken symmetry state we have to introduce the nonzero expectation value $m = \langle \phi \rangle$ of the order parameter. We shall consider the symmetry broken in the 1-direction, i.e., $\langle \phi_1 \rangle \neq 0$ in the n-component model and call this the longitudinal direction. The rest n-1 directions will be called the transverse directions and $\langle \phi_i \rangle = 0$ for i = 2,3, ... n. The fluctuating longitudinal field is $\psi_L = \langle \phi_1 - \langle \phi_1 \rangle \rangle = \phi_1 - m$ and the magnitude of the transverse field will be denoted by $\psi_T{}^2 = \sum_{i=2}^{n} \phi_i^2$. For $\tau < 0 (T < T_c)$, the free energy of Eq. (1.17) can be written in terms ψ_L and $\psi_T{}^2$ as

$$F = \int d^D x \{ |\tau| \psi_L^2 + \tfrac{1}{2} (\nabla \psi_L)^2 + \tfrac{1}{2} (\nabla \psi_T)^2$$

$$+ u m \psi_L (\psi_L^2 + \psi_T^2) + \frac{u}{4} (\psi_L^2 + \psi_T^2)^2 \}, \tag{2.20}$$

where all constant terms have been dropped and to the lowest order m has been chosen to keep $\langle \psi_L \rangle = 0$, i.e., remove linear terms in ψ_L in F, which leads to

$$m^2 = | \tau | / u \tag{2.21}$$

This gives the mean-field value (1/2)) of the exponent β. Setting $\langle \psi_L \rangle = 0$ at the next order leads to the value of β quoted in Eq. (1.24). We find that the free energy of Eq. (3.1) introduces a three point coupling whose coupling constant is of strength $u^{1/2}$. Also note that it is the longitudinal field which has a mass, the transverse modes are massless and are known as the Goldstone modes. They have rather profound effects on correlation functions. It is because of the Goldstone modes that the longitudinal correlation function diverges at long wavelengths. Their presence affects the shape of the correlation function of the square of the order parameter that we are considering here. In terms of the fields ψ_L and ψ_T the desired correlation function can be written as (it is the correlation of fluctuations in ϕ^2)

$$\begin{aligned} G_2(x_{12}) &= \langle(\phi^2(x_2) + 2m\,\psi_L(x_2))\,(\phi^2(x_1) + 2m\,\psi_L(x_1))\rangle \\ &= \langle\phi^2(x_2)\,\phi^2(x_1)\rangle + 4m\,\langle\psi_L(x_2)\,\phi^2(x_1)\rangle \\ &\qquad + 4m^2\,\langle\psi_L(x_2)\,\psi_L(x_1)\rangle \\ &= \langle\phi^2(x_2)\,\phi^2(x_1)\rangle + 4m\,\langle\psi_L(x_2)(\psi_L^2(x_1) + \psi_T^2(x_1))\rangle \\ &\qquad + 4m^2\,\langle\psi_L(x_2)\,\psi_L(x_1)\rangle \qquad (2.22) \end{aligned}$$

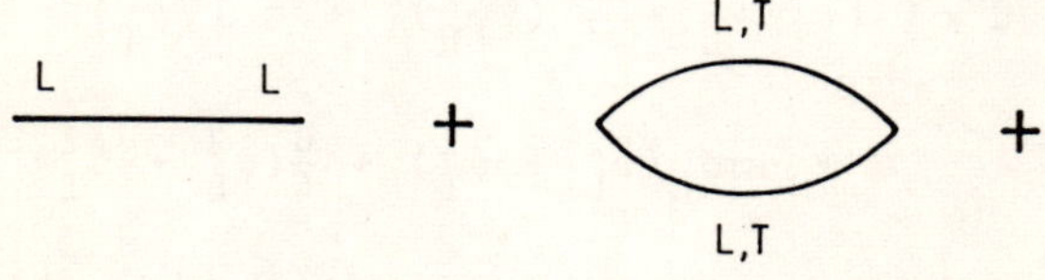

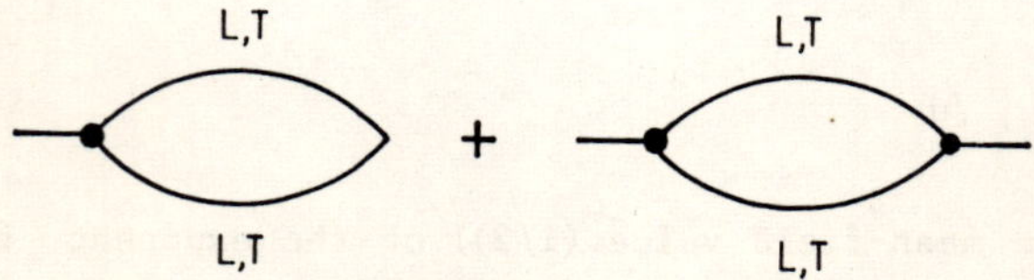

Fig. 2 : One loop graphs for specific heat ($T < T_c$)

The diagrammatic expansion of the resulting $C(k, \kappa)$ is given in Fig. 2. The loop with two longitudinal lines has its value at D=4 is given by Eq. (2.11) with $\kappa^2 = 2\,um^2$. The loop with two transverse lines has the value $\ln \Lambda/k + 1/2$ at D=4. Using the proper combinatorial factors associated with the diagrams we can write down

$$C(k, \kappa) \propto - \frac{4m^2}{(k^2 + 2u\,m^2)} + 2\left(1 - \frac{6u\,m^2}{k^2 + 2u\,m^2}\right)^2 \times$$

$$\times \left[\tfrac{1}{2}\ln \frac{\Lambda^2}{2u\,m^2} - \tfrac{1}{2} + B(y)\right] +$$

$$+ 2(n-1)\left(1 - \frac{2u\,m^2}{k^2 + 2u\,m^2}\right)^2 \left(\ln \frac{\Lambda}{k} + \tfrac{1}{2}\right), \qquad (2.23)$$

where $y^2 = k^2/2\,um^2$. After a little algebra this expression can be written as

$$C(k, \kappa) \propto \frac{n}{2} \ln \frac{\Lambda^2}{2u\,m^2} + \frac{2m^2 u\left(1 - \varepsilon \frac{n+2}{n+8} \ln \frac{\Lambda^2}{2u\,m^2}\right)}{u\left(k^2 + 2u\,m^2\left(1 - \frac{\varepsilon}{2} \ln \frac{\Lambda^2}{2u\,m^2}\right)\right)} +$$

$$+ (n-1)\left(\frac{k^2}{k^2 + 2u\,m^2}\right)^2 \left(-\ln y + \tfrac{1}{2}\right) +$$

$$+ \left(\frac{k^2 - 4u\,m^2}{k^2 + 2u\,m^2}\right)^2 \times (1 + B(y)). \qquad (2.24)$$

Noting that $\frac{\nu}{\beta} = 1 + \varepsilon/2$ and $\frac{\alpha}{\nu} = \varepsilon \frac{4-n}{n+8}$, we can to $O(1)$, write

$$C(k, \kappa) \propto n \ln \frac{\Lambda}{\kappa} + \frac{\tilde{C}_o\, \kappa^{-\alpha/\nu}}{1 + x^2} + (n-1) \frac{x^4}{(1 + x^2)^2}\left(\tfrac{1}{2} - \ln x\right) +$$

$$+ \left(\frac{x^2 - 2}{x^2 + 1}\right)^2 \left[-\tfrac{1}{2} + B(x)\right] \qquad (2.25)$$

where $x = k/\kappa$ and $\tilde{C}_o$ is a constant. We now take the limit of $\alpha/\nu \to 0$, which is the case of interest. Since our computation is exact near D=4, it would be appropriate to take n=4 to get the above limit. This weights the contribution from the Goldstone modes three times as heavily as the contrribution from the longitudinal mode. A particularly simple scaling function is obtained if we assign all the weight to the Goldstone modes and in this approximation

$$C(k,\kappa) = A\left[\ln\frac{\Lambda}{\kappa} + \frac{\tilde{C}_o}{1+x^2} + \frac{x^4}{(1+x^2)^2}(1 - \ln x)\right] \qquad (2.26)$$

where the charge in the addition to the logarthm has been necessary to ensure that the same lambda point value ($\kappa = 0$) is achieved from above and below T_λ.

In comparing with experiments the constants A, Λ , $\tilde{C}_o$ and the amplitudes associated with the correlation length above and below T_λ are to be determined. The thermodynnamic specific heat for $T > T_\lambda$ which can be written as $C_p = \nu \tilde{A} \ln t_o/t$ fixes the constants A and Λ / κ_o, where $\kappa = \kappa_o t^{2/3}$, t being the reduced temperature $(T - T_\lambda)/T_\lambda$. The constant κ_o is determined from the light scattering data for $T > T_\lambda$. For scattering at a given k, we determine the reduced temperature t at which the intensity levels off. This temperature corresponds to $\kappa = k/a$, and that yields κ_o. The ratio of κ_o to κ'_o the amplitude for $T < T_\lambda$ is obtained[27)] from ultrasonics data. This number together with the thermodynamic specific heat for $T < T_\lambda$ fixes the remaining constant $\tilde{C}_o$. With the constants so determined one can compare experimental data for the total scattering intensity[30] above and below T_λ with the expression in Eq. (2.16) and Eq. (3.7) respectively. A more sensitive probe is the quantity[31)] $(\gamma-1)$, which is the ratio of the Rayleigh-line intensity to the Brillouin-line intensity. The comparison at a pressure of 23.1 atmospheres is shown in Fig. 3.

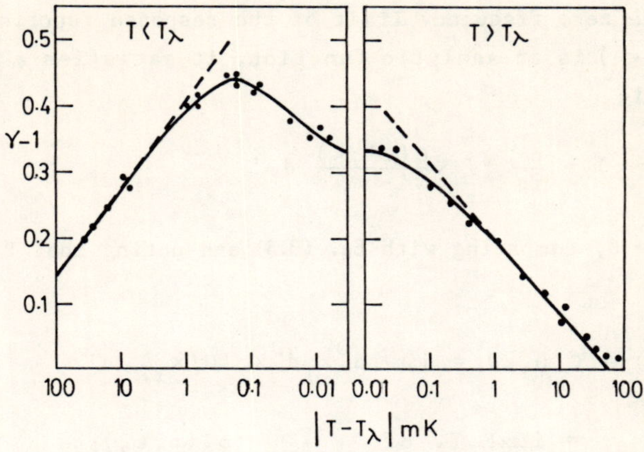

Fig. 3 : Comparison of measured light scattering intensity with the theory. The dashed lines are the asymptotic thermodynamic forms. The data are from Tarvin et al[31])

3. DYNAMIC SCALING FUNCTION

We now turn to the frequency dependent specific heat. It must be related to the fourier transform of the time dependent correlation function $G_2(x_{12}, t_{12})$ defined as

$$G_2(x_{12}, t_{12}) = \langle \phi^2(x_2, t_2)\phi^2(x_1, t_1)\rangle_c \tag{3.1}$$

The thermodynamic specific heat is clearly the integral of the equal time ($t_{12} = 0$) limit of Eq. (3.1). The fourier transform of $G_2(x_{12}, t_{12})$ is defined as

$$G_2(x_{12}, \omega) = \frac{1}{2(\pi)} \int_{-\infty}^{\infty} G_2(x_{12}, t_{12}) e^{i\omega t_{12}} dt_{12} \tag{3.2}$$

The thermodynamic specific heat $C(\kappa)$ is clearly

$$C(\kappa) \propto \int d^Dx_1 d^Dx_2 \int_{-\infty}^{\infty} G_2(x_{12}, \omega)\, \frac{d\omega}{\pi} \tag{3.3}$$

which is the zero frequency limit of the response function $C(\omega, \kappa)$. Since $C(\omega, \kappa)$ is an analytic function, it satisfies a dispersion relation and

$$C(\omega, \kappa) = \frac{P}{i\pi} \int_{-\infty}^{\infty} \frac{C(\omega', \kappa)}{\omega' - \omega}\, d\omega' \tag{3.4}$$

Setting $\omega = 0$, comparing with Eq. (3.3) and noting that $P \int_{-\infty}^{\infty} \frac{d\omega}{\omega} = 0$, we obtain

$$C(\omega, \kappa) - C(0, \kappa) = i\omega \int d^Dx_1 d^Dx_2\, G(x_{12}, \omega)$$

$$= i\omega \times \text{F.T. of} \sum_{p,q,i,j} \phi_i(p, t_2)\phi_i$$

$$(-p, t_2)\phi_j(q, t_1)\phi_j(-q, t_1) \tag{3.5}$$

Thus the computation of the time dependent correlation function in Eq. (3.5) is the basic task of the theory.

Before proceeding any further we need to specify the equation of motion for the order parameter ϕ_i. The relaxational part of the equation of motion for $\phi_i(k)$ is a Langevin equation of the form

$$\dot{\phi}_i(k) = -\Gamma \frac{\delta F}{\delta \phi_i} + N \tag{3.6}$$

for the non-conserved order parameter (liquid helium) and

$$\dot{\phi}_i(k) = -\Gamma k^2 \frac{\delta F}{\delta \phi_i} + N \tag{3.7}$$

for the conserved order parameter as in classical fluids and magnets. N is a Gaussian noise that satisfies

$$\langle N(x,t)N(x',t')\rangle = \Gamma\, \delta(x - x')\, \delta(t - t') \tag{3.8}$$

for the nonconserved case and

$$\langle N(x,t)\, N(x', t')\rangle = \Gamma \quad k^2 \delta(x-x') \quad \delta(t-t') \tag{3.9}$$

for the conserved case. The correlations expressed in Eqs. (3.8) and (3.9) ensure that in spite of the dissipative nature of the equation of motion, the free energy does not change with time and the definition of correiation function remains intact. We note that

$$\frac{\delta F}{\delta \phi_i} = (\kappa^2+k^2)\phi_i(k) + u \sum_{j,k_1,k_2} \phi_j(k_1)\phi_j(k_2)\phi_{i'}(k-k_1-k_2) \tag{3.10}$$

In the absence of the four point coupling u, the correlation function $\langle \phi_i(k, t_2) \; \phi_i(-k, t_1)\rangle$ will decay as $\exp(-\Gamma(\kappa^2 + k^2)t_{21})$ or $\exp(- k^2(k^2 + \kappa^2)t_{21})$ as the case may be. We have not said any thing about Γ as yet. The Onsagar coefficient Γ, because of reversible mode coupling terms (not shown here, but discussed elsewhere in this volume), will in general be temperature and wave number dependent. This dependence comes because of the self enrgy insertions of the form shown in Fig. 1(d). We will assume that all such effects have been taken into account and $\Gamma(k,\kappa)$ is a known function. Thus it will not be necessary to treat graphs such as Fig.1(d) separately once the correct $\Gamma(k,\kappa)$ is used in the propagators in the graph shown in Fig. (1a).

We now illustrate the use of perturbation theory to obtain the time dependent correlation function using the model of Eq. (3.6) which when combined with Eq. (3.10) yields

$$\dot{\phi}_i(k) = -\Gamma(k^2+\kappa^2)\phi_i(k) - u\Gamma \sum_{j,k_1,k_2} \phi_j(k_1)\phi_j(k_2)\phi_i(k-k_1-k_2) + N \tag{3.11}$$

We expand $\phi_i(k)$ in powers of u as

$$\phi_i(k) = \phi_i^{(0)}(k) + u\,\phi_i^{(1)}(k) + O(u^2) \tag{3.12}$$

Inserting this expansion in Eq. (3.11) and equating terms with equal powers of u on either side we obtain

$$\dot{\phi}_i^{(o)}(k) = -\Gamma(k^2+\kappa^2)\,\phi_i^{(o)}(k) + N \qquad (3.13)$$

$$\dot{\phi}^{(1)}(k) = -\Gamma(k^2+\kappa^2)\,\phi_i^{(1)}(k) - \Gamma \sum_{j,k_1,k_2} \phi_j^{(o)}(k_1)\phi_j^{(o)}(k_2)\,\phi_i^{(o)}(k,-k_1-k_2) \qquad (3.14)$$

and so on. The correlation function $G_2(t_{21},\kappa)$ can be expanded as

$$\begin{aligned} G_2(t_{21},\kappa) &= \sum_{p,q,i,j} \langle \phi_i(p,t_2)\phi_i(-p,t_2)\,\phi_j(q,t_1)\,\phi_j(-q,t_1)\rangle \\ &= \sum_{p,q,j,i} \langle \phi_i^{(o)}(p,t_2)\,\phi_i^{(o)}(-p,t_2)\,\phi_j^{(o)}(q,t_1)\,\phi_j^{(o)}(-q,t_1)\rangle \\ &\quad + u\,\{\sum_{p,q,i,j} \phi_i^{(1)}(p,t_2)\phi_i^{(o)}(-p,t_2)\phi_j^{(o)}(q,t_1)\phi_j^{(o)}(-q,t_1)\rangle + 3 \text{ similar terms }\} + 0(u^2) \end{aligned} \qquad (3.15)$$

We note that Eq. (3.13) can be readily solved to yield

$$\langle \phi_i(k,t_2)\,\phi_i(-k,t_1)\rangle = (k^2+\kappa^2)^{-1}\exp(-\gamma(k,\kappa)t_{21}) \qquad (3.16)$$

$\gamma(k,\kappa)=\Gamma(k,\kappa)(k^2+\kappa^2)$ is the relaxation rate. Working with Eq.(3.15) to 0(1), we immediately see that the pairwise factorisation of the fields corresponding to Fig. (1a) leads to

$$G_2(t_{21},\kappa) = 2n \int \frac{d^D p}{(p^2+\kappa^2)^2}\, e^{-\gamma(p,k)t_{21}} + 0(u) \qquad (3.17)$$

The Fourier transform and the subsequent conversions to the response function via Eq. (3.5) gives

$$C(\omega,\kappa) \propto 2n \int \frac{d^D p}{(p^2+\kappa^2)^2} \frac{\gamma(p,\kappa)}{-i\omega+\gamma(p,\kappa)} + O(u) \qquad (3.18)$$

To compute the O(u) terms we first express $\phi_i^{(1)}(k,t)$ in terms of $\phi_i^{(o)}$ at earlier times by solving Eq. (3.14) using standard Green's function techniques. Insertion of this $\phi_i^{(1)}$ in the second term of Eq. (3.15) and pairwise factoring of the $\phi_i^{(o)}$ fields keeping track of all possible time orderings leads after some algebra to the expected result

$$C(\omega,\kappa) \propto \frac{1}{C_D} \int \frac{d^D p}{(p^2+\kappa^2)^2} \frac{2\gamma(p,\kappa)}{-i\omega+(p,\kappa)} - u(n+2) \times$$

$$\times \left(\frac{1}{C_D} \int \frac{d^D p}{(p^2+\kappa^2)^2} \frac{2\gamma(p,\kappa)}{-i\omega+2\gamma(p,\kappa)}\right)^2 + O(u^2) \qquad (3.19)$$

corresponding to Figs. (1a) and (1b). The structure of this result remains unchanged for the conservative order parameter of Eq. (3.7). The difference from the case of the non-conserved order parameter lies in the fact that the relaxation rate $\gamma(k,\kappa)$ is now $\gamma(k,\kappa) = \Gamma(k,\kappa)\, k^2(k^2 + \kappa^2)$ with the appropriate Onsagar coefficient $\Gamma(k,\kappa)$.

The discussion of the scaling function closely parallels the discussion of Section 2. We shall evaluate the integrals associated with the first two terms of Eq. (3.19) and exponentiate the answer to the form

$$C(\omega,\kappa) \propto A(\kappa^{-\alpha/\nu}\, e^{\alpha/\nu\, L(\Omega)} - 1)\, \frac{\nu}{\alpha} + B. \qquad (3.20)$$

Here $\Omega = \omega/\gamma(k,\kappa)$ and $L(\Omega)$ satisfies the boundary condition

$$L(\Omega) = \begin{cases} 0 \text{ for } \Omega = 0 \\ -1/z_o (\ln \Omega - \text{const}) \text{ for } \Omega >> 1 \end{cases} \tag{3.21}$$

The two systems of interest are the liquid vapor or the binary liquid (n=1, conserved order parameter) and superfluid He^4 (n=2, non-conserved order parameter). We treat them separately.

Conserved order parameter: The relaxation rate $\gamma(k, \kappa)$ is given by the Kawasaki function[13)]. A convenient approximation is to write

$$\gamma(k, \kappa) = \Gamma_o k^2(k^2 + \kappa^2)^{D/2 - 1} \tag{3.22}$$

We will generalise this relaxation rate to

$$\gamma(k, \kappa) = \Gamma_o k^2(k^2 + \kappa^2)^{D'/2 - 1} \tag{3.23}$$

and write the integral corresponding to the first term of Eq. (3.19) as

$$I_{DD'}(\omega,\kappa) = \frac{1}{C_D} \int \frac{d^D p}{(p^2+\kappa^2)^2} \frac{2\Gamma_o p^2(p^2+\kappa^2)^{D'/2-1}}{-i\omega+2\Gamma_o p^2(p^2+\kappa^2)^{D'/2-1}} \tag{3.24}$$

The approximation that corresponds to Eq. (2.12) for the static case now requires the evaluation of $I_{DD'}$ at $D=D' = 4$. The resulting $L(\Omega)$ is

$$L(\Omega) = \frac{1}{z} \ln z - \ln z+2+ \frac{2}{4z-1}(\frac{1}{z}-3)\tan^{-1}\sqrt{4z-1}+O(\varepsilon) \tag{3.25}$$

where $z = -i\Omega$. We note that this shape of $L(\Omega)$ is correct to $O(\varepsilon)$; for the scaled frequency however one must use the exact result

$$\Omega = \omega/2\Gamma_o \kappa^D \tag{3.26}$$

where the constant Γ_o can be determined from light scattering experiments. In writing Eq. (3.26) we have ignored the small viscosity anomaly which would change D to D - z_η where[13] z_η= 0.054 for D = 3. In this approximation, the constant addition to ln Ω for $\Omega >> 1$ in Eq. (3.21) is found to be 2 and if this addition is written as ln a, then

$$a = e^2 + O(\varepsilon) \tag{3.27}$$

A different approach is to hold D´ = 3, and carry out the ε -expansion in the phase space factor alone. Denoting the resulting approximation as $L_{43}(\Omega)$, we find

$$L_{43}(\Omega) = -\frac{z}{2}\int_0^\infty \frac{x\,dx}{(1+x)^2}\cdot\frac{1}{z+x(1+x)^{\frac{1}{2}}} \tag{3.28}$$

The O(ε) correction can be carried out as indicated following Eq.(2.13).

Yet another approach involves evaluating $I_{DD'}$, at D=D´ = 3 and exponentiating as in Eq. (2.17) for the static scaling function. However analytic expressions are not available for D=D´ = 3. One can evaluate the two cases D=3, D´=4 and D=3, D´=2 and interpolate between them. The two functions are very close to each other numerically, which makes this a reliable approach.

We close with a discussion of the constant '$a_{DD'}$'. A two term ε -expansion shows a substantial reduction in the value of ´a´ from the one shown in Eq. (3.27). The value obtained from I_{33} is

$$a_{33} = (8/3\sqrt{3})^3 \simeq 3.65 \tag{3.29}$$

This value needs further correction from

i. use of the correct Kawasaki function[13] for $\Gamma(k, \kappa)$

ii. the vertex correction coming from the reversible terms in the equation of motion.

The Kawasaki correction is simple and leads to a 18% increase in the value of 'a'. The vertex correction arising from the coupling to the transverse velocity field shown in Fig. 1(c), is more complicated. The effect has been calculated in an ε expansion framework. The final outcome is[23)]

$$a \simeq 2.7 \tag{3.30}$$

Using this value to redefine the frequency scale and then using the function form of Eq. (3.25) yields a good representation of the dependent specific heat. An empirical fit to the ultrasonic attenuation data gives a form for $L(\Omega)$ which can be written as

$$L_{emp.}(z) = \frac{2z^2}{1+z^2} + z^2 \frac{3-z^2}{(1+z^2)^2} \ln z + \frac{\sqrt{2\pi}\; z^{3/2}}{(1+z^2)^2}(1 - 2z - z^2) + \frac{\pi}{2}\frac{-z+3z^3}{(1+z^2)^2} \tag{3.31}$$

Here once again the frequency scale has to be redefined to be consistent with $a = 2.7$.

<u>Non-conserved order parameter</u>: The system of interest is liquid ^{4}He near the lambda transition. The relaxation rate of the order parameter has a considerably simplified form because of the rather special features associated with the dynamics of the lambda transition. We shall not discuss them here but merely state that because of these special effects the relaxation rate can be written as $\gamma(k, \kappa) = \Gamma_o(k^2 + \kappa^2)$, where Γ_o is a constant. The critical behavior of Γ_o sets in at temperature very close to T_λ and for the present purpose can be ignored[25)]. In the ε-expansion framework the scaling function is

$$L(\Omega) = -(1 + 1/z)\ln(1 + z) + 1 + O(\varepsilon) \tag{3.32}$$

where $z = -i\,\Omega$ and $\Omega = \omega/2\Gamma_o\kappa^2$.

$T < T_\lambda$: This becomes interesting because of the presence of the Goldstone modes and the correlation function has to be written as the sum of three separate terms exactly as in Eq. (2.22). The equation of motion leads to the graphs shown in Fig. 2 and the approximations that led to Eq. (2.26) for the static scaling function now lead to

$$C(\omega,\kappa) = A\left[z_o \ln\frac{\Lambda}{\kappa} + \frac{C_o}{1+z} - \frac{z^2}{(1+z)^2}\ln z + 1\right] \quad (3.33)$$

This function was used to explain the ultrasonic attenuation[27) in ^4He for $T < T_\lambda$.

REFERENCES

1. For a recent review see Fisher, M.E. in "Critical Phenomena, Ed. Hahne, F.J.W., Springer-Verlag (1983).
2. Wilson, K.G. and Fisher, M.E., Phys. Rev. Lett. 28, 240 (1972).
3. Ferrell, R.A. and Scalapino, D.J., Phys. Rev. Lett. 34, 200 (1975).
4. Bray, A.J., Phys. Rev. Lett. 36, 285 (1976) and Phys. Rev. B14, 1248 (1976).
5. Chang, R.F., Burstyn, H., Sengers, J.V. and Bray, A.J., Phys. Rev. Lett. 37, 1481.
6. Chang, R.F., Burstyn, H. and Sengers, J.V., Phys. Rev. A19, 866 (1979).
7. Bray, A.J., J. Phys. C12, 1777 (1979).
8. Kroll, D.M., Z. Phys., B31, 309 (1978).
9. Nicoll, J.F., Phys. Rev. B20, 4527 (1979).
10. A review of the basics is given in Hohenberg, P.C. and Halperin, Rev. Mod. Phys. 49, 435 (1977).
11. Ferrell, R.A., Menyhard, N., Schmidt, H., Scchwabl, F. and Szepfalusy, P., Ann. Phys. (N.Y.), 47, 565 (1968).
12. Halperin, B.I. and Hohenberg, P.C., Phys. Rev. 177, 952 (1969).
13. Kawasaki, K., Ann. Phys. (N.Y.) 61, 1 (1970).
14. Ferrell, R.A., Phys. Rev. Lett. 24, 1169 (1970).

15. Ferrell, R.A. and Bhattacharjee, J.K., Phys. Rev. Lett. 42, 1638 (1979).
16. Dohm, V. and Folk, R., Phys. Rev. Lett. 46, 349 (1981), Z. Phys. B41, 251 (1981), B45, 129 (1981).
17. Ferrell, R.A. and Bhattacharjee, J.K., Phys. Rev. B28, 121 (1983).
18. Resibois, P. and Piette, C., Phys. Rev. Lett., 24, 514 (1970).
19. Bhattacharjee, J.K., Ph.D. Thesis, Univ. of Maryland, (1979).
20. Freedman, R. and Mazenko, G.F., Phys. Rev. B13, 4967 (1976).
21. Bhattacharjee, J.K. and Ferrell, R.A., Phys. Rev. A24, 1643 (1981).
22. Ferrell, R.A. and Bhattacharjee, J.K., Phys. Rev. B24, 4095 (1981).
23. Ferrell, R.A., Bhattacharjee, J.K., Phys. Rev. A31, (1985).
24. Bhattacharjee, J.K. and Ferrell, R.A., Phys. Lett. 86A, 109 (1981).
25. Ferrell, R.A. and Bhattacharjee, J.K., Phys. Rev. Lett. 44, 403 (1980).
26. Ferrell, R.A. and Bhattacharjee, J.K., Phys. Rev. B23, 2434 (1981).
27. Fixman, M., J. Chem. Phys. 36, 1961 (1962).
28. Mistura, L. in "Proceedings of Varenna Summer School in Critical Phenomena", Ed. Green, M.S. (Academic, N.Y. 1971).
29. Singaas, A. and Ahlers, G., Phys. Rev. B (1984).
30. Tagg, R.P., Ph.D. Thesis, M.I.T. (1984).
31. Tarvin, J., Vidal, F. and Greytak, T.J., Phys. Rev. B15, 4193 (1977).

METAL INSULATOR TRANSITION IN DISORDERED SYSTEMS

T.V. Ramakrishnan

Department of Physics
Banaras Hindu University
Varanasi-221005
INDIA

ABSTRACT

Recent progress in the field of disordered electronic systems is summarized, focussing primarily on their behaviour near the metal to insulator transition caused by disorder (Anderson transition). The effect of interactions between electrons is briefly discussed.

1. An electron subject to a random potential can have a spatially localized wavefunction[1], depending on its energy, the degree of disorder, the dimensionality of the system, etc.. Mott pointed out[2],[3] that this implies that with increasing disorder, the states near the Fermi energy can change from extended to localized, so that the system undergoes a metal to insulator transition. In the last five years, a detailed theoretical picture of this transition and of the disordered metal has emerged. A number of experiments exploring this regime provide support for this picture; several characteristic interaction effects have been discovered and identified as well. It now appears that there are several classes of metal insulator transitions depending on physical conditions and relative importance of different kinds of interactions. Work on this difficult problem is just beginning and the present status is summarized here.(See Lee and Ramakrishnan[4] for a recent review).

We first introduce the basic ideas of Anderson localization, the Anderson transition, minimum metallic conductivity, and Mott transition (Section I). We then discuss the scaling theory of localization, its predictions, and experimental resuslts (Section 2). Finally, in Section 3, the problem of the metal insulator transition in a disordered interacting electron gas is formulated, special solutions described and compared with experiment.

1.1 Anderson Localization

The localization concept of Anderson was in contradiction to the then prevalent belief that no qualitative change occurs in the nature of states with increasing disorder; electrons always diffuse, but more slowly as randomness increases. Anderson considered a simple tight binding model.

$$H = \sum_{i,j} t_{ij} a_i^+ a_j + \sum_i \varepsilon_i n_i \qquad (1)$$

where the sites i are on the lattice, t_{ij} is the nearest neighbour hopping energy, and ε_i is the energy of the electron at site i. The latter is assumed to vary randomly and independently from site to site in the range $-\frac{W}{2} < \varepsilon_i < \frac{W}{2}$, with equal probability. For small fluctuations W (i.e. for $W << zt_{ij}$ where z is the number of nearest neighbours) the system is a weakly disordered metal. For <u>large</u> fluctuations W(i.e. $W >> zt_{ij}$) Anderson showed that the states are localized, by considering the motion of an electron initially at the origin. Such an electron can escape out by admixing its state with another. However, states nearby in space are unlikely to be close in energy if ε_i fluctuates widely. Typically this admixture is of order (t/W), so that a perturbation theory in hopping converges for $(t/W) << 1$. A sophisticated version of this argument was presented by Anderson[1)] who showed that an electron of energy E has a nonvanishing residence probability for $(zt/W) < N_c$ where N_c is a number of order unity, depending on the electron energy and the dimensionality as well as type of the system lattice.

For strong enough disorder, states are exponentially localized, i.e. the envelope of the wave function decreases exponentially with scale ξ called the localization length. With decreasing randomness, ξ increases till it diverges at a critical value. For strong disorder, it is of the order of an atomic length.

1.2 Anderson Transition :

For a given system, i.e. a given disorder, electronic states in a certain range of energies are localized, and other states are extended. The boundary separating extended and localized states was termed the mobility edge by Mott. At zero temperature, the available states are occupied in order of increasing energy. If the last occupied state is in the region of localized states, the lowlying excited states are spatially localized and the system is an insulator. On the other hand, if the Fermi energy is in the region of extended states, the system is a disordered metal. Thus, as the disorder increases, or as the electron density decreases (or with a combination of the two) the Fermi energy crosses the mobility edge (Fig.1) and the system undergoes a transition from a metallic to an insulating state.

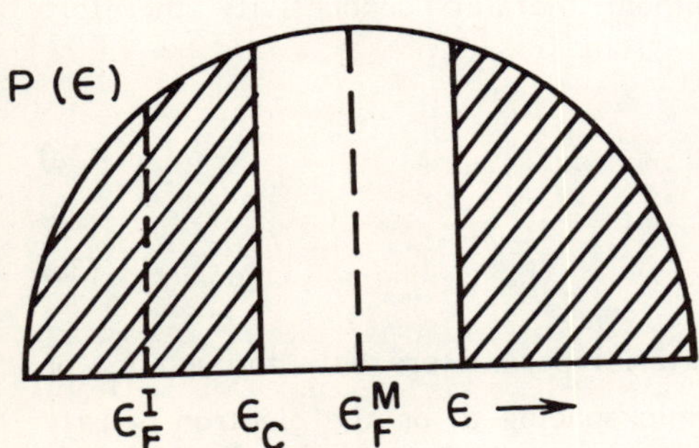

Fig.1 Density of electronic states $\rho(\varepsilon)$ vs. energy ε. The mobility edge ε_c, localized states (shaded), and Fermi energies E_F^M and E_F^I for metallic and insulating phases are shown.

Fig.2 Zero temperature electrical conductivity of a disordered metal plotted as a function of its Fermi energy E_F. The Mott prediction of a minimum metallic conductivity ($\sigma_{min.}$) is shown (full line) as well as the scaling theory prediction (dotted line).

This disorder induced metal to insulator transition was termed the Anderson transition by Mott, who was the first to investigate its characteristics. We now outline some of these.

1.3 Disordered Metal and the Minimum Metallic Conductivity :

Consider approaching the Anderson transition from the metallic side. For a disordered metal the d.c. conductivity σ is given in standard weak scattering transport theory by the formula

$$\sigma = \frac{ne^2\tau}{m} = \left(\frac{e^2}{\hbar}\right)\left(\frac{n}{k_F^2}\right) k_F l \tag{2}$$

where n is the electron density, τ the relaxation time, and l the mean free path due to collision with scatterers. Ioffe and Regel[5)] pointed out that since scatterers have a spacing $\sim k_F^{-1}$, mean free paths $l < k_F^{-1}$ are difficult to visualize. Mott argued[6),3)] that therefore the condition $k_F l = 1$ sets a lower limit on the mean free path or conductivity. For greater disorder, the mode of transport must change, and the system cannot remain metallic. The minimum metallic conductivity therefore has the following values :

$$\sigma^{3d}_{min} = (1/3\pi^2)(e^2/\hbar)\, k_F \tag{3a}$$

$$\sigma^{2d}_{min.} = (e^2/2\pi\hbar) \tag{3b}$$

We notice that σ^{3d} depends on material parameters only through k_F or equivalently the inverse of interatomic spacing a, or the electron density n $(k_F \simeq (3\pi^2 n)^{1/3})$. It has a quantum conductance scale $(e^2/\hbar\pi^2)$; $\sigma^{3d}_{min.}$ is the zero temperature conductivity of a degenerate electron gas. Extending the idea to disordered films, the minimum conductivity is seen to be universal, independent of material parameters. It has the dimensions of conductance. The associated universal resistance is about 30,000 ohms.

There exists a large body of experimental evidence supporting the idea that the conditions Eq.(3) mark a change of transport regime (Fig.2). The evidence comes from low to high temperature conductivity measurements on both sides of the metal insulator transition, but not too close to it. However, recent experimental work done close to the metal insulator transition at very low temperatures shows clearly conductivities that are much smaller than the Mott value.

As the transition is approached from the localized side, the localization length diverges. The polarizability and hence the dielectric constant depend on the square of the localization length, so that the dielectric constant diverges as ξ^2. Such a divergence has been observed in phosphorus doped silicon near the metal insulator phase transition.

Considered as a zero temperature phase transition, localization is asymmetric; it is discontinuous from the metallic side and continuous from the insulating side. This is due to the asymmetric way in which quantum interference is treated. Just on the insulating side, the interference between components scattered repeatedly back and forth by the random potential is such that the state has exponentially small amplitude beyond the localization length. However, for disorder just less than critical, it is assumed this interference does not affect physical properties such as conductivity which are calculated in the single collision approximation. In the scaling theory of localization[7)], one considers the evolution of this quantum interference as a function of scale size, and finds a continuous transition (Section 2).

1.4 Mott Transition :

A system without disorder can undergo a metal to insulator transition due to electron correlation. This was recognized as early as 1937, when de Boer and Verwey[8)] described several insulating systems which ought to be metallic according to one electron band theory because they have an odd number of electrons per unit cell. A simple hypothetical model used by Mott[9),10)] to describe this transition is a lattice of hydrogen atoms. Consider the structure to be fixed and the lattice constant a to be the only variable. For small lattice constant, the electronic states on adjacent sites overlap strongly, and the system is a good metal with a half filled conduction band (one electron per unit cell). For large lattice constant a however, one manifestly has a lattice of neutral hydrogen atoms, i.e. an insulator. In this phase, to produce an electronic excitation, i.e. to remove an electron from one hydrogen atom and to put it on another atom costs an energy U = I-A. Here I is the ionization energy and A is the electron affinity of a hydrogen atom. Thus electrons are not mobile; every electron is localized at one or the other lattice site. In a model where the long range coulomb attraction between an extra electron and a hole is ignored, and only the above short range correlation term U is considered, there is most probably a second order transition from metal to insulator for $U \simeq zt$ (the bandwidth). The effects of long range coulomb interactions, of orbital degeneracy, of coupling to the lattice, of deviation from half filling etc. have not been pinned down yet though plausible arguments exist. More relevant here is the effect of disorder. At least one system, Si : P is approximated well as a disordered correlated electron system. We shall briefly review its transition behaviour in Section 3, and shall find that it does not fit well into any one of the types investigated so far.

2. SCALING THEORY OF LOCALIZATION

Abrahams, Anderson, Licciardello and Ramakrishnan[7] considered how quantum mechanical interference effects for an electron wave in a random medium evolve as a function of scale size L of the system. The basic microscopic length scale is the mean free path l beyond which electron motion is diffusive . The first question is:- what is the measure of quantum interference? Abrahams et al argue , following ideas due to Thouless[11] that it is the electrical conductance g(L) defined on length scales L > l. It is small if disorder is large (strong interference), and is large if disorder is small. A small change in the length scale L leads to an increase in interference (or effective disorder or g(L)) by an amount which depends on the effective disorder at that length scale. One can thus plausibly argue for a single parameter scaling theory, namely that

$$(\frac{dg(L)}{g(L)})/(\frac{dL}{L}) = \frac{d \ln g(L)}{d \ln L} = \beta (g) \qquad (4)$$

Such a scaling theory can also be derived more rigorously, e.g. from many body perturbation theory, or from a mapping of the localization probelm into a field theory (see below). Now if $L \gg l$, the scaling function does not depend on the details of the system, but only on g which has a universal scale $g_c = (e^2/\hbar \pi^2)$, and on dimensionality d. $\beta(g)$ can be constructed from a knowledge of limiting forms and perturbative corrections, and used to make predictions about macroscopic electrical behaviour (e.g. g(L)) given the microscopic state of disorder (i.e. g(l)).

For large conductance, the system is a good metal, current flow is classical and Ohm's law, i.e. the relation $g = \sigma L^{d-2}$ is valid for a hypercube of extent L. This leads to

$$\beta (g) = (d-2) \qquad (5)$$

for $g \gg 1$ (in units of $g_c = (e^2/h \pi^2)$). For very small conductance, i.e. $g \ll 1$, states near the Fermi energy being exponentially localized, conduction from one end of the sample to the other is by hopping on to states with exponentially small amplitude exp $(-L/\xi)$ across the system boundary. Thus $g(L) \sim \exp(-L/\xi)$ so that

$$\beta(g) = \ln g \qquad (6)$$

for $g \ll 1$.

For large g, it was shown by Abrahams et al. that a particular set of processes contributes a negative term $(-1/g)$ to $\beta(g)$. The process is interference between 'forward' motion and backscattering in the field of the random potential. It is illustrated in Fig.3. An electron can go from r to r' in many ways. The total amplitude or propagator is a properly weighted integral over all possible paths. Two of these paths are shown in the figure. In one (full line) the path returns to the some point (or within a distance l) and goes on to r'. In the other (dotted line) the loop path is traversed in the opposite direction. If the system is invariant under time reversal, the phase changes or the action are equal so that the in square of the amplitude, there is constructive interference. The probability that an electron returns to the same point is enhanced because of this interference. This is the basic localizing process. It reduces conductivity more effectively in lower dimensions. In one dimension, the interference is perfect since only such paths exist. In two dimensions, there are paths nearby which do not exactly backtrack; consequently the localization length is exponential in the mean free path, not linear as in one dimension.

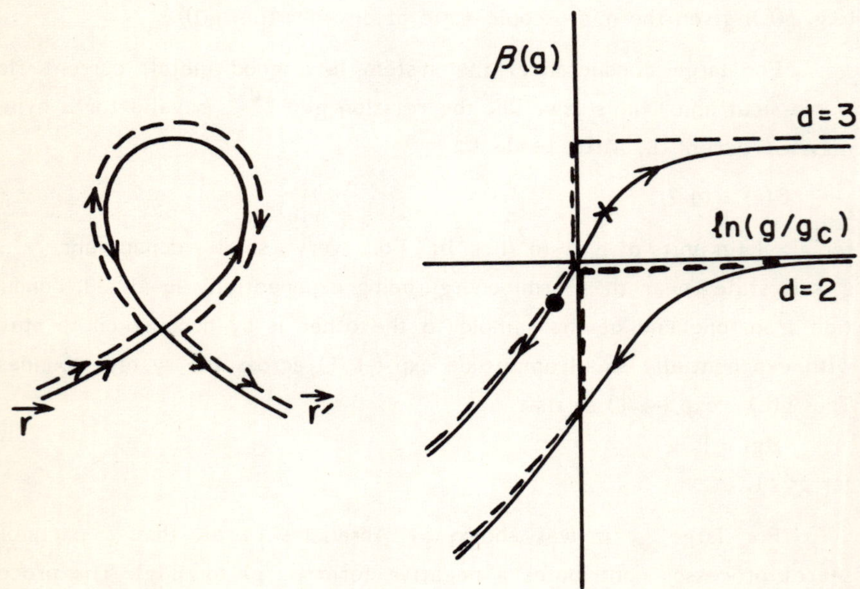

Fig.3. Two strongly interfering paths for an electron propagating from r to r'.

Fig.4. The scaling curve β(g) for conductance in two and three dimensions. Dotted lines show the behaviour expected from the minimum metallic conductivity concept.

A field theory formulation of the localization problem is due to Wegner[12] (See Ref.4 for a summary). The action functional describing amplitudes for various paths is akin to the free energy functional. Wegner showed that the quantity of interest, namely the diffusion constant, is analogous to the effective spin wave stiffness of a fixed length Heisenberg spin model. The fixed length is related to the electron density of states. One is interested in the effective spin wave stiffness at large length scales, starting from a certain bare value for it at short length scales. The renormalization of stiffness occurs because of fluctuations which leave the local density of states unaltered. The mapping shows clearly that only one scale parameter is relevant. Effects of perturbations class of the field theory, e.g. an external magnetic field which breaks time reversal invariance, can be calculated. Perturbative analysis of the $\beta(g)$ function using the field theory shows that to order $(1/g^3)$ there are no new terms.

2.1 Predictions of Scaling Theory :

The scaling curve $\beta(g)$ can be used to calculate, by integration, the conductance g at a macroscopic length scale L given the value g_o at the microscopic scale or lower length cutoff l. There are two regimes. If $g_o > g_c$, i.e. the intrinsic disorder is less than the critical value for which $\beta(g) = 0$ (see Fig.4) one starts on the positive branch of the $\beta(g)$ curves, e.g. at the point marked by the cross, g increases with length scale and one ends up in the Ohm's law regime for which $g(L) = \sigma L$. The conductivity σ is given by

$$\sigma = (e^2/\hbar\pi^2\xi_>) = (e^2/\hbar\pi^2 l)\left(\frac{g_c - g_o}{g_c}\right)^{-\nu} \tag{7}$$

where ν is the slope of the $\beta(g)$ curve near $\beta(g)=0$. The first term is the small length scale or Mott conductivity, and the second factor describes its reduction due to scale dependent interference. Due to the latter, the conductivity goes smoothly to zero at the metal insulator transition.

The Mott $\sigma_{min.}$ approximation ignores this effect.

The other regime for conductance occurs for disorder larger than critical, i.e. $g_o < g_c$ (e.g. a point marked by the circle). One then has $\beta(g) < 0$, scaling reduces conductance and for large length scales one goes over to an exponentially localized behaviour, i.e.

$$g(L) \sim g_o \exp(-L/\xi_<) \tag{8}$$

with $\xi_<$ diverging as $\{(g_c - g_o)/g_c\}^{\nu}$. The localization transition is thus a continuous transition characterized by a single correlation length diverging with an exponent ν. On the extended side, the correlation length describes how the conductivity goes to zero. On the localized side, again close to the transition, the correlation length is simply the localization scale. The perturbative estimate of the universal exponent ν is unity, i.e. $\nu = 1$. In the field theory, ν can be calculated in an expansion in powers of $\varepsilon = (d-2)$, and is found to be equal to ε. The point $\beta(g) = 0$ is an unstable fixed point of the one parameter scaling transformation.

The above results are for zero temperature. Temperature is a relevant perturbation since inelastic processes terminate quantum interference by introducing random phases. If the inelastic scattering mean free path is L_{in}, this is the effective upper length scale, not the system size. As temperature decreases L_{in}, increases, so that quantum interference or localization effects are more clearly seen. Conversely, experiments done at not very low temperatures effectively probe small length scales, i.e. systems exhibit a Mott minimum metallic conductivity like behaviour. A microscopic analysis of the inelastic scattering process confirms these ideas; it leads to a mass in the particle particle channel for electron scattering. Since due to localization effects the conductivity decreases with increasing scale size i.e. with increasing L_{in} or decreasing temperature, one has, because of incipient localization, a conductivity that decreases with decreasing temperature; this is 'nonmetallic' behaviour.

An external magnetic field is an interesting probe of the localization process. It affects the phases of the backscattering paths (Fig.3) oppositely, so that interference is not as strong in presence of the magnetic field. An external magnetic field thus leads to a negative magnetoresistance[13]. It has a characteristic H dependence for large fields in three dimensions, and a magnetic field scale H such that the typical orbit with area L_{in}^2 encloses a flux quantum ϕ i.e. $L_{in}^2 H \sim (hc/2e)$. Further, in a geometry where the flux enclosed by the electron paths can be precisely controlled, one expects the conductivity to return to the same value when the magnetic field induced phase difference changes by $2n\pi$[14]. This oscillatory conductivity behaviour has been seen experimentally[15], and is a dramatic confirmation of the reality of the backscattering process.

Another qualitatively new prediction of scaling theory is the instability of the metallic state in two dimensions. From the scaling curve for d=2 (Fig.4) we see that $\beta(g)$ is negative so that on increasing the length scale, g (or equivalently in 2d, the conductivity σ) decreases, finally tending to exponentially localized behaviour. The characteristic length scale for localization is $\xi_{loc.} \sim l \exp(\pi k_F l)$ i.e. exponential in the inverse disorder, whereas in one dimension, $\xi_{loc.} \sim l$. (The difference is due to the fact that in d = 1 one has exact backscattering). Thus at T=0, for a disordered film there is no truly metallic state. At nonzero temperatures, the tendency towards localization manifests itself in a logarithmic increase of resistivity with decreasing temperature.

3. INTERACTING DISORDERED ELECTRON GAS

3.1 Perturbative Corrections :

At about the same time as the scaling theory of localization was proposed, Altshuler and Aronov[16] pointed out that in a disordered electron gas, interactions lead to characteristically new effects for equilibrium and transport properties. These arise crudely from the fact that diffusing electrons stay around each other longer, so that interactions between them are stronger and are characteristically retarded. A simple uncertainty principle argument based on the $t^{-3/2}$ decay of the probability for an electron to stay put leads to an energy shift of $E^{3/2}$ for an electronic excitation with energy E. This gives rise to a density of states correction $E^{1/2}$. There are other associated effects, all becoming more prominent with disorder. The net result is that the consequences of disorder and interaction are strongly coupled, so that the metal insulator transition in a disordered system is rather complex.

We first briefly describe interaction effects to leading order in perturbation theory. Consider for example the energy shift of an electron due to interaction[16),17)]. Denoting the eigenstates in the absence of intearaction by $\Psi_m(r)$, the energy shift of a particular state m to lowest order in interaction v(r-r') is

$$\Sigma_m = -\sum_{\text{occupied}} \iint d\vec{r}\, d\vec{r}'\, \Psi_m^*(r)\Psi_n^*(r')\Psi_m(r')\Psi_n(r)\, v(r-r') \tag{9}$$

where the summation is over all occupied states n. The energy of a particle with E will be shifted by an amount Σ_E where

$$\Sigma_E = (1/N(o))\ \delta(E-E_m)\Sigma_m = -\int_{-\infty}^{o} dE'\ F(E,E';\ r,r')\ v(r-r')\ d\vec{r}\ d\vec{r}' \tag{10}$$

where F(E,E', r,r') is just the amplitude for a density fluctuation with energy (E-E') to diffuse from r to r'. This is known and leads to

$$\Sigma_E = \left(\frac{dn}{d\mu}\right)\int_{-\infty}^{o} dE' \frac{d\vec{q}}{(2\pi)^d} \frac{Dq^2}{(E-E')^2 + (Dq^2)^2} v(q) \tag{11}$$

The change in density of states is proportional to $\delta\Sigma_E/\delta E$ and goes as

$$[\delta N(E)/N(o)] = [2\pi^2(dn/d\mu)D]^{-1}[E/2D]^{1/2} \tag{12}$$

namely as $\sqrt{E}$, with a coefficient which depends on $D^{-3/2}$, i.e. on the conductivity. The interaction strength is absent from Eq.(12) since it is the dynamically screened coulomb potential. The same process leads to a logarithmically decreasing density of states for two dimensional systems. Consequences for resistivity, magnetoresistance etc. have also been worked out (see Ref.4 for a review) using many body perturbation theory.

3.2 Field Theory :

Finkelshtein[18),19)] has constructed a field theory for the interacting Fermion problem. The standard technique of describing Fermions by Grassmann fields $\Psi^{\sigma}(\vec{r},n)$ is used where σ is the spin index, $\vec{r}$ the coordinate and n the frequency index, $(2n+1)\, i\pi\, k_B\, T = i\varepsilon_n$. This nonzero frequency variable describes quantum fluctuations. The partition function of the system is given by

$$z_M = \int D\overline{\Psi}\, D\Psi \; \exp\{ S_d(\overline{\Psi},\Psi) + S_{e-e}(\overline{\Psi},\Psi) \} \tag{13}$$

where S_d is the action associated with pure disorder and S_{e-e} includes all electron electron interactions. S_d has the following form :

$$S_d(\overline{\Psi},\Psi) = \int dr \; \{ \sum_{n,i,\sigma} \overline{\Psi}^{i\sigma}_n (i\varepsilon_n + 1/2\Delta + \mu)\Psi^{i\sigma}_n + \frac{1}{4\pi N(o)\tau} \sum_{n,i,\sigma} (\overline{\Psi}^{i\sigma}_n \Psi^{i\sigma}_n)^2 \}$$

where $1/2\,\Delta$ is the kinetic energy operator $(+\hbar^2\, \nabla^2/2m)$ with $\hbar=m=1$, μ is the chemical potential, and the quartic term is due to electron scattering by a zero range random potential with a Gaussian strength distribution. On integrating over the Gaussian, one has the quartic term whose coefficient $(4\pi N(o))^{-1}$ is a measure of the strength of the random potential. Here τ is the relaxation time, σ is the spin component, and i is the replica index (i=1 to M). The limit of electrons moving in a random potential corresponds to $M \to 0$.

At this stage, since different frequency components are uncoupled (scattering is from a static one body potential) one has a classical field theory which is the starting point of Wegner's work[12]. The interaction part S_{e-e} is given by

$$S_{e-e}(\bar{\Psi},\Psi) = - \frac{T}{2N(o)} \iiint \frac{d^d k \, d^d p_1 \, d^d p_2}{(2\pi)^{3d}} \times$$

$$\sum_{\substack{n_1,n_2,m \\ i,\sigma,\sigma'}} \Gamma \, \bar{\Psi}^{i\sigma}_{n_1}(p_1) \, \bar{\Psi}^{i\sigma'}_{n_2}(p_2) \Psi^{i\sigma}_{n_2-m}(p_2-k) \, \Psi^{i\sigma}_{n_1+m}(p_1+k)$$

$$+ \Gamma_2 \bar{\Psi}^{i\sigma}_{n_1}(p_1) \, \bar{\Psi}^{i\sigma'}_{n_2}(p_2) \bar{\Psi}^{i\sigma'}_{n_1+m}(p_1+k) \Psi^{i\sigma}_{n_2-m}(p_2-k)$$

$$+ \Gamma_3 \bar{\Psi}^{i\sigma}_{n_1}(p_1) \, \bar{\Psi}^{i\sigma'}_{-n_1+m}(-p_1+k) \, \Psi^{i\sigma'}_{-n_2+m}(-p_2+k) \Psi^{i\sigma}_{n_2}(p_2) \qquad (15)$$

The three terms described inEq.(15) correspond to direct (Γ), exchange (Γ_2) and electron electron (Γ_3) channels. The first is spin singlet while the second has both spin singlet and triplet terms. The coefficients Γ, Γ_2, Γ_3 are the small momentum transfer limits of the scattering amplitude (e.g. in a microscopic theory starting with the bare coulomb interaction and including screening). These are the bare short length and time scale coupling constants of the field theory. One is interested in their renormalization as the effect of fluctuations with $q < q_c \simeq l^{-1}$ and $\omega < \tau^{-1}$ is included. The renormalization is due to effective higher order interaction via diffusing electron - hole and electron - electron excitations.

The general problem has three coupling constants and a bare diffusion constant D to begin with, and has not been solved. Several special cases have been considered to lowest order in D^{-1}. We mention two here.

i) Singlet Only: In the presence of scattering by magnetic impurities, the triplet and particle particle channels are massive (the corresponding long wave ength excitations relax with a finite lifetime). The variation of Γ_s and of t (inverse conductance in dimensionless units) with scale $\lambda (\simeq q_c)$ is given by

$$- \frac{d\Gamma_s}{d\ln \lambda} = - \frac{1}{2} t\Gamma_s \quad , \quad - \frac{dt}{d\ln \lambda} = - \frac{\varepsilon t}{2} + t^2 \tag{16}$$

The equations (16) have a fixed point solution $\Gamma_s^* = 0$, and $t^* = \varepsilon/2$. The conductivity goes to zero with an exponent $(1/\varepsilon)$ so that in three dimensions $(\varepsilon = 1)$ the exponent is unity, like in the pure localization problem. The density of states goes to zero at the metal insulator transition, in contrast to the pure localization case where it is nonsingular.

ii) No Particle Particle Scattering: In this case[19)], one has two coupling constants, spin triplet and singlet. It turn out that on scaling, the effective spin triplet coupling constant becomes very lage, and diverges at a finite length scale l_c. At this length scale density diffusion $D(l_c)$ is nonzero, but spin diffusion vanishes. The spin susceptibility diverges (all these results are for T=0). One thus has a clump of frozen magnetization on this length scale, i.e. a local moment. Unfortunately, the scaling approach (perturbative in the coupling strength) breaks down when coupling constants are large, and the above resuslts for a spin localization transition (Mott transiton?) are only suggestive.

This brief description suggests that there is a rich variety of types of metal insulator transitions. Because of the technical complexity of the problems only a few special cases have been explored. However, the entire approach so far is perturbative whereas transitions like the Mott transition are due to strong local correlation. Band filling or commensurate electron number plays no special role in these basically wide band theories.

3.3 Experimental Results :

We now briefly summarize the experimental situation (see also Ref.4). A large number of experiments done at low temperatures on relatively weakly disordered systems have verified many scaling theory predictions, such as logarithmic temperature dependence of resistance in films, negative magnetoresistance, oscillatory magnetic field dependence of resistivity, and effects of other relevant perturbations such as spin orbit scattering, magnetic impurity scattering etc.. It turns out that interactions in a disordered systems gives rise to comparable effects. In some cases, by measuring a number of properties,one can disentangle these two effects and evaluate their sizes. We mention below some experimental results near the metal insulator transition.

Low temperature experiments on a number of systems (see Ref.20 for a recent summary) show that the conductivity transition is smooth, and that the conductivity goes to zero on the metallic side with an exponent of either unity or half. The former is found in systems such as Nb_xSi_{1-x} (metallic for $x > 0.12$ and insulating for $x < 0.12$) Au_xSi_{1-x}, doped compensated semiconductors, and Gd_3S_4 with vacancies driven through the metal insulator transition by a magnetic field. The exponent of half is so far restricted to doped uncompensated semiconductors, e.g. Si:P. A crossover from the latter to the former behaviour with compensation has also been noted. In all these systems, conductivity values much less than the Mott $\sigma_{min.}$ have been observed, e.g. twenty times less.

Anderson localization has an exponent of unity for the conductivity (Section 2.2). The interacting disordered electron gas also has a metal insulator transition with a conductivity exponent of unity when the spin triplet and electron electron channels are suppressed and only the singlet channel is effective , e.g. due to a magnetic field or magnetic impurity scattering (Section 3.2). However, effects such as critical density of states dip in Nb_xSi_{1-x} (dip going as $(E/\Delta)^{1/2}$ with Δ proportional to σ^2), point to the presence and importance of interaction effects so that the question is :- how does the localization without interaction behaviour survive? The conductivity transition in Gd_3S_4 in presence of a strong magnetic field may be an example of the interacting spin singlet case.

Perhaps the most widely investigated and least understood system is Si:P. It was first thought to be a case of Mott transition, since the shallow levels are hydrogenic and there is one electron per phosphorus atom. Correlation effects ought therefore to be increasingly important with decreasing density and finally lead to an insulating phase. Mott argued[9),10)] that excitonic effects, i.e. long range coulomb force of attraction between electron hole excitations, lead to a first order transition. The transition is however actually continuous. The system also is intrinsically disordered, with the hopping energy t_{ij} having a wide distribution of values due to random location of phosphorus atoms. If we neglect correlations altogether the transition can be thought of as one of localization. The conductivity exponent and other features of transition, e.g. slowing down of the spin diffusion, enhancement of spin susceptibility and specific heat, are not consistent with this. One thus has a puzzle, namely a Mott-Anderson transition with features not described yet in any of the theories developed. An ad hoc explanation is to attribute the exponent of (1/2) to Anderson localization in the presence of magnetic impurity scattering, and argue that slowed down spin fluctuations act as local spins. This approach does not describe how other properties such as susceptibility enhancement can be analyzed. The difference between predictions and theory in this system is a fairly accurate indicator of our understanding of metal insulator transitions in disordered materials.

REFERENCES

1. Anderson, P.W., Phys. Rev. 109, 1492 (1958).
2. Mott, N.F., J. Noncryst. Solids 1, 1 (1968).
3. See also the book Mott, N.F., and Davis, E.A., Electronic Processes in Non-Crystalline Materials, 2nd Edition (Clarendon, Oxford, 1979).
4. Lee, P.A., and Ramakrishnan, T.V., "Disordered Electronic Systems", Rev. Mod. Phys. April 1985.
5. Ioffe, A.F., and Regel, A.R., Progress in Semiconductors, 4, 237 (1937).
6. Mott, N.F., in Electronic and Structural Properties of Amorphous Semiconductors (eds. P.G. Le Comber and J. Mort, Academic Press, London, 1973) p.1.
7. Abrahams, E., Anderson, P.W., Licciardello, D.C. and Ramakrishnan, T.V., Phys. Rev. Lett. 42, 673 (1979).
8. de Boer, J.H., and Verwey, E.J.W., Proc. Phys. Soc. A 49, 59 (1937).
9. Mott, N.F.; Proc. Phys. Soc. A 62, 416 (1949).
10. See also Mott, N.F., Metal Insulator Transitions (Taylor and Francis, London, 1974).
11. Thouless, D.J., in Ill Condensed Matter (ed.G.Toulouse and R.Balian, North Holland, New York, 1979).
12. Wegner, F.J., Z. Phys. B 35, 207 (1979).
13. Hikami, S., Larkin, A.I., and Nagaoka, Y., Prog. Theor. Phys. 63, 707 (1980).
14. Altshuler, B.L., Aronov, A.G. and Spivak, B.Z. JETP Lett. 33, 94, (1981).
15. Sharvin, D.Yu. and Sharvin, Yu.V., JETP Lett. 34, 285 (1981).
16. Altshuler, B.L. and Aronov, A.G., Sov. Phys. JETP 50, 968 (1979).

17. Abrahams, E.A., Anderson, P.W., Lee, P.A. and Ramakrishnan, T.V., Phys. Rev. B 24, 6783 (1982).

18. Finkelshtein, A.M., Sov. Phys. JETP 57, 97 (1983).

19. Finkelshtein, A.M., Z. Phys. B 56, 189 (1984).
Castellani, C. et al , Phys. Rev. B 30, 1596 (1984).

20. G.A. Thomas, in Proceedings of International Conference on Localization, Interactions and Transport Phenomena in Impure Metals (Braunschweig, 1984).

WETTING AND LAYERING TRANSITIONS: AN OVERVIEW

Rahul Pandit

Department of Physics
Indian Institute of Science
Bangalore 560 012
INDIA

ABSTRACT

Recent theories of wetting transitions in binary-fluid mixtures and wetting and layering transitions in the adsorption of gases on substrates are reviewed. The systematics of these transitions and the relationship between them is outlined. Some open questions are also listed.

In this talk I shall give a brief overview of certain surface phase transitions*: wetting, prewetting, critical wetting, and layering. I shall restrict myself to a discussion of equilibrium phenomena. This is an introductory talk, not a comprehensive review. For such reviews see the articles of Pandit, Schick, and Wortis[5)], de Gennes[6)], and Sullivan and Telo da Gama[7)]. For general introductions see the articles by Sullivan[8)], Moldover and Schmidt[9)], and Hauge[10)].

This talk is organised as follows: Section I describes wetting transitions in (1) fluid mixtures and (2) the adsorption of gases on substrates. Section II discusses the essential physics of these transitions, their mean-field theory, and the importance of certain fluctua-

---

*For a review of surface phase transitions see Ref.1. For related interfacial phase transitions see Refs.2-4.

tions neglected by this mean-field theory. Section III reviews some recent experiments. Section IV is devoted to critical wetting, the fluctuations that accompany this transition, and the role of long-range forces in suppressing it (in most cases). Section V summarises some recent theoretical results. Section VI concludes this talk with a brief discussion of some unanswered questions in this field.

I. INTRODUCTION

A. Fluid Mixtures

Consider two fluid phases α (say, vapour) and β (say, liquid) above a substrate or wall ω. Let all thermodynamic parameters (e.g., T, the temperature, and μ, the difference of the chemical potentials of α and β) be such that the phases α and β coexist in the bulk, i.e., far away from the substrate ω. The substrate ω can be (a) incompletely wet (Fig. 1a), (b) completely wet by β (Fig. 1b), or (c) completely wet by α. (If β is a liquid and α its vapour, then wetting by α is referred to as drying.)

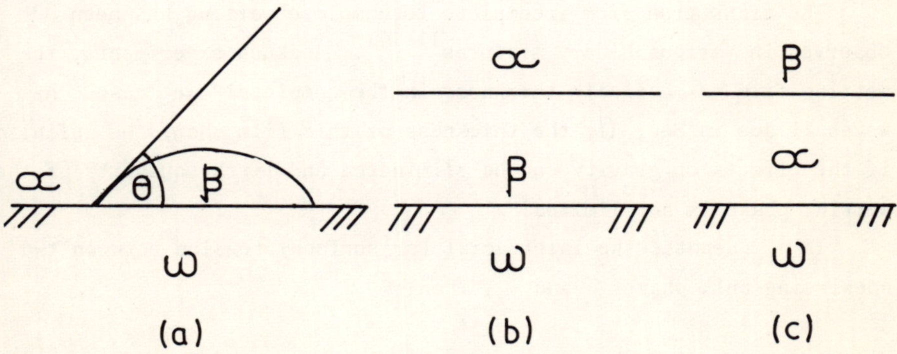

Figure 1. A substrate ω that is (a) incompletely wet, (b) completely wet by β, and (c) completely wet by α. In (a) θ is the contact angle.

If ω is incompletely wet, then as T approaches the critical temperature T_c at which α and β mix* either β or α completely wets ω at a wetting temperature $T_w < T_c$. In some systems, the role of the substrate ω is played by a third phase γ and the incompletely wet configuration is as shown in Fig.2; the completely wet configurations are like Figs. 1b and c.

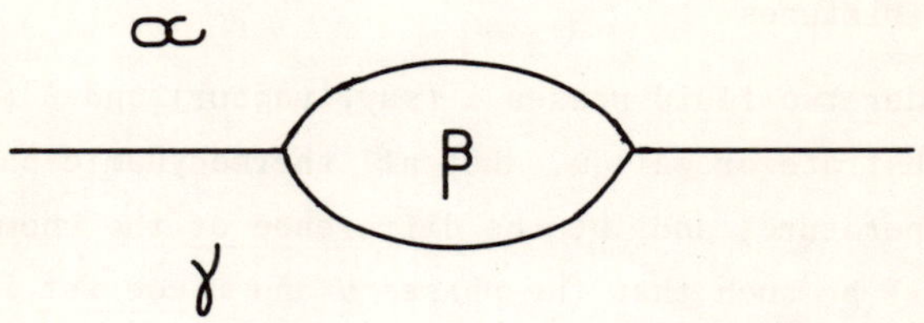

Figure 2. A lens of the β phase at the α/γ interface between the phases α and γ.

The transition from incomplete to complete wetting has been observed in various binary mixtures[11-16]. In such experiments, the wetting film has a finite thickness in the completely wet case. As we shall see in Sec. II, the thickness of this film should be infinite if the effects of gravity can be eliminated and if the quantity of the wetting phase is not limited.

If $\sigma_{\lambda\nu}$ denotes the interfacial (or surface) tension between two coexisting bulk phases λ and ν, then

$$|\sigma_{\alpha\omega} - \sigma_{\beta\omega}| < \sigma_{\alpha\beta} \quad , \tag{1a}$$

when ω is incompletely wet, and

---

*For ease of presentation, I shall restrict the discussion to mixtures with an upper consolute point.

$$\sigma_{\alpha\omega} - \sigma_{\beta\omega} = \sigma_{\alpha\beta} \quad , \tag{1b}$$

when ω is completely wet by β . (The sign of the left-hand side of Eq.(1b) is reversed if α wets ω). Mechanical equilibrium of the droplet in Fig. 1a requires[17)]

$$\sigma_{\alpha\omega} - \sigma_{\beta\omega} = \sigma_{\alpha\beta} \cos\Theta \ , \tag{1c}$$

where Θ is the contact angle. In the completely wet phase $\Theta = 0$ or $\Theta = \pi$.

B. Adsorption

When a solid substrate like graphite is placed in contact with a gas like Kr, some gas is adsorbed* on the substrate. A convenient measure of the adsorption is the excess surface density

$$n_s(T, \mu) = \int_0^\infty (n(z) - n_b)dz, \tag{2}$$

where n_b is the bulk density in the absence of the substrate and $n(z)$ the density at a distance z perpendicularly above the substrate at $z = 0$. An experimental way of estimating n_s is described in Sec. III.

Typical adsorption isotherms are shown in Fig. 3. These are plots of n_s vs. the chemical potential μ(or the pressure P) at a constant temperature. There are two qualitatively different types of isotherms⁺: class-I isotherms in which n_s diverges as $\mu \to \mu_o(T)$, the chemical potential at which the adatom gas coexists[#] with a dense

---

*I shall only consider physisorption on inert substrates.

⁺I follow the classification of Ref.5. For different (though similar) classification schemes see Refs. 18 and 19.

#In all the experiments that I am aware of, $\mu_o(T)$ is approached from the gas phase ($\mu < \mu_o(T)$). Drying transitions can occur on the high-density side of the coexistence curve. However, this region is not easily accessible to experiments.

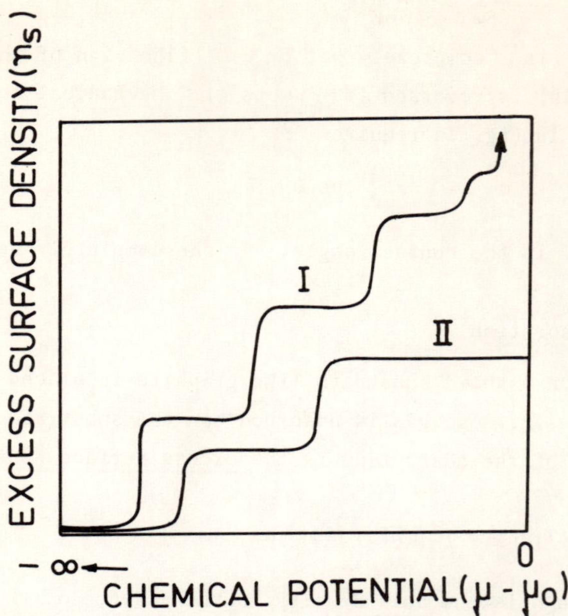

Figure 3. A schematic plot of adsorption isotherms of classes I and II. Sharp steps occur at layering transitions.

phase, either solid* or liquid; and class-II isotherms in which n_s goes to a finite constant as $\mu \to \mu_o(T)$. Such isotherms have been observed in many systems.[18,20,21] For example, the adsorption of Kr on graphite[22] yields a class-I isotherm at 77.3K; a class-II isotherm is obtained[23] when CH_3Cl is adsorbed on Cd at 143.6K. The adsorption of C_2H_4 on graphite yields especially interesting results:[24] at low temperatures ($\simeq$ 77K) the isotherms are of class II, but at high temperatures ($\simeq$ 106K) the isotherms are of class I. This passage from a low-temperature, low-adsorption phase to a high-temperature, high-adsorption phase is a manifestation of the wetting transition, which I have discussed in the context of fluid mixtures. The incompletely wet configuration of Fig. 1a corresponds to a low-adsorption phase; unadsorbed, but condensed, adatoms form droplets or

---

*If the dense bulk phase is a solid, n_s might not diverge as discussed in Sec. V.

crystallites on the substrate. The completely-wet configuration of Fig. 1b corresponds to a high-adsorption phase: an infinitely thick film of the adsorbate is formed on the substrate as the excess surface density n_s diverges.

C. Layering, Prewetting, and Critical Wetting

So far I have concentrated on phenomena that occur near the coexistence boundary $\mu = \mu_o(T)$. The sharp steps in the adsorption isotherms of Fig. 3 indicate that there are some phase transitions below this coexistence boundary. We shall see that these steps arise because of first-order <u>layering</u> transitions in which a layer of adatoms is adsorbed on the substrate*. There can be an infinite sequence of such transitions.

In fluid mixtures there are no layering transitions; however, they can be replaced by one first-order <u>prewetting</u> transition at which a thin film of fluid on a substrate suddenly becomes a thick film. At bulk coexistence ($\mu = \mu_o(T)$), the thick film becomes infinitely thick and the prewetting transition becomes a first-order <u>wetting</u> transition. The wetting transition can be continuous. In this case it is called a <u>critical wetting</u> transition. Critical wetting does not have a prewetting transition associated with it.

In the next section I shall outline the theory of wetting, prewetting, critical-wetting, and layering transitions, examine the relationships between them, and give an overview of the systematics of these transitions. In Sec. III I shall discuss some experiments that have tried to verify this theory and the degree to which it has been verified.

---

*This is only approximately true at finite temperatures, but strictly true at T = 0 (if the layering transition occurs at T = 0).

II. WETTING TRANSITIONS: THEORY

A. Essential Physics

Two competing interactions* are responsible for the wetting transition: The adatom-adatom interaction V and the adatom-substrate interaction U. Both of these are attractive, in most cases, beyond nearest-neighbour distances. For simple fluids (like rare gases), V can be taken to be of the Lennard-Jones form

$$V(r) = v\,\{4((\frac{a}{r})^{12} - (\frac{a}{r})^{6})\}\ , \qquad (3a)$$

where r is the separation between two adatoms, and a and v are parameters the vary from system to system. If the adatoms interact with the substrate atoms via a Lennard-Jones potential, then a lateral average over a semi-infinite, planar substrate yields an adatom-substrate potential of the form

$$U(z) = u\,\{\frac{4\pi}{45}(n_W a_W^3)\,((\frac{a_W}{z})^9 - \frac{15}{2}(\frac{a_W}{z})^3)\}\ , \qquad (3b)$$

where z is the distance between the adatom and the surface of the substrate, n_W is the number density of atoms in the substrate, and a_W and u are parameters that characterise a given adatom-substrate system. In Eqs.(3a) and (3b) the prefactors v and u measure the overall strengths of the potentials.

Most realistic potentials are long-ranged, i.e., they decay algebraically with distance as in Eqs.(3a) and (3b). I shall also discuss short-range potentials which decay exponentially (or even faster than exponentially) with distance.

Large values of v lead to a large interfacial tension $\sigma_{\alpha\beta}$ at low temperatures. Large values of $\sigma_{\alpha\beta}$ favour incomplete wetting (Fig. 1a)

---

* The following discussion is couched in the language of adsorption. The extension to fluid mixtures is straightforward. The crucial ingredients for a wetting transition are two coexisting, bulk phases α and β and a substrate ω which has a preferential attraction for one of these phases.

because the interfacial free energy is minimised by the formation of droplets of β (liquid) in α (vapour). On the other hand, large values of u favour the adsorption of thick films, i.e. complete wetting (Fig. 1b). As the temperature is increased, $\sigma_{\alpha\beta}$ decreases*. So, even if $\sigma_{\alpha\beta}(T=0)$ is large enough to force incomplete wetting at low temperatures ($T < T_W$), at high temperatures ($T_W<T<T_c$) the adatom-substrate interaction can win over $\sigma_{\alpha\beta}(T)$ and force complete wetting.⁺

From the preceding paragraph it is clear that the systematics of wetting and adsorption depends crucially on the ratio u/v. This also emerges from the theory I outline below. The principal results are: (1) For strong-substrate systems ($u/v \gg 1$) the substrate is completely wet at all temperatures, i.e. $T_W = 0$; for intermediate-substrate systems ($u/v \simeq 1$) there is a wetting transition at $T = T_W$ ($0<T_W<T_c$); for weak-substrate systems ($u/v \ll 1$) there is a drying transition on the high-density side of the coexistence boundary $\mu = \mu_o(T)$. (2) Class-II isotherms are obtained for $0 < T < T_W$ and class-I isotherms for $T_W < T < T_c$. (3) For class-I isotherms, the precise way in which n_s diverges as $\mu \to \mu_o(T)$ depends on the functional forms of V and U (Sec. IID). (4) Class-I isotherms can have an infinite sequence of

---

*In d dimensions $\sigma_{\alpha\beta}$ vanishes as[2,3] $|T-T_c|^{(d-1)\nu}$ as $T \to T_c^-$, where T_c is the critical temperature at which α and β mix and ν is the critical exponent which characterises the divergence of the bulk correlation length at T_c.

⁺If we assume that U couples to $\Delta\rho$, the density difference between the liquid β and the vapour α, then, as T increases, $\Delta\rho$ must decrease more slowly than $\sigma_{\alpha\beta}$ for the above argument to be true. Cahn[25] used such an argument to show that a substrate is completely wet (or completely dry) near T_c. For a critique of Cahn's argument, see Ref.5, p.5125. For an extension of Cahn's argument to systems with long-range potentials, see Ref. 26.

sharp steps (Fig. 3), a few sharp steps followed by a continuous increase of n_s, or just a continuous increase of n_s as $\mu \to \mu_o(T)$; which one of these situations obtains depends on whether the interface between the phases α and β is <u>smooth</u> or <u>rough</u>[27] as explained below.

B. Surface Statistical Mechanics

All <u>bulk</u> thermodynamic properties of a system follow from its <u>bulk</u> free energy. Similarly, all <u>surface</u> thermodynamic properties of a system follow from its <u>surface</u> free energy. In particular, the surface free energy is nonanalytic at a surface phase boundary where a surface phase transition, such as wetting, occurs. This subsection contains precise definitions of surface free energies, surface phases, and surface phase transitions. Readers familiar with these definitions should go on to the next one.

Consider a gas of adatoms* in a region Ω of volume $V(\Omega)$ and bounded by inert, planar walls⁺(substrates) of area $S(\Omega)$. The total free energy of this system is

$$F_{V,S}(\Omega,T,\mu,u,v) = -k_B T \ln(\mathrm{Tr}\,\{e^{-\beta(\mathcal{H} - \mu N)}\}) , \tag{4}$$

where $\mathcal{H}$ is the Hamiltonian of the system, N the number of adatoms, $\beta = 1/k_B T$, and k_B the Boltzmann constant. I use the grand canonical ensemble. This free energy depends on the temperature T, the chemical potential μ, the adatom-adatom and adatom-substrate potentials v and u, and the size and shape of Ω . The <u>bulk free energy</u> per unit volume is obtained by taking the thermodynamic limit

---

*Again the discussion is couched in the language of adsorption, but the definitions are easily extended to other systems[28,29].

⁺I assume that all walls have the same crystallographic orientation, say (1,0,0). Walls with different orientations have different surface free energies.

$$f_b(T,\mu, v) \equiv \lim_{V(\Omega)\to\infty} \frac{F_{V,S}(\Omega,T,\mu,u,v)}{V(\Omega)} . \tag{5}$$

For reasonable potentials* and regions+ the limit f_b exists,[28] is independent of the size and shape of Ω, and does not depend on the properties of the walls, i.e. u. All bulk thermodynamic functions can be obtained from f_b. For example, the bulk particle density n_b is given by

$$n_b = -\partial f_b/\partial\mu . \tag{6}$$

<u>Bulk phase boundaries</u> in <u>bulk phase diagrams</u> are the loci of non-analyticities of f_b in the space of parameters T, μ, and v. In this space, <u>bulk phases</u> are the regions where f_b is an analytic function of its arguments.

<u>The surface free energy</u> per unit surface area f_s is obtained by subtracting the total bulk free energy $V(\Omega)f_b$ from $F_{V,S}$ and taking the thermodynamic limit

$$f_s(T,\mu,v,u) \equiv \lim_{\substack{V(\Omega)\to\infty \\ S(\Omega)\to\infty}} \frac{F_{V,S}(\Omega,T,\mu,u,v)-Vf_b}{S(\Omega)} , \tag{7}$$

where the size of Ω is taken to infinity with a fixed shape. Again, for reasonable potentials and regions Ω, f_s can be proved to exist.[29] Note that f_s is a function of both adatom-adatom and adatom-substrate potentials. All surface thermodynamic functions can be obtained from f_s. For example, the excess surface density n_s is given by

$$n_s = -\partial f_s/\partial\mu . \tag{8}$$

The loci of nonanalyticities of f_s in the space of parameters T,μ,u,

---

* Reasonable potentials are those which are bounded below, have a sufficiently hard core to prevent collapse, and do not decay too slowly at large distances.

\+ Reasonable regions are those for which the surface to volume ratio goes to zero in the thermodynamic limit.

and v are the surface phase boundaries of the surface phase diagram. In this space surface phases are regions where f_s is an analytic function of its arguments.

The surface regions are in intimate contact with the bulk, so all bulk phase boundaries appear in the surface phase diagram. In addition to these boundaries, there are surface phase boundaries which arise because of phase transitions in regions near surfaces. Such phase boundaries do not appear in bulk phase diagrams because the surface-to-volume ratio vanishes in the thermodynamic limit (5) (if the limit exists).

In the following sections, I shall let v set the scale of energies and plot surface phase diagrams in the (T,μ) plane for different values of u/v. (I shall use magnetic variables below.)

C. Continuum Models

The wetting transition was predicted on theoretical grounds by Ebner and Saam[30)] and, independently, by Cahn.[25)] Ebner and Saam used a density-functional theory, Cahn a Landau-Ginzburg theory. Both these theories neglected fluctuations which are important when the wetting transition is continuous. These fluctuations were first studied by Lipowsky, Kroll, and Zia[31)] and by Brézin, Halperin, and Leibler.[32,33)]

I discuss a version of Cahn's theory[34,35)] here. In Sec. IV I discuss the effect of fluctuations, as treated by Brézin, Halperin, and Leibler.[33)] For uniformity of presentation, I use the notations of Brézin, Halperin, and Leibler. Unfortunately, this necessitates a change to magnetic variables from the fluid variables I have been using so far. The exact equivalences between these variables are given in the context of the lattice model discussed in the next subsection. For the discussion here the following equivalences are enough: large positive (negative) values of the bulk chemical potential μ correspond to large positive (negative) values of the bulk magnetic field H; the density n corresponds to the magnetization M; the high-density (low-density) phase corresponds to the up-spin or

$M > 0$ (down-spin or $M<0$) phase*; the z-dependent adatom-substrate potential (3b) corresponds to an inhomogeneous magnetic field $H(z)$ which is positive if $U(z)$ is sufficiently attractive (Sec. II D); the adatom-adatom potential (3a) corresponds to a two-spin exchange coupling which is ferromagnetic if $V(r)$ is attractive; the coexistence boundary $\mu = \mu_o(T)$ corresponds to $H=0$, $T<T_c$, where T_c is the Curie temperature.

I restrict the discussion to short-ranged adatom-adatom potentials[25,34,35] for the moment. For other continuum models see Refs. 19,30, and 36-48.

Consider the Landau-Ginzburg free-energy functional

$$F_s = \int_{z>0} dz\, d\rho^{d-1} \{\tfrac{1}{2}(\nabla\phi)^2 - \tfrac{\tau}{2}(\phi^2 - M^2) + \tfrac{g}{4!}(\phi^4 - M^4) - (h + h(z))(\phi - M)\} + \int d\rho^{d-1} \{\tfrac{1}{2} c\, \phi^2(0, \underset{\sim}{\rho}) - h_1 \phi(0, \underset{\sim}{\rho}) - (\tfrac{1}{2}cM^2 - h_1 M)\}, \quad (9)$$

where d is the dimension, $\phi(z, \rho_1, \ldots, \rho_{d-1})$ is a one-component order parameter, $\tau \equiv -(T-T_c)/T_c \geq 0$, T_c is the bulk critical temperature, $h \equiv H/k_B T$, $h(z) \equiv H(z)/k_B T$, M is the equilibrium value of the bulk magnetization, c is a surface temperature⁺ ($c>0$), $h_1 \equiv H_1/k_B T$ is a surface magnetic field, and $g>0$. The integration is restricted to the semi-infinite half-space $z>0$, $-\infty < \rho_1, \ldots, \rho_{d-1} < \infty$: the substrate lies at $z = 0$, the coordinate z is perpendicular to the substrate, and $\rho_1, \ldots, \rho_{d-1}$ are parallel to it. The model (9) can either be written down on phenomenological grounds[25] or it can be obtained from a microscopic Ising model.[34]

---

*This theory is too simple to account for solid-liquid-gas triple points. See Sec.V for a discussion of bulk triple points and wetting transitions.

⁺I shall only consider $c>0$. This is the case for the physical situations described here. For an extension to other values of c, see Nakanishi and Fisher.[35]

In the mean-field approximation, there are no fluctuations in the bulk magnetization and its equilibrium value $M \to -(6\tau/g)^{\frac{1}{2}}$ as $H \to 0^-$ (the down-spin or low-density side of coexistence). The M-dependent terms in Eq.(9) are put in to subtract the (mean-field) bulk free energy from the total free energy (cf. Eq.(7)).

In the mean-field approximation, the order parameter ϕ depends only on z; all other fluctuations are suppressed. By minimising F_s with respect to ϕ (i.e., by setting $\delta F_s/\delta\phi = 0$), we get

$$-\frac{d^2\phi}{dz^2} - \tau\phi + \frac{g\phi^3}{6} = h + h(z) \tag{10}$$

and the boundary condition

$$\left.\frac{d\phi}{dz}\right|_{z\to 0^+} = c\,\phi\,(z=0) - h_1 \quad ; \tag{11a}$$

in addition, ϕ must approach the bulk magnetization far away from the substrate, so

$$\phi(z = \infty) = M. \tag{11b}$$

The equilibrium, mean-field profile $\phi_c(z)$ is obtained by solving Eq.(10) subject to the boundary conditions (11) and then picking the solution which gives the global minimum of F_S. For arbitrary h(z), Eq.(10) has to be solved numerically; however, for h(z) = 0 but $h_1 > 0$, a qualitative analysis shows why a wetting transition must occur. I give this qualitative analysis below, but, before that, I summarise the main results.

For $T < T_c$ and $H \lesssim 0$, two profiles compete for the global minimum of F_S : The profile ϕ_{cII} drops rapidly* from its value at z=0 to its bulk value M (Fig.4a); i.e., the adsorbed film is thin or, equivalently, the α/β interface is pinned to the substrate ω . In the profile ϕ_{cI}, the order parameter is substantially greater than M over a large region $0 < z < l$ (T,H) (Fig.4b); i.e., the adsorbed film is thick. In this case $l \to \infty$ as $H \to 0^-$: the β/α interface is unpinned from the

*Unless $T \to T_c$ and the bulk correlation length ξ_b diverges. See the caption of Fig.4.

substrate ω . If $h(z)$ and h_1 are sufficiently large and positive (i.e., the adatom-substrate potential is sufficiently attractive), then, infinitesimally below coexistence ($H = 0$, $0 < T < T_c$), the thin-film solution ϕ_{cII} is the equilibrium profile at low temperatures ($0 \leqslant T \leqslant T_W$). The thick-film solution ϕ_{cI} is the equilibrium

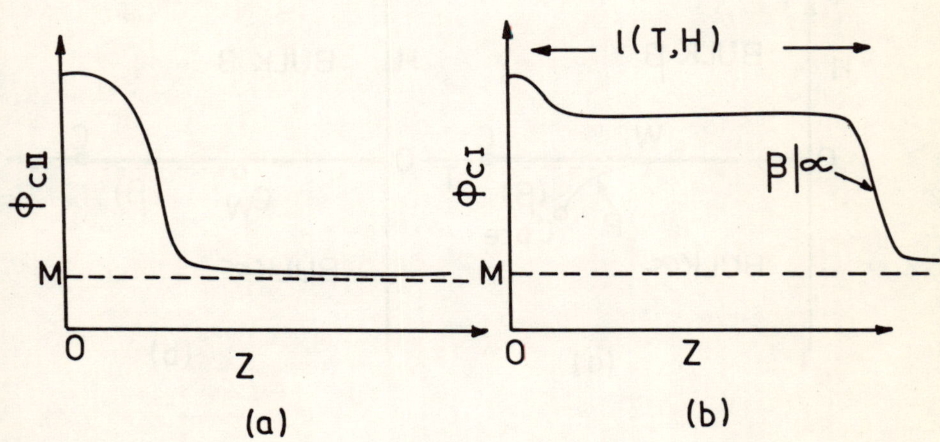

Figure 4. The order-parameter profiles corresponding to (a) a thin, adsorbed film and (b) a thick, adsorbed film. The nonuniform region near $z = 0$ has a width of the order of $\max(\xi_b, r_o)$, where ξ_b is the bulk correlation length and r_o the range over which $h(z)$ differs appreciably from zero. The nonuniform region near the β/α interface in (b) has a width of the order of ξ_b. In case (b), the film thickness $l(T,H) \to \infty$ as $H \to 0^-$ (see text).

profile at high temperatures ($T_W \leqslant T < T_c$). Thus, there is a wetting transition at T_W (unless $T_W = 0$). Adsorption isotherms are of class II for $0 < T < T_W$ and of class I for $T_W < T < T_c$.

The wetting temperature T_W decreases with increasing overall strengths of $h(z)$ and h_1. The wetting transition can be first-order, as at W in Fig. 5a, or continuous, as at the critical wetting point* C_W in Fig. 5b. If it is first-order, then a first-order prewetting boundary P emanates from W, moves away tangentially[10,44] from the coexistence boundary H = 0, and ends in a <u>prewetting critical point</u> C_{pre} (Fig. 5a).

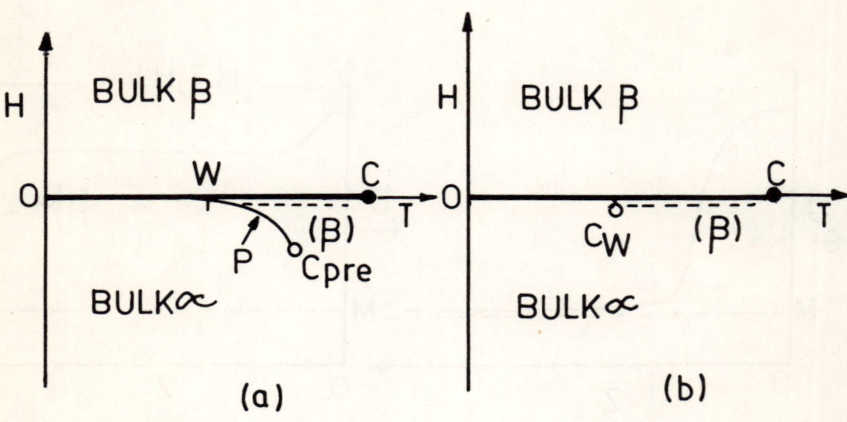

Figure 5. Schematic surface phase diagrams showing (a) first-order wetting W, first-order prewetting P, and a prewetting critical point C_{pre}; and (b) critical wetting C_W. The bold line is the bulk phase boundary separating the bulk phases β(up-spin or liquid) and α(down-spin or gas). The dots indicate the region of the bulk phase boundary where the substrate is completely wet by a film of β(denoted by (β)). C is the bulk critical point.

I give below a qualitative analysis of Eqs. (10) and (11) which shows why a wetting transition occurs. Readers who are not interested in this should skip over the rest of this subsection.

---

*Critical wetting was discovered by Sullivan[19] in the context of another continuum model.

If $h(z) = 0$, the solution of Eq. (10) can be reduced to quadratures. It is useful to portray the solutions of Eq. (10) in the phase space $(\frac{d\phi}{dz}, \phi)$ as shown in Fig. 6. The only solutions of interest to us are the ones which satisfy both the boundary conditions (11a) and (11b). The boundary condition (11b) can be satisfied only if the

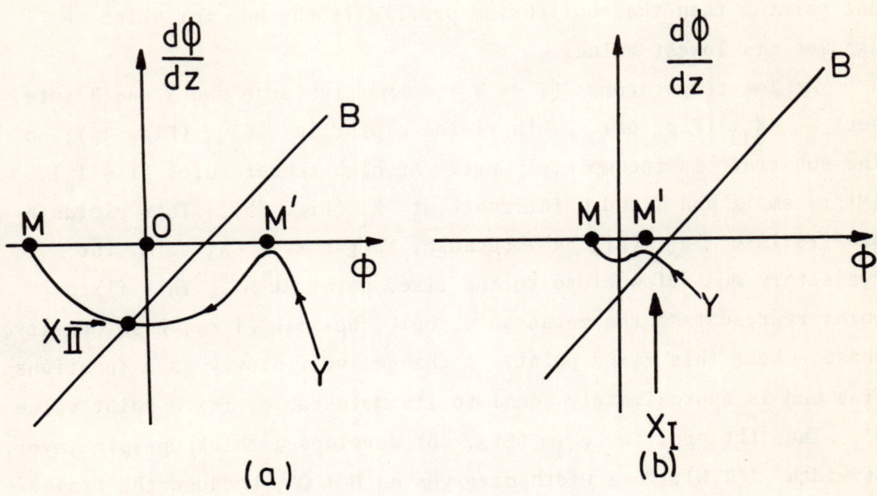

Figure 6. A schematic representation of the solution of Eq.(10) in the phase space $(\frac{d\phi}{dz}, \phi)$, with $h(z) = 0$, $h \lesssim 0$, for (a) low temperatures $(0 < T < T_W)$ and (b) high temperatures $(T_W < T < T_c)$. The trajectory Y is the only one which satisfies the boundary conditions (11). The boundary condition (11a) appears as the line B. M, O, and M' are fixed points corresponding to thermodynamically stable, unstable, and metastable phases, respectively. $M \to -(6\tau/g)^{\frac{1}{2}}$ as $h \to 0^-$. $h_1 > 0$ in these figures.

phase trajectory* flows into the fixed point M (i.e., $\phi = M, \frac{d\phi}{dz} = 0$). There are only two such trajectories, but only one of them (Y in Fig. 6) satisfies the boundary condition (11a). The line B represents this boundary condition ($h_1 > 0$ in Fig.6). A possible profile ϕ_c is obtained by starting (at z = 0) at the intersection of Y and B and following Y to the fixed point M. If Y and B intersect at more than one point+, then the equilibrium profile is the one for which F_s assumes its lowest value.

At low temperatures ($T \ll T_c$), $|M|$ is large and Y and B intersect at X_{II} (Fig. 6a). This yields a profile ϕ_{cII} (Fig. 4a), so the substrate is incompletely wet. At high temperatures ($T \lessapprox T_c$), $|M|$ is small and Y and B intersect at X_I (Fig. 6b). This yields a profile like ϕ_{cI} (Fig. 4b) because, to get from X_I to M, the trajectory must flow close to the fixed point at M'. This fixed point represents** the metastable, bulk, up-spin (i.e., high-density) phase. Near this fixed point, ϕ changes very slowly as a functions of z and is approximately equal to its metastable, fixed-point value M'. Thus the profile ϕ_{cI} (Fig. 4b) develops a thick, up-spin layer of width $l(T,H)$; this width diverges as $H \to 0^-$, because the trajectory Y flows very close to M' in this limit, and the substrate is completely wet.

---

*I use the language of dynamical systems. A phase trajectory represents a solution of Eq.(10). A fixed point is obtained by setting $\frac{d^2\phi}{dz^2} = \frac{d\phi}{dz} = 0$. For details see Pandit and Wortis[34]. For a slightly different way of viewing this dynamical analogy see Widom[49].

+Y and B intersect either at one point or at three.

**The fixed points are obtained by setting $\frac{d^2\phi}{dz^2} = 0$ in Eq.(10) and and solving for ϕ. The solution $\phi = M \to -(6\tau/g)^{1/2}$ (as $H \to 0^-$) is the global minimum of the bulk free energy. $\phi = M' \simeq +(6\tau/g)^{1/2}$ and $\phi \simeq 0$ are, respectively, metastable and unstable (thermodynamically not mechanically).

The crossover from an incompletely-wet substrate at low temperatures $(0 < T < T_W)$ to a completely-wet substrate at high temperatures $(T_W < T < T_c)$ can occur via a first-order or a continuous transition. A first-order transition is obtained if Y and B intersect at three points as Y evolves from its low-temperature form (Fig. 6a) to its high-temperature form (Fig. 6b). This happens for h_1 large and positive. If Y and B always intersect at one point, the wetting transition is continuous at coexistence. This happens for h_1 small but positive. The crossover from first-order to continuous wetting occurs at a tricritical wetting point. First-order and critical prewetting can also be studied by the method given above. Drying transitions occur on the high-density side of the coexistence boundary $(H \to 0^+, T < T_c)$ if $h_1 < 0$. For details and extensions of the analysis given here see Pandit and Wortis[34)] and Nakanishi and Fisher[35)].

I have discussed what I consider to be the simplest model for the wetting transition. Many other continuum models have been studied[19,30,36-48)]. Both long-range[30,36-43)] and short-range[37-39,& 44-48)] potentials have been used. As we shall see later, long-range potentials can suppress critical wetting and can force the first-order prewetting boundary (P in Fig. 5a) to lie very close to the bulk coexistence boundary[40)]. Foiles and Ashcroft[36)] and Tarazona and Evans[37)] have shown that long-range correlations develop parallel to the β/α interface (Fig.4b) in the completely-wet phase. Lipowsky and Speth[45-47)] have studied an interesting class of models in which the wetting layer is a disordered (paramagnetic) phase. Some numerical simulations have also been carried out[50)].

D. Lattice Models

In the continuum model (9) there is no sequence of layering transitions because it is impossible to distinguish one fluid layer from another. I now describe a lattice-gas model which exhibits layering transitions. The imposed lattice structure makes it possible to distinguish one layer from another for $T < T_R$, the roughening temperature (see below).

Consider a semi-infinite, lattice-gas model described by occupation-number variables $n_i = 0,1$ at the sites i. Let the atoms of this gas (adatoms) interact via pairwise forces v_{ij}, and let the substrate be represented by a single-particle potential u_i acting on the adatoms. Thus,

$$\mathcal{H} - \mu N = \sum_{\langle i,j \rangle} v_{ij} n_i n_j + \sum_i (u_i - \mu) n_i , \tag{12}$$

where $\mathcal{H}$ is the Hamiltonian, μ the chemical potential, and $\langle i,j \rangle$ are distinct pairs of sites. The potential u_i is uniform in directions parallel to the substrate whose surface is taken to be at z = 0 (cf. Eq.(9)). Both v_{ij} and u_i are attractive; the hard-core part (of, say, Eqs. (3a)) is accounted for by the constraint n_i=0,1. This lattice-gas model was first introduced by de Oliveira and Griffiths[51)] to study layering transitions in strong-substrate systems. It has been studied subsequently by Ebner[52,53)], Pandit, Schick, and Wortis[5)], Kim and Landau[54)], Dietrich and Schick[55)], Ebner, Saam, and Sen[56)], and Binder and Landau[57)]. The systematics of layering transitions that emerges from these studies is outlined below.

For ease of comparison with model (9), it is necessary to change to magnetic variables. The model (12) is equivalent to the semi-infinite Ising model (upto a constant term)

$$\mathcal{H}_I = - \sum_{\langle ij \rangle} J_{ij} S_i S_j - \sum_i (H + H_i) S_i , \quad S_i = \pm 1, \tag{13}$$

with

$$S_i = 2n_i - 1, \tag{14a}$$

$$H = \tfrac{1}{2} (\mu - \tfrac{1}{2} \sum_j v_{ij}) , \tag{14b}$$

$$H_i = \tfrac{1}{2} (-u_i + \tfrac{1}{2} \sum_j{}' v_{ij}) , \tag{14c}$$

and

$$J_{ij} = -\tfrac{1}{4} v_{ij} , \tag{14d}$$

where the prime on the sum indicates that only "virtual" sites in the region $z<0$ are to be counted*. Equations (14a) - (14d) make precise the correspondence between fluid and magnetic variables.

The analogue of the order parameter $\phi(z, \rho)$ (Eq.(9)) is the magnetization of the ith site $M_i = \langle S_i \rangle$, where the angular brackets indicate a thermal average. In the mean-field approximation M_i is taken to be uniform in directions parallel to the substrate and the analogues of Eqs. (10) and (11) are⁺

$$M_1 = \tanh\left(\beta \{(H + H_1) + 2(d-1)\, \mathbf{J}M_1 + \mathbf{J}M_2\} \right) , \tag{15a}$$

$$M_k = \tanh\left(\beta \{(H + H_k) + 2(d-1)\, \mathbf{J}M_k + \mathbf{J}(M_{k-1} + M_{k+1})\} \right), \; k > 2, \tag{15b}$$

and

$$M = M_b ,$$

where the bulk magnetization M_b satisfies the Curie-Weiss self-consistent equation[34)]

$$M_b = \tanh\left(\beta \{H + 2d\, \mathbf{J}M_b\} \right), \tag{16}$$

and d is the dimension. The infinite number of equations (15a) and (15b) can be truncated to a manageable number N (N = 300 in a recent calculation[56)]), M_N set equal to M_b, and this set of N coupled, nonlinear equations solved by standard numerical methods[34,52)]. If many solutions exist, then the one which is the global minimum of the surface free energy** is the equilibrium solution.

---

*In the absence of the substrate, adatoms in the region $z < 0$ attract adatoms in the region $z > 0$. The substrate does not allow for such an interaction and leads to an effective repulsion

$$\tfrac{1}{2} \sum_j{}' v_{ij} \quad \text{(Eq.(14c))}.$$

+For simplicity I give the expressions for nearest-neighbour, adatom-adatom interactions **J**.

**The mean-field surface free energy is given by a discrete version of Eq.(9). See Ref.34.

The models (12) and (13) have surface phase diagrams that depend crucially on the relative strengths of the adatom-substrate and adatom-adatom potentials (i.e., on the ratio u/v) and, to a lesser extent, on the forms of these potentials. For the systematics of these phase diagrams see Ref. 5.

Figures 7a and 7b show possible surface phase diagrams for strong-substrate and intermediate-substrate systems, respectively. Instead of one prewetting transition P (Fig.5a), an infinite sequence of first-order layering transitions $\{L_n\}$ is obtained. Roughly speaking, one layer is

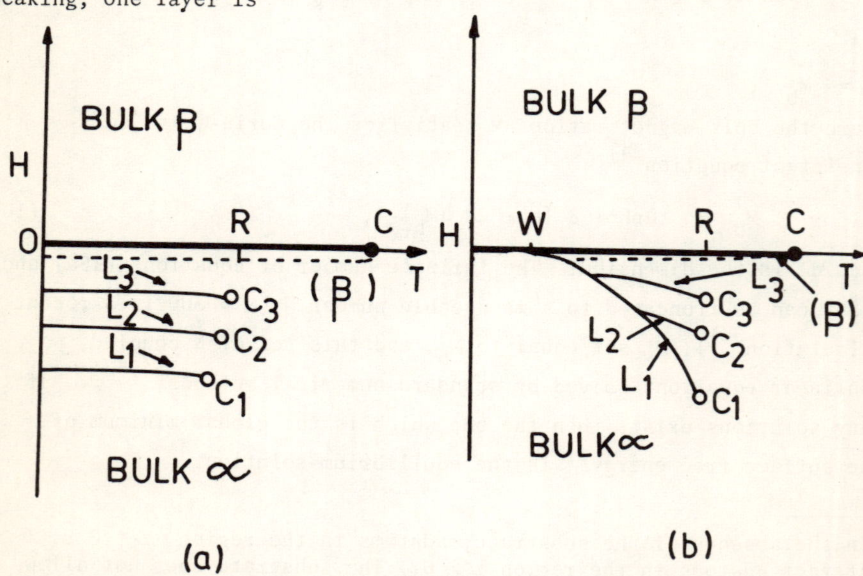

Figure 7. Possible surface phase diagrams for (a) strong-substrate systems (u/v >> 1) and (b) intermediate-substrate system (u/v ≃1). Layering phase boundaries L_n (only L_1, L_2, and L_3 are shown) and critical points C_n, first-order wetting W, and the roughening point R are shown. (Note that $\lim_{n\to\infty} T_{C_n} = T_R$.) The bold line separates the bulk phases β(up-spin) and α(down-spin). The dots indicate the region of the bulk phase boundary where the substrate is completely wet by a film of β (denoted by (β)). C is the bulk critical point.

adsorbed or desorbed when a layering transition occurs*; this leads to sharp steps in adsorption isotherms (Fig.3). The first-order layering phase boundaries terminate in <u>layering critical points</u> C_n. For a d-dimensional model, these critical points lie in the universality class of the (d-1) - dimensional Ising model[51,58,59]. These layering critical points approach the roughening point R (see below) asymptotically, i.e., $T_{C_n} \to T_R$ as $n \to \infty$. The wetting transition W is the accumulation point of the layering boundaries L_n on the bulk coexistence boundary.

In the mean-field approximation (Eqs. (15) and (16)) there is no roughening point R and the layering critical points C_n converge on the critical point C. In their pioneering paper, de Oliveira and Griffiths[51] argued that this mean-field result is qualitatively wrong. Before presenting their argument, it is necessary to define the roughening point.

An interface between two coexisting phases can be either <u>smooth</u> or <u>rough</u>[27]. It is smooth (rough) if the mean-square fluctuations in its position about the position of the zero-temperature interface are bounded (unbounded). For $d \leq 3$, interfaces between fluid phases are rough at any finite temperature[60]. However, solid-fluid or lattice-model interfaces can be smooth at low temperatures ($0 \leq T < T_R$) and rough at high temperatures ($T_R \leq T < T_c$). A <u>roughening transition</u>[27] occurs at $T = T_R$. For the d-dimensional Ising model, $T_R=0$ for $d = 2$, $0 < T_R < T_c$ for $d = 3$ (numerical calculations give $T_R \simeq T_c/2$), and $T_R = T_c$ for $d = 4$.

The de Oliveira - Griffiths argument proceeds as follows : If a large number of layers have been adsorbed on the substrate, then the interface between the adsorbed layers and the adatom gas is like an Ising-model interface. Thus, in d = 3, it is smooth for $0 \leq T < T_R$

---

*This is strictly true at T = 0 (if the layering transition exists at T = 0). Also, by a suitable choice of potentials it is possible to obtain layering transitions in which more than one layer (say 2) is adsorbed or desorbed at T = 0.

and rough for $T_R \leqslant T < T_c$. If the interface is rough, then it is impossible to distinguish one layer from another. Thus the layering transitions L_n cannot extend beyond R, and T_{C_n} must tend to T_R as $n \to \infty$. This plausible argument* has been verified by a variational theory[58)] which incorporates roughening, Monte Carlo simulations[53,54,57)], and renormalization-group calculations[59,61,62)]. These renormalization-group calculations predict that $(T_R - T_{C_n}) \sim 1/(\ln(n))^2$ as $n \to \infty$, for long-range adatom-substrate potentials.

A simple extension[5)] of the de Oliveira - Griffiths argument implies that if $T_R < T_W$, then wetting in lattice models is similar to wetting in fluids (Figs. 5a and 5b). Numerical simulations[53,57)], renormalization-group calculations[62)], and exact solutions of two-dimensional models[63)] support this argument. In two-dimensional models, only critical wetting (or drying) is possible: the substrate is one-dimensional, so it cannot have ordered layers on it at any finite temperature; the surface free energy (7) can be nonanalytic only when the bulk free energy (5) is nonanalytic, i.e., at the bulk coexistence boundary.

If the adatom-substrate potential is very weak, H_1 can be negative (Eq. (14b)), and drying transitions are obtained on the dense side ($H > 0$) of the bulk coexistence boundary. An infinite sequence of layering transitions on this dense side is possible for the model (12), but unphysical : this model overemphasizes the importance of hard-core repulsion, which should be unimportant for drying because low-density films are involved[5)]. For other limitations of the lattice-gas model (12) see Ref.5 and Sec.V.

The surface phase diagrams for model (12) depend on the forms of U and V (Eq.(3)). This dependence is discussed in Refs. 5,7, and 62. The way in which the excess surface density n_s (Eq.(2)) diverges at the coexistence boundary (for $T_W < T < T_c$) depends simply on the asymptotic form of U far from the substrate: For

*This argument is plausible only as $n \to \infty$. Low-order (small n) layering phase boundaries can extend beyond T_R.

short-range adatom-substrate potentials n_s diverges as $|\ln|\mu - \mu_o||$ (or $|\ln|H||$ in magnetic variables); for long-range adatom-substrate potentials n_s diverges as $|\mu - \mu_o|^{-\frac{1}{m}}$ if $U(z) \sim z^{-m}$ as $z \to \infty$. Most experiments[18,20], give m =3, which is consistent with the form of the potential (3b).

III. SOME RECENT EXPERIMENTS

A. Fluid Mixtures

Heady and Cahn[64] first noted a thin layer of a dense liquid sandwiched between a vapour and another less dense liquid. In the terminology of this paper, the dense liquid had completely wet the interface between the vapour and less dense liquid. (The formation of an infinitely thick wetting film was inhibited by gravity, which was not included in the model discussed in Sec. II). Moldover and Cahn[65] first observed a wetting transition from an incompletely-wet to a completely-wet interface. They concluded that this transition was first-order, but, after Sullivan's discovery[19] of critical wetting, their conclusion was questioned.

The wetting transition has now been observed in many binary-fluid mixtures[11-16]. Both direct and indirect observations have been used : The thickness of wetting films have been measured by ellipsometric methods[13,15,16]. In binary-fluid mixtures with closely matched densities, large, pendent drops* of the heavier fluid have been observed at the interface of vapour and the lighter fluid[14]. A discontinuity in $d\sigma\alpha\omega/dt$ has been observed at $T=T_W$;[12] such a discontinuity has been predicted by Cahn[25] for a first-order wetting transition. The clearest demonstration of a first-order wetting transition in binary-fluid mixtures has been given by Schmidt and Moldover[16]: For a binary mixture of isopropanol ($i\text{-}C_3H_7OH$) and perfluoromethylcyclohexane (C_7F_{14}), they have determined T_W (= 38.0 ± 0.01 C) and measured the thickness of the film above and

---

*In some cases, these large, pendent drops are metastable configurations of the system and the equilibrium wetting film is quite thin (a few hundred Angstroms) because of gravity.

below T_W.

Unfortunately, none of these experiments shows critical wetting ; nor do they show any evidence of prewetting, which is surprising. Tarazona and Evans[40] suggest that the way out of this dilemma lies in the quantitative treatment of wetting transitions with realistic potentials (Eqs. (3a) and (3b)). When they use parameters that are appropriate for the adsorption of Ar on solid CO_2, they find that the prewetting phase boundary lies very close to the bulk coexistence boundary. If this is true for all binary fluid mixtures, then prewetting will be hard to observe.

There are some puzzling experiments[65] which find substrates incompletely wet very close to T_c. Unless $(T_W - T_c)$ is too small to be observed, these experiments contradict the prediction[25] that a substrate is completely wet or completely dry near T_c.

B. Adsorption

Multilayer adsorption has been studied for many years[5,18-21] and by a variety of techniques ranging from electron diffraction to thermal and vapour-pressure measurements. In particular, adsorption isotherms have been catalogued for many adsorbate-substrate systems. However, in most cases, sufficient data have not been collected over wide ranges of temperature, so it has not been possible to obtain the complete surface phase diagram for any adsorbate-substrate system. This is principally because multilayer-adsorption experiments are hard to do, especially at temperatures below the triple point of the adsorbate : equilibration times are long, there are difficulties with capillary condensation[20], and the excess surface density n_s has to be measured very accurately. In spite of these difficulties, there has been a lot of experimental work in this field in recent years, stimulated, at least partially, by the theoretical work described in Sec. II. These experiments have in turn indicated certain limitations of the models of Sec. II and have stimulated new theoretical work, which is summarised in Sec. V.

The experiments of Menaucourt, Thomy, and Duval[24] first

indicated that graphite is incompletely wet by C_2H_4 at low temperatures (≃77 K, i.e., well below the triple point of C_2H_4) and completely wet by C_2H_4 at high temperatures (≃106 K, i.e., just above the triple point of C_2H_4). Recent x-ray[66] and neutron-diffraction[67] experiments support this result : they show a sequence of three layering transitions as the triple point is approached along the bulk gas-solid phase boundary. It is still not clear whether there is an infinite sequence of layering transitions or just a few of them followed by a critical wetting point at the triple point (Sec.V), but the latter seems to be the most likely possibility. Similar data have been obtained for the adsorption of C_2H_4 on PbI_2.[68]

Reflection high-energy electron diffraction (RHEED) has been used to study the adsorption of CF_4 on graphite[69]. A first-order wetting transition is found at 37 K, which is well below the triple point of CF_4 at 89.5 K. The resolution of this experiment is not good enough for the observation of separate layering transitions. A RHEED study[70] of the adsorption of Ne, N_2, Xe, and Ar on graphite has yielded unexpected results : Xe and Ar completely wet graphite, but Ne and N_2 do not. If the criterion of Sec.II is used, Ne and N_2 should wet graphite more strongly than Xe and Ar : with graphite as the substrate, u/v is larger for Ne and N_2 than for Xe and Ar. (For a graphite substrate, u/v = 9.2, 7.1, 3.7, and 2 for Ne, N_2, Xe, and Ar, respectively[70].) This enigma has been resolved recently[71-73] (Sec.V).

Thermal and vapour-pressure measurements have been used to identify wetting and layering transitions in many systems[74-76]. In the methane-graphite system, Hamilton and Goodstein[76] have found six layering transitions and the critical points they end in. If these critical points are extrapolated to the bulk coexistence boundary, they indicate a roughening point at 78 K, which is well below the triple point of methane*.

*Note that the roughening temperature for a gas-solid interface has to be less than or equal to the triple-point temperature, because the gas-liquid interface is always rough.

Ramesh, Zhang, Torso, and Maynard[77)] have used fourth sound to study the multilayer adsorption of ^{4}He on graphite. They have measured the layering-critical-point temperatures T_{C_n} for $4 < n < 8$. An extrapolation to $n = \infty$ yields a roughening point at 1.17 ± 0.03K, which is in very good agreement with the experimental value[78)] of the roughening temperature of the (0001) surface of bulk hcp ^{4}He crystals.

Krim, Dash, and Suzanne[79)] and Gregory and Bartosch[80)] have developed an elegant method of measuring adsorption isotherms. In this method, the substrate is an oscillator whose frequency depends on the amount of material adsorbed on it. (This provides a direct way of estimating the excess surface density n_s.) Adsorption isotherms are obtained by monitoring this frequency as a function of the chemical potential. Krim, Dash, and Suzanne[79)] have studied the adsorption of Ar, Kr, Xe, N_2, O_2, CH_4, and C_2H_6 on the (111) surface of a gold film deposited on a quartz crystal (the oscillator). They find that for all these gases, there is a critical-wetting transition at the solid-liquid-gas triple point. This technique has also been used to study the approach to critical wetting (away from the triple point) in the adsorption of ^{4}He on silver[81)].

IV. CRITICAL WETTING

Critical wetting has been the subject of a lot of recent theoretical work. Lipowsky, Kroll, and Zia[31)] and Brézin, Halperin, and Leibler[32,33)] have studied the fluctuations neglected by the mean-field theory of Sec. II ; and many authors[31,32,37-39,42,43,55,56,62,82-87)] have discussed the role of long-range forces in suppressing critical wetting. This suppression has been controversial, but recent work[55,56,87)] has settled this controversy by showing that a careful choice of the parameters in long-range potentials can lead to critical wetting. I summarise below the effects of fluctuations and long-range forces on critical wetting.

My treatment of fluctuations follows that of Brézin, Halperin, and Leibler[33)]. From the qualitative analysis of Eq. (10) (Sec. IIC) it follows that, in the mean-field approximation, the model (9) has a critical-wetting transition if

$$c \geqslant \sqrt{2\,\tau} \quad . \tag{17}$$

The equality holds at the critical-wetting temperature defined by

$$\tau_W \equiv 1 - \frac{T_W}{T_c} = \frac{gh_1^2}{6c^2} \quad . \tag{18}$$

For $T < T_W$, an explicit solution of Eq. (10) yields the equilibrium profile

$$\phi_c(z) = - M \tanh \left(\sqrt{\frac{\tau}{2}} \; (z-z_o) \right) \quad , \tag{19}$$

where z_o is determined by the boundary condition (11a). The thickness of the film is $\simeq z_o$. As $T \to T_W^-$, z_o diverges continuously as

$$z_o \sim \frac{1}{\sqrt{2}\;\tau_W} \ln \left(\frac{1}{T_W - T} \right) \quad . \tag{20}$$

Finally, the surface free energy per unit area can be put in the form (for $|T_W - T| << 1$)

$$\frac{F_s}{L^{d-1}} \approx \frac{-3}{2g} \left(\frac{T_W - T}{T_c} \right)^{2-\alpha_W} \frac{1}{c - \sqrt{2}\;\tau_W} \quad , \tag{21}$$

with the specific-heat exponent $\alpha_W = 0$.

To study the fluctuations of the order parameter about the mean-field profile $\phi_c(z)$ we write

$$\phi(z, \underset{\sim}{\rho}) = \phi_c(z) + \chi(z, \underset{\sim}{\rho}) \tag{22}$$

and expand F_s about $\phi_c(z)$ to second order in χ :

$$F_s = F_s(\phi_c) + \tfrac{1}{2} \int \chi(z_1, \underset{\sim}{\rho}_1) \;\; \chi(z_2, \underset{\sim}{\rho}_2) \;\; A^{(2)}(z_1, z_2, \;\; \underset{\sim}{\rho}_1 - \underset{\sim}{\rho}_2) + O(\chi^3) \, , \tag{23a}$$

where

$$A^{(2)} = \delta(z_1-z_2)\ \delta^{(d-1)}(\underset{\sim}{\rho}_1 - \underset{\sim}{\rho}_2)\ \Big(- \frac{\partial^2}{\partial z_1^2} - \nabla_{||}^2 - \frac{3\ \tau}{\cosh^2(\sqrt{\frac{\tau}{2}}(z-z_o))} + 2\ \tau \Big) \ , \tag{23b}$$

and the subscript $||$ refers to the directions parallel to the substrate. The correlation function $G(\vec{X}_1, \vec{X}_2) = <\phi(\vec{X}_1)\ \phi(\vec{X}_2)> - <\phi(\vec{X}_1)>\ <\phi(\vec{X}_2)> = (A^{(2)})^{-1}$ is evaluated by obtaining the eigenmodes and eigenvalues of $A^{(2)}$. These follow[33] from a Fourier transform with respect to the (d-1) variables $\underset{\sim}{\rho}$ parallel to the substrate and the solution of an ordinary differential equation for an eigenfunction $\psi_n(z)$. The lowest eigenmode with eigenvalue

$$\varepsilon_o(q_{||}) \sim q_{||}^2 + \frac{1}{\xi_o^2}\ (T_W - T)^2 \tag{24}$$

($q_{||}^2$ comes from the Fourier transform and ξ_o depends on c, T_c, and T_W) dominates the correlation function whose Fourier transform is

$$\tilde{G}(z,z'\ ;\ q_{||}) \underset{\sim}{\approx} \frac{\psi_o(z)\ \psi_o(z')}{\varepsilon_o(q_{||})} \quad . \tag{25}$$

An inverse Fourier transform yields an exponentially decaying correlation function with a correlation length

$$\xi_W = \xi_o\ (T_W - T)^{-\nu_W} \quad , \tag{26}$$

in directions parallel to the substrate. The exponent $\nu_W = 1$ in contrast to the value $\nu = \frac{1}{2}$ that is obtained from Gaussian fluctuations at an ordinary critical point.

If an interface exists in an infinite, translationally-invariant system, then the position of this interface (z_o in Eq.(19)) is arbitrary. As a result of this, the capillary waves (low $q_{||}$ modes which lead to distortions of the interface) are gapless. In the model (9), the substrate at z = 0 breaks this translational invariance and the interface is pinned to the substrate. However, at the wetting transition, the interface is unpinned from the substrate, translational invariance is restored, and the gapless modes reappear.

The upper-critical dimension d_{uc} below which non-Gaussian fluctuations affect the values of exponents can be shown[31,33] to be 3. The analysis is technical, so I do not include it here. Note, though, that the hyperscaling relation[88] $(d-1)\,\nu_W = 2 - \alpha_W$ is satisfied if $d = d_{uc} = 3$ and ν_W and α_W assume their Gaussian values of 1 and 0, respectively*.

For $d = d_{uc}$ there are mild, logarithmic corrections to the power-law divergences predicted by mean-field theory for an ordinary critical point (Ref. 88, p.88). However, for a critical-wetting point, the corrections to mean-field exponents in $d = d_{uc}$ are not at all mild. In addition, these exponents are not universal : they depend in a complicated way on the dimensionless parameter

$$\omega_W \equiv k_B T_W/(4\pi \xi_b^2 \sigma), \tag{27}$$

where ξ_b is the bulk correlation length and σ is the interfacial tension of the free interface. The exponent ν_W is given by[32)]

$$\nu_W = (1 - \omega_W)^{-1}, \qquad 0 < \omega_W < \tfrac{1}{2}, \tag{28a}$$

$$\nu_W = (\sqrt{2} - \sqrt{\omega_W})^{-2}, \qquad \tfrac{1}{2} < \omega_W < 2, \tag{28b}$$

and $\nu_W = \infty$ for $\omega_W > 2$ because ξ_W diverges exponentially as $(T - T_W) \to 0^-$. There are logarithmic corrections to the power-law divergence of ξ_W implied by Eqs. (26) and (28b).

The calculations which lead to Eqs. (27) and (28) use an effective free energy

$$F_{eff} = \int d^{d-1}\rho \left(\frac{\sigma}{2} \left| \nabla_\rho z(\underset{\sim}{\rho}) \right|^2 + V_{eff}(z(\underset{\sim}{\rho})) \right), \tag{29}$$

---

*The usual hyperscaling relation is $d\nu = 2 - \alpha$. I use (d-1) because the substrate has dimension (d-1). Also, it is well known[88] that Gaussian exponents satisfy the hyperscaling relation if $d = d_{uc}$.

where $z(\underset{\sim}{\rho})$ is the separation between the interface and the substrate at the point $\underset{\sim}{\rho}$ on the (d-1)-dimensional substrate (Fig. 8a) and $V_{eff}(z(\underset{\sim}{\rho}))$ is an effective potential. Such a model does not allow for the overhang excitations shown in Fig.8b ; this is a good approximation [3,89,90] if T_W is much less than T_c, for such overhangs cost a lot of energy and are formed very infrequently at low temperatures ($T \ll T_c$). For the functional form of V_{eff}, the details of the calculations which lead to Eqs. (27) and (28), and some extensions see Refs. 32 and 91.

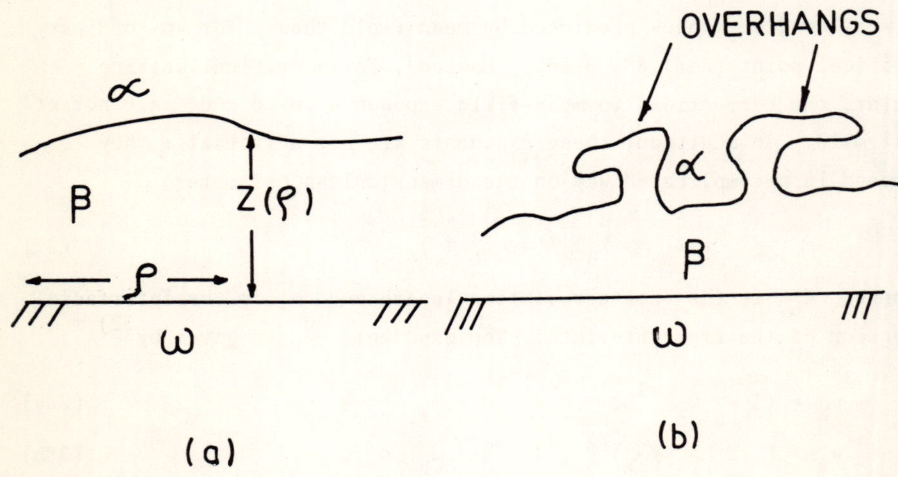

Figure 8. Schematic plots of fluctuating α/β interfaces above a substrate ω in two dimensions. The interface in (b) has two overhangs. In (a), the height $z(\underset{\sim}{\rho})$ of the interface above the substrate at $\underset{\sim}{\rho}$ is uniquely defined.

So far I have discussed critical wetting in the presence of short-range forces. I now summarise the effects of long-range forces on critical wetting. Unless the parameters that characterise long-range potentials are chosen very carefully, critical wetting becomes first-order wetting*. Let T(z) denote the potential felt by an adatom above a substrate made of the same adatoms and let T(z) and U(z), the usual adatom-substrate potential, have the following forms for large z :

$$T(z) \approx -\frac{t_3}{z^3} - \frac{t_4}{z^4} + \cdot \cdot \cdot \cdot \cdot , \tag{30a}$$

and

$$U(z) \approx -\frac{u_3}{z^3} - \frac{u_4}{z^4} + \cdot \cdot \cdot \cdot \cdot , \tag{30b}$$

with t_3 and u_3 positive. Critical wetting occurs if

$$\tfrac{1}{2} < \frac{u_3}{t_3} < 1 \tag{31a}$$

and

$$t_4/t_3 < u_4/u_3 \quad . \tag{31b}$$

Thus, both the leading and next-to-leading terms in T(z) and U(z) have to be controlled to obtain critical wetting[55,56]. At this critical wetting transition, the excess surface density n_s diverges as $(T_W - T)^{-1}$ as $T \to T_W^-$. (This has been seen in a recent experiments[81]). Multicritical wetting can also be obtained[56]. For most of these transitions, the upper-critical dimension $d_{uc} < 3$, so mean-field exponents suffice in d = 3.

V. A SUMMARY OF RECENT THEORETICAL RESULTS

I have already discussed some recent theoretical results. I list some others below.

The models discussed in Secs. II and IV have only two bulk phases: a high-density phase (solid** or liquid) or a low-density phase

---

*This change from critical to first-order wetting emerges simply from an effective model like Eq.(29). See Ref. 6.

**The imposed lattice structure in a lattice-gas model makes the dense phase solid-like, but this is certainly not like a real solid.

(gas). Most simple substances have at least three bulk phases - solid, liquid, and gas - and experiments [24,66-68,74,79] indicate that layering and wetting transitions often occur near such bulk triple points. Lattice-gas[92] and phenomenological[93] models have been developed for such triple-point wetting and a variety of possible surface phase diagrams have been suggested. However, only one of them has been observed so far[68].

Some experiments[70] have indicated that infinitely thick films are not formed for strong-substrate systems if the wetting film is crystalline. Lattice-gas[71] and phenomenological[72,73] theories indicate that this arises because of the mismatch between the crystal structure of the substrate and that of the solid film. This mismatch leads to strains which limit the thickness of the wetting film. The free-energy cost of this mismatch increases with the strength of the adatom-substrate potential. Thus, the stronger the substrate, the thinner the adsorbed film. This is contrary to the theoretical result of Sec. II, but agrees with experiments[70]. Note that above the triple point the wetting film is liquid-like and there is no mismatch; so we should expect that a strong-substrate system should become completely wet at the triple point. This has been observed in some systems[66-68,79].

The interplay between in-plane ordering in layers and thick-film adsorption has been studied by lattice-gas models[71,94]. The gross systematics of multilayer adsorption discussed in Sec. II are not altered, but new surface phases of laterally ordered layers appear. Dewetting phenomena like the ones described in the previous paragraph also occur[71].

Wetting in finite geometries (slabs and strips) has been studied recently[95,96]. This should pave the way for an understanding of capillary condensation, recent experiments[97], and numerical simulations[98].

Wetting on a disordered substrate has also been studied recently[99]. This study suggests that critical wetting is unaffected by quenched disorder. However, first-order wetting in a pure, three-dimensional system becomes continuous by the addition of an infinitesimal amount of substrate disorder.

In a recent paper[100] the adsorption of molecules has been studied. Reentrant wetting transitions are predicted by this study.

VI. OUTLOOK

A lot of work has been done on the equilibrium statistical mechanics of wetting and layering transitions. I have outlined some of this in Secs. II, IV, and V : Possible surface phase diagrams have been suggested for different adsorbate-substrate and binary-fluid systems ; novel critical and multicritical points have been predicted. A systematic experimental study of these phenomena has been initiated (Sec. III), but a lot remains to be done : Prewetting and critical wetting have not been observed in binary-fluid mixtures. Detailed phase diagrams showing multilayer adsorption have not been mapped out, though much more is known now than a few years ago. It should be possible to observe prewetting and critical wetting by plating substrates with different materials. The order-parameter profile of critical, binary-liquid mixtures near a substrate is changed by such substrate treatment[101]. Also, the microbalance techniques of Refs. 79 and 80 should expedite the mapping-out of the phase diagrams discussed in Secs. II and V.

Though equilibrium wetting phenomena are understood fairly well at the level of the models discussed here, some unanswered questions remain in complex systems. An example is the wetting of grain boundaries by a liquid[102,103], which is of importance in metallurgy. There are reports[104] of a discontinuous change in the dihedral angle at the wetting transition at a liquid-solid-solid trijunction. This is in disagreement with theoretical expectations[102,103].

The dynamics of wetting phenomena might prove to be a fertile area of research. It has been studied much less than the statics, both theoretically and experimentally, but some work has been done recently. Surface spinodal decomposition[105], the growth of a liquid layer[106], and the dynamics of critical wetting[107] (for the case of a nonconserved order parameter in an Ising-like model, i.e., model A in the terminology of Hohenberg and Halperin[108]) have been studied. The dynamics of a spreading droplet[109] and contact-angle hysteresis[110] have been studied. For a review of the latter see the article by de Gennes[6]. A (nonequilibrium) wetting layer in a binary-fluid mixture has been observed to undergo a Rayleigh-Taylor instability[111].

ACKNOWLEDGEMENTS

I would like to thank C. Bartosch, S. Chatterjee, C. Ebner, M.E. Fisher, C. Franck, E.S.R. Gopal, S. Gregory, S. Guha, D. Huse, H. Nakanishi, V. Privman, C. Rottman, M. Schick, V. Vani, B. Widom, and especially M. Wortis for many discussions on wetting and multilayer adsorption. I would also like to thank the University Grants Commission for financial assistance and Y.A. Rakma and M.V. Manjula for typing this article.

REFERENCES

1. Binder, K., "Critical Behaviour at Surfaces", in Phase Transitions and Critical Phenomena, Vol. 8, edited by C. Domb and J.L. Lebowitz (Academic, New York, 1983) pp. 1-144.

2. Rowlinson, J.S., and Widom, B., Molecular Theory of Capillarity (Clarendon, Oxford, 1984).

3. Jasnow, D., Rep. Prog. Phys. 47, 1059 (1984).

4. Widom, B., "Phase Equilibrium and Interfacial Structure", Cornell University preprint (1984) and Centenary Lecture of the Royal Society of Chemistry (1984).

5. Pandit, R., Schick, M., and Wortis, M., Phys. Rev. B 26, 5112 (1982). For a short summary see Wortis, M., Pandit, R., and Schick, M., in Melting Localization and Chaos, edited by R.K. Kalia and P. Vashishta (North-Holland, New York, 1982).

6. de Gennes, P.G., "Wetting : Statics and Dynamics", Collège de France preprint (1984), to appear in Rev. Mod. Phys.

7. Sullivan, D.E., and Telo da Gama, M.M., to appear in Fluid Interfacial Phenomena, edited by C.A. Croxton (Wiley, New York, 1985).

8. Sullivan, D.E., Comments Solid State Phys. 11, 17 (1983).

9. Moldover, M.R., and Schmidt, J.W., in Proceedings of the Conference on Fronts, Interfaces and Patterns, Los Alamos National Laboratory (1983), to be published.

10. Hauge, E.H., "The Wetting Transition", to appear in Fundamental Problems in Statistical Mechanics VI, edited by E.G.D. Cohen (North-Holland, Amsterdam, 1985); see also M. Wortis's article in the same volume.

11. Moldover, M.R., and Cahn, J.W., Science 207, 1073 (1980).

12. Pohl, D.W., and Goldburg, W.I., Phys.Rev.Lett. 48, 1111 (1982).

13. Kwon, O'D., Beaglehole, D., Webb, W.W., Widom, B., Schmidt, J.W., Cahn, J.W., Moldover, M.R., and Stephenson, B., Phys.Rev.Lett. 48, 185 (1982).

14. Vani, V., Guha, S., Gopal, E.S.R., Madhusudhana Rao, S., Phys. Lett. 99A, 441 (1983).

15. Beaglehole, D., J.Phys.Chem. 87, 4749 (1983) and references cited therein.

16. Schmidt, J.W., and Moldover, M.R., J.Chem.Phys. 79, 379 (1983).

17. Adamson, A.W., Physical Chemistry of Surfaces (Wiley, New York, 1976).

18. Dash, J.G., Phys.Rev. B 15, 3136 (1977); J.Phys. (Paris) Colloq. 38. C4-201 (1977).

19. Sullivan,D.E., Phys.Rev. B 20 , 3991 (1979) ; J.Chem.Phys. 74, 2604 (1981) ; Faraday Symp. Chem.Soc. 16, 191 (1981).

20. Dash, J.G., Films on Solid Surfaces (Academic, New York, 1975); Steele, W.A., The Interaction of Gases with Solid Surfaces (Pergamon, Oxford, 1974).

21. Thomy, A., Duval, X., and Regnier, J., Surf.Sci. Rep. 1, 1 (1981)

22. Thomy, A., and Duval, X., J. Chim.Phys. Physicochim. Biol. 66, 1966 (1969); 67, 286 (1970); 67, 1101 (1970).

23. Génot, B., thése docteur, Université de Nancy (1974).

24. Menancourt, J., Thomy, A., and Duval, X., J.Phys. (Paris), Colloq. 38 C4-195 (1977). See also, Menancourt, J., thése docteur, Université de Nancy (1977) pp 86-90.

25. Cahn, J. W., J.Chem.Phys. 66, 3667 (1977).

26. Nightingale, M.P., and Indekeu, J.O., "The Effect of Criticality on Wetting Layers", University of Rhode Island preprint (1984), to be published.

27. Weeks, J.D., in Ordering in Strongly Fluctuating Condensed Matter Systems, edited by T. Riste (Plenum, New York, 1980) p 293.

28. Griffiths, R.B., in Phase Transitions and Critical Phenomena, Vol.1, edited by C. Domb and M.S. Green (Academic, New York, 1972) p.7.

29. Fisher, M.E., and Caginalp, G., Commun.Math.Phys. 56, 11 (1977); 65, 247 (1979).

30. Ebner,C., & Saam, W.F., Phys.Rev.Lett. 38, 1486 (1977).

31. Lipowsky, R., Kroll, D.M., and Zia, R.K.P., Phys.Rev. B 27, 4499 (1983) ; Kroll, D.M., and Lipowsky, R., Phys.Rev. B 26, 5289 (1982).

32. Brézin, E., Halperin, B.I., and Leibler, S., Phys.Rev.Lett. 50, 1387 (1983).

33. Brézin, E., Halperin, B.I., and Leibler, S., J.Phys. (Paris) 44, 775 (1983).

34. Pandit, R., and Wortis, M., Phys. Rev. B 25, 3226 (1982).

35. Nakanishi, H., and Fisher, M.E., Phys.Rev.Lett. 49, 1565 (1982).

36. Foiles, S.M., and Ashcroft, N.W., Phys.Rev. B 25, 1366 (1982).

37. Tarazona, P., and Evans, R., Molec.Phys. 47, 1033 (1982); 48, 799 (1983); J.Chem.Phys. 80, 587 (1984).

38. Telo da Gama, M. M., and Evans, R., Molec.Phys. 48, 251 (1983); 48, 687 (1983).

39. Tarazona, P., Telo da Gama, M. M., and Evans, R., Molec.Phys. 49, 283 (1983); 49, 301 (1983).

40. Evans, R., and Tarazona, P., Phys.Rev. A 28, 1864 (1983).

41. Tarazona, P., and Evans, R., Molec.Phys. 52, 847 (1984).

42. Teletzke, G.F., Scriven, L.E., and Davis, H.T., J.Chem.Phys. 77, 5794 (1982); 78, 1431 (1983).

43. Benner, Jr., R.E., Teletzke, G.F., Scriven, L.E., and Davis, H.J., J.Chem.Phys. 80, 589 (1984).

44. Hauge, E.H., and Schick, M., Phys.Rev.B 27, 4288 (1983).

45. Lipowsky, R., Phys.Rev.Lett. 49, 1575 (1982); Z.Phys. B 51, 165 (1983); J.Appl.Phys. 55, 2485 (1984).

46. Lipowsky, R., Z.Phys.B. 55, 335 (1984); 55, 345 (1984).

47. Lipowsky, R., and Speth, W., Phys.Rev. B 28, 3983 (1983).

48. Costas, M.E., Varea, C., and Robledo, A., Phys.Rev. Lett. 51, 2394 (1983).

49. Widom, B., Faraday Symp. Chem.Soc. 16, 7 (1981).

50. Sullivan, D.E., Levesque, D., and Weis, J., J.Chem.Phys. 72, 1170 (1980).

51. de Oliveira, M.J., and Griffiths, R.B., Suf.Sci. 71, 687 (1978).

52. Ebner, C., Phys.Rev. A 22, 2776 (1980).

53. Ebner, C., Phys.Rev. A 23, 1925 (1981).

54. Kim, I.M., and Landau, D.P., Surf.Sci. 110, 415 (1981).

55. Dietrich, S., and Schick, M., University of Washington, Seattle, preprint (1984), to be published.

56. Ebner, C., Saam, W.F., and Sen, A.K., Ohio State University preprint (1984), to be published.

57. Binder, K., and Landau, D.P., Universität Mainz preprint (1984), to be published.

58. Weeks, J.D., Phys.Rev. B 26, 3998 (1982); and in Phase Transformation in Solids, edited by T. Tsakalakos (Elsevier, New York, 1984).

59. Huse, D.A., Phys.Rev. B 30, 1371 (1984).

60. Buff, F.P., Lovett, R.A., and Stillinger, F.H., Phys.Rev.Lett. 15, 621 (1965); Robert, M., Phys.Rev.Lett., 54, 444 (1985).

61. Saam, W.F., Surf.Sci. 125, 253 (1983).

62. Nightingale, M.P., Saam, W.F., and Schick, M., Phys.Rev.B 30, 3830 (1984).

63. Abraham, D.B., Phys.Rev.Lett. 44, 1165 (1980); Burkhardt, T.W., J.Phys.A. 14, L63 (1981); Chui, S.T., and Weeks, J.D., Phys.Rev. B 23, 2438 (1981); van Leeuwen, J.M.J., and Hilhorst, H.J., Physica 107A, 319 (1981); Chalker, J.T., J.Phys. A 14, 2431

(1981); Vallade, M., and Lajzerowicz, J., J.Phys.(Paris) 42, 1505 (1981); Chui, S.T., and Ma, K.B., Phys.Rev.B 28, 2555 (1983); Kroll, D.M., and Lipowsky, R., Phys.Rev.B 28, 5273 (1983); Wolff, W.F., and Švrakić, N.M., J. Phys. A 17, 3383 (1984); Fisher, M.E., J.Stat. Phys. 34, 667 (1984).

64. Heady, R.B., and Cahn, J.W., J.Chem.Phys. 58, 896 (1973).

65. Ross, S., and Kornbrekke, R.E., J.Coll. and Int.Sci., 99, 223 (1984).

66. Sutton, M., Mochrie, S.G.J., and Birgeneau, R.J., Phys.Rev.Lett. 51, 407 (1983); Mochrie, S.G.J., Sutton, M., Birgeneau, R.J., Moncton, D.E., and Horn, P.M., Phys.Rev. B 30, 263 (1984).

67. Satija, S.K., Passell, L., Eckart, J., Ellenson, W., and Patterson, H., Phys.Rev.Lett. 51, 411 (1983).

68. Bassignana, I.C., and Lahrer, Y., Surf.Sci. 147, 48 (1984).

69. Suzanne, J., Seguin, J.L., Bienfait, M., and Lerner, E., Phys. Rev.Lett. 52, 637 (1984).

70. Seguin, J.L., Suzanne, J., Bienfait, M., Dash, J.G., and Venables, J.A., Phys.Rev.Lett. 51, 122 (1983); Bienfait, M., Seguin, J.L., Suzanne, J., Lerner, E., Krim, J., and Dash.J.G., Phys.Rev. B 29, 983 (1984).

71. Ebner, C., Rottman, C., and Wortis, M., Phys.Rev.B 28, 4186 (1983).

72. Huse, D.A., Phys.Rev.B 29, 6985 (1984).

73. Gittes, F.T., and Schick, M., Phys.Rev. B30, 209 (1984).

74. Awschalom, D.D., Lewis, G.N., and Gregory, S., Phys.Rev.Lett. 51, 581 (1983).

75. Specovius, J., and Findenegg, G.H., Ber. Bunsenges. Phys.Chem. 82, 174 (1978); 84, 690 (1980); 84, 696 (1980); Findenegg, G.H., and Löring, R., J.Chem.Phys. 81, 3270 (1984).

76. Hamilton, J.J., and Goodstein, D.D., Phys.Rev.B 28, 3838 (1983).

77. Ramesh, S., Zhang, Q., Torso, G., and Maynard, J.D., Phys.Rev. Lett. 52, 2375 (1984).

78. Wolf, P., Balibar, S., and Gallet, F., Phys.Rev.Lett. 51, 1366 (1983); Lipson, S.G., Physica (Utrecht) 109 & 110 B, 1805 (1982).

79. Krim, J., Dash, J.G., and Suzanne, J., Phys.Rev.Lett.52, 640 (1984).

80. Bartosch, C.E., and Gregory, S., Cornell University preprint (1985), to be published.

81. Mignone, A.D., Krim, J., Dash, J.G., and Suzanne, J., University of Washington, Seattle, preprint (1984), to be published.

82. de Gennes, P.G., C.R. Acad.Sci.Paris (II) 297, 9 (1983).

83. Nightingale, M.P., Saam, W.F., and Schick, M., Phys.Rev.Lett. 51, 1275 (1983).

84. Privman, V., J.Chem.Phys. 81, 2463 (1984).

85. Lipowsky, R., Phys.Rev.Lett. 52, 1429 (1984).

86. Dietrich, S., Nightingale, M.P., and Schick, M., University of Washington, Seattle,preprint (1984), to be published.

87. Kroll, D.M., and Meister, T.F., Institut fur Festkörperforschung der Kernforschungsanlage, Jülich,preprint (1984), to be published.

88. Pfeuty, P., and Toulouse, G., Introduction to the Renormalization Group and to Critical Phenomena (Wiley, New York, 1977) pp.22-28.

89. Zia, R.K.P., in Proceedings of the Scottish Universities Summer School in Physics, edited by K. Bowler and A. McKane (1983), to be published.

90. Huse, D.A., van Saarloos, W., and Weeks, J.D., AT&T Bell Laboratories preprint (1984), to be published.

91. Fisher, D.S., and Huse, D.A., AT&T Bell Laboratories preprint (1984), to be published.

92. Ebner, C., Phys.Rev.B 28, 2890 (1983).

93. Pandit, R., and Fisher, M.E., Phys.Rev.Lett. 51, 1772 (1983).

94. Kennedy, S.J., and Walker, J.S., Phys.Rev. B 30, 1498 (1984).

95. Evans, R., and Tarazona, P., Phys.Rev.Lett. 52, 557 (1984).

96. Lipowsky, R., and Gompper, G., Phys.Rev.B 29, 5213 (1984).

97. Moldover, M.R., and Gammon, R.W., J.Chem.Phys. 80, 528 (1984); see also Legait, B., and de Gennes, P.G., J.Phys (Paris) Lett. 45, L-647 (1984).

98. Lane, J.E., and Spurling, T.H., Aust.J.Chem. 33, 231 (1980); 34, 1529 (1981).

99. Forgács, G., Orland, H., and Schick, M., C.E.N. Saclay preprint (1984), to be published.

100. Sluckin, T.J., and Chalker, J.T., J.Phys.(Paris) Lett. 45, L-863 (1984).

101. Dixon, J.A., Schlossman, M., Wu, X-L., and Franck, C., Cornell University preprint (1984), to be published.

102. Cahn, J.W., J.Phys.(Paris) Colloq. C6-199 (1982).

103. Rottman, C., Ohio State University preprint (1984), to be published.

104. Glicksman, M.E., and Vold, C.L., Surf.Sci. 31, 50 (1972).

105. Nakanishi, H., and Pincus, P., J.Chem.Phys. 79, 997 (1983).

106. Meister, T., and Müller - Krumbhaar, H., Phys.Rev.Lett. 51, 1780 (1983); Z.Phys. B55, 111 (1984).

107. Forgács, G., Orland, H., and Schick, M., C.E.N. Saclay preprint (1984), to be published.

108. Hohenberg, P.C., and Halperin, B.I., Rev.Mod.Phys. 49, 435 (1977).

109. de Gennes, P.G., Compt. Rend., Acad.Sci.Paris, 298II, 111(1984); 298II, 475 (1984).

110. Joanny, J.F., and de Gennes, P.G., J.Chem.Phys. 81, 552 (1984).

111. Chatterjee, S., Vani, V., Guha, S., & Gopal, E.S.R., Indian Institute of Science, Bangalore, preprint (1985), to be published.

ELECTRODYNAMICS OF LIGHT RAYS

E.C.G. Sudarshan
Institute of Mathematical Sciences
Madras, India

There are many ways to deal with light. There are two descriptions - light as a pencil of rays and light as a wave. A ray of light can be reflected from a mirror or refracted through a thin lens - the adjective "thin" meaning that a ray emerges as a ray after passing through a thin lens. On the other hand a wave passes through a thin lens and emerges with a quadratic shift of phase i.e., an incoming spherical wave front in a particular direction after passing through a thin lens changes its curvature and axial direction. The question now is whether these two pictures are reconcilable; whether waves can be described in terms of rays and vice versa. It is quite obvious that the correspondence between a single ray and a wave does not hold; a wave will diffract while a ray will proceed along a straight line. So the correspondence has to be between pencils of rays which have an angular spead and light waves. For the time being we will talk of only classical waves and also ignore the phenomena of polarization - we deal with scalar propagating waves. We could correct the wave and ray pictures approximating waves of short wavelength and long propagation length by rays. This is the geometric optics approximation; but we wish to make an exact correspondence rather than an approximate one.

In elementary non-relativistic physics, we take the wave front of a plane wave as $e^{ik.x}$ where k is a wave number, x is the distance. For a spherical lens the thickness s can be calculated to yield a

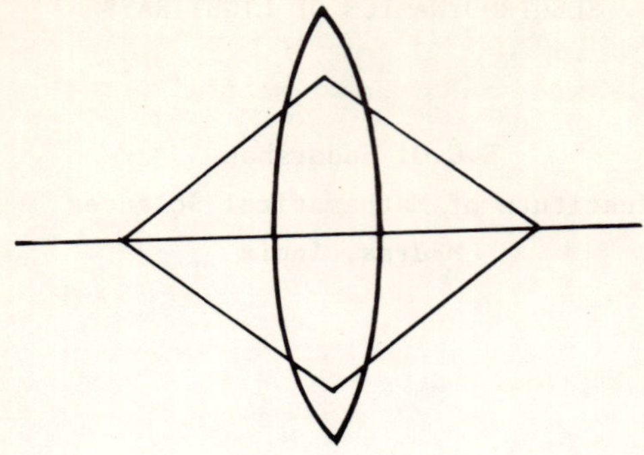

geometric thickness proportional to $(x^2 + y^2)$ and therefore, an optical thickness of the form $s \sim g(x^2 + y^2)/2$ where $g \sim f^{-1}$ is the inverse of focal length. On meeting such a convex lens the ray should be bent inwards, towards the focus. The deflection of rays is proportional to the thickness of lens and the phase shift of the wave front depends on s and depends quadratically on x and y. In a sense, the energy of the advancing wave contains a term proportional to k^2 and if the wave is of the above form, then the deflection could be understood. We need to derive the above from wave optics and as mentioned before, we are dealing with "non-relativistic" waves. In quantum mechanics, we usually consider waves whose phase advances in time, but here we are dealing with those waves where phases propagate in space rather than in time since we are dealing with stationary system.

We must have a rigorous theory for this sort of stationary waves propagating only in space and not in time. We must make them relativistic, incorporate the fact that they are transverse in nature and lastly include the photon nature of light so that we must have a second quantised theory as well. Ultimately all the above ingredients must produce the ray description of light.

Before going deeply into the theory, let us look at a single ray propagating in space. It does not propagate in time, but the propagation is described by a time like parameter. This single ray can be parametrised by two parameters x and p, viz. position and momentum. The propagation of the ray in vaccum can be described by the following equation involving the "transfer matrix".

$$\begin{bmatrix} x^1 \\ p^1 \end{bmatrix} = \begin{bmatrix} 1 & D \\ 0 & 1 \end{bmatrix} \begin{bmatrix} x \\ p \end{bmatrix}$$

where D is the distance.

As the wave propaagates x increasees, but p remains unchanged in case of deflection by a lens, the deflection of the ray is not uniform but proportional to the distance and in this case, the transfer matrix is given by

$$\begin{bmatrix} x^1 \\ p^1 \end{bmatrix} = \begin{bmatrix} 1 & 0 \\ -g & 1 \end{bmatrix} \begin{bmatrix} x \\ p \end{bmatrix}$$

Only the direction of ray changes at the point of intersection of the ray and lens, but the position does not change. This is because the lens is very thin. The two matrices defined above are special cases of more general matrices namely the metaplectic matrices belonging to Sp (2, R) in two dimensions. These have the property that $S^T KS = K$, where $K = \begin{pmatrix} 0 & -1 \\ 1 & 0 \end{pmatrix}$, where S^T is the transpose of the matrix S. In classical mechanics this corresponds to canonical transformations in two variables. The interpretation in this case is that the incoming light is retransmitted and redistributed, but the total radiance in over all directions and areas is the same before and after meeting the lens. In other words, light is neither created nor destroyed.

We observe that products of any two matrices satisfying $S^T KS = K$ also have the same property. We can arrange a combination of lenses with alternate free propagation and lenses as follows:

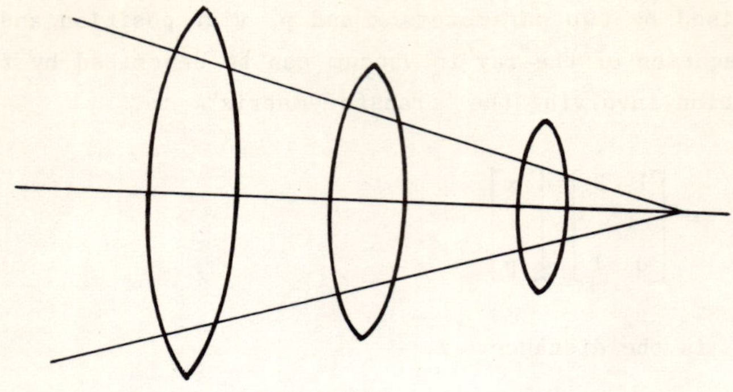

The resulting 2 x 2 matrix would have a very complicated structure but would still follow the metaplectic property $S^T KS = K$. Because of the group which we have deduced, the whole question of ray propagation can be rephrased in a group theoretical form. In the above, we have used classical considerations. But quantum mechanically, x and p, being noncommuting variables, can not be measured simultaneously. In quantum theory in terms of wave optics we can not expect to write down the above 2 x 2 matrices.

Pencils of rays correspond to bilinears in wave amplitudes and not to amplitudes themselves, since the energy of a wave is proportional to the square of its amplitude. Therefore, if we take pencils of rays, we define the density function W(x, p), which should be related to bilinear quantities in the wave amplitudes. A simple method of doing this would be to write

$$W(x,p) = \int A^* (x + 1/2\, y)\, A(x - 1/2\, y)\, e^{ip\cdot y}$$

where x,y are neighbouring vectors and we have Fourier transformed the bilinear quantity A^*A. Here A is the wave amplitude. If A is a plane wave, then A(x + 1/2 y) is also a plane wave, so the phases are $e^{-ik(x + 1/2\, y)}$ and $e^{ik(x - 1/2\, y)}$ which yields e^{iky}. Talking the Fourier transform of e^{iky} with $e^{ip\cdot y}$ we get $\delta(p-k)$, a delta function in p. So the ray density function is a uniform distribution in x, which is necessary because we are considering plane wave propagation; the plane wave exists over all space. On the other hand, if we localised the wave with a $\delta(x)$, then the wave would stay at one point, but now the moments are arbitrary. In the general case, the "Wolf function" W(x,p) representing ray density is real. But if W(x,p) is < 0. then the correspondence between pencil of ray-waves can not be physically realized, and so we assume it to be positive definite for the present.

It is easily seen that no pencil of rays can be too narrow since x and p are noncommuting variables. If we localise the pencil of rays to a narrow region x, the momenta increase indefinitely and vice versa. So, in this sense, a correspondence between pencils of light rays and wave amplitudes is established via the Wolf function W(x,p). In fact it describes a "general" light rays i.e., coherent or not, polarized light, laser light and all the other varieties of light.

At this point we notice that we have too many variables. In general we have x_1, x_2, z, t, i.e., three spatital variables and time. Instead, we can do with one variable which describes temporal development and two variables to describe two transverse direction. So from x_1, x_2, z, t, we wish to go to x_1, x_2, τ where τ is a time like variable density propagation of wave front. Define $\tau = 1/2(t + z)$ and $\sigma = t - z$. The corresponding four momentum p_1, p_2, p_3, p_0 obeys $p_1^2 + p_2^2 + p_3^2 = p_0^2$ in natural units. Rewriting

$$p_1^2 + p_2^2 = (p_0 + p_3)(p_0 - p_3) .$$

The conjugate vaariables corresponding to τ and σ are now

$$p_\tau = p_0 - p_3 , \; p_\sigma = 1/2(p_0 + p_3)$$

Here $p_\sigma = 1/2(p_0 + p_3)$, $p_\tau = p_0 - p_3$. For axial rays i.e., rays parallel to the axis, the Z component of p, p_z is greater than p_1, p_2, the transverse momenta, which implies that p_τ is less than p_σ. If we deal with monochromatic waves only, frequency (ν) is constant. Instead, define $M = 1/2(p_0 + p_3)$ = constant, where M is the average of the frequency and the third component of momentum. So we may write $(p_1^2 + p_2^2)/2M = p_\tau$. This is a "non-relativistic" relation.

If we now have a wavefront $\psi_M(x_1, x_2, \tau)$, then a non-relativistic Schrodinger equation can be immediately written down:

$$-i \frac{\partial}{\partial\tau} \psi_M (x_1, x_2, \tau) = (p_1^2 + p_2^2)/2M \ \psi_M$$

One calls the wave described by ψ_M as a "henochromatic" wave to distinguish it from monochromatic waves. The above is the wave-mechanical description of light rays appropriate to wave optics.

The non-relativistic Schrodinger equation is usually associated with Galilian group which contains translation along various directions, rotations amongst these direction, boosts (velocity changes) along these directions, and time translations. For a particle, the mechanical momentum is the generator of the Galilian group. So the boost generator divided by the mass is some sort of "coordinate". The same treatment can be generalised to a lens. An ideal lens in this picture can be thought of as a harmonic oscillator potential, quadratic in the distance from the axis.

Supposing we had an arrangement of coaxial lenses, e.g., a microscope or a telescope; how do we describe such a system and what would happen to the pencils of rays ? This may be visualised as free propagation, encounter with an instantaneous harmonic oscillator potential, subsequent free propagation and so on. Let us find out what would happen if we had a series FLFL ... where F denotes free propagation and L a lens. Free propagation is described by the matrix

$$F(D) = \begin{pmatrix} 1 & D \\ 0 & 1 \end{pmatrix} ; D > 0.$$

Propagation through a lens of optical power is given by

$$L(g) = \begin{pmatrix} 1 & 0 \\ -g & 1 \end{pmatrix} \ g \gtrless 0$$

The general product of the matrices would be of the form $S = \begin{pmatrix} a & b \\ c & d \end{pmatrix}$ which necessarily has det S = 1, since aall individual matrices have determinant equal to 1. We now consider whether such a general matrix can be obtained by a combination of free propagation and lenses. There are only three free parameters. So FLF or LFL are the minimum number required. However, since D's are constrained to be positive definite, all general matrices can not be so constructed. For example, magnification without inversion is not possible using only FLF or LFL. If M is positive it is a plain magnifier and if M is negative it inverts. Also, we can not have inverse propagation i.e., propagation through a negative distance using only three of the elements FLF or LFL, but with six of them such as FLFLFL, any general ray transfer matrix S can be realized. The above discussion is also applicable to cylindrical and circular lenses.

For the case of non-coaxial lenses, the matrices need to be extended to the type

$$\begin{bmatrix} a & b & e \\ c & d & f \\ 0 & 0 & 1 \end{bmatrix}$$

with same kind of group structure. The above is an inhomogenous extension, because previously we had $S \sim g(x^2 + y^2)/2$, which was quaadratic in x, but now we have $S \sim 1/2\, g(x - a)^2$ which gives rise to a term linear in x, i.e., the displacement.

We consider now the case of anisotropic lenses. We now have to consider the group Sp(4, R), which consists of all real 4 x 4 matrices S which satisfy $S^T K S = K$ with

$$K = \begin{bmatrix} 1 & 0 & 0 & 0 \\ 0 & 1 & 0 & 0 \\ 0 & 0 & -1 & 0 \\ 0 & 0 & 0 & -1 \end{bmatrix}$$

All of them have unit determinant. The free propagation matrix is still of the same form as in the isotropic case:

$$F(D) = \begin{pmatrix} 1 & D \times 1 \\ 0 & 1 \end{pmatrix}$$

except that the entries now refer to 2 x 2 matrices which is isotropic. But for a lens, the matrix L now becomes

$$L = \begin{pmatrix} 1 & 0 \\ G & 1 \end{pmatrix}$$

where G in general is a 2 x 2 matrix which is anisotropic. The question now is whether one can construct systems composed only of lenses and free propagation as in the isotropic case. We recognise that the ray transfer matrix S belongs to the group Sp(4, R) (which corresponds to canonical transformation in two pairs of variables) which is a subgroup of SL(4, R). Both F and L belong to Sp(4, R) and combination of F′s and L′s will not realise anything outside the group. It is possible to show that this is in fact the largest allowed ray transfer matrix group: there is no possibility of going beyond Sp(4, R). It can also be shown that every such matrix can be realized using a sufficient number of thin lenses and free propagations.

We now turn to the Quantum Electrodynamic aspect of the subject. For an ordinary proragating wave in space, the phase change is proportional to the optical path traversed for a monochromatic wave. For the Wolf function W(x, p) which describes pencils of rays (i.e., ray density) of given position and momenta the following simple relation holds:

$$W(x, p) = W'(x'p') ,$$

where x′, p′ are the new coordinates and moments. The new ray density is obtained very simply from the old ray density; in fact it just transforms as a "scalar field". We can regard classical optical configurations as coherent states of a quantized field. Coherent states are eigenstates of the operator x + ip, which though not Hermitian, is bounded and is non-normal, (does not commute with its adjoint). Still it has a complete set, actually an over-complete set of eigenstates. The ground state of the harmonic oscillator is the one where x + ip or x + d/dx acts on it to give zero eigenvalue.

This eigenstate is $\exp(-x^2/2)$. One can construct a complete set of eigenstates from the lowest eigenstate, they are in general not mutually orthogonal. Because there are so many of them we can so choose them that any density matrix can be made diagonal. This is called "diagonal coherent state representation". Every state can be expressed as a weighted average of coherent states; every coherent state in second quantised theory, the results are identical with classical theory. So we now have a second quantised theory of classical electromagnetic field corresponding to thin lenses.

It must be mentioned here that the optical power g which was assumed to be a constant for each lens can be generalized to be a function of henochromatic frequency.

If we do not restrict ourselves to these thin lenses with quadratic phase shifts, but include thick lenses with quadratic phase shifts or more, then the transformation $W(x, p) \to W'(x'p')$ will change from a point transformation to an integral transformation.

In the treatment so far wave optical treatment of ray optics was given, which did not yield any new effects not found in geometric ray optics. We now extend the treatment and show some new non-trivial effects associated with the quantum electrodynamical description of ray optics.

Before we go into the details, we discuss the Van Cittert-Zernicke theorem in coherence optics. If we consider ordinary light streaming from a bulb for example, we would find that light emanating from any two points on the bulb are independent of each other i.e., they are incoherent in the sense that a two slit experiment with these two sources will not produce interference. We can introduce a two point correlation function as follows:

$$\Gamma\,(x_1\;x_2,\;y_1\;y_2) = \langle\, A^*(x_1\;x_2)\;A(y_1\;y_2)\,\rangle\;,$$

where Γ is the two point correlation function for two points $(x_1\;x_2,\;y_1\;y_2)$ and which is equal to the ensemble average of the quantity

$A(x_1\ x_2)\ A(y_1\ y_2)$. Here A's are wave amplitudes. If the two points $(x_1\ x_2)$ and $(y_1\ y_2)$ are not correlated whenever $x_1\ x_2$ do not coincide with $y_1\ y_2$, then

$$\langle A^*(x_1\ x_2)\ A(y_1\ y_2) \rangle = \delta(x_1 - y_1)\ \delta(x_2 - y_2)\ I(x_1, y_1)$$

where $I(x_1, y_1)$ is some intensity function of x_1, y_1. So the two point correlation function for such incoherent illumination can be specified once the intensity is known. This is the simplest case (i.e., uncorrelated case) and can be thought of as a primary source of illumination. The ensemble average over amplitudes described above is a reflection of the coherence between two sources of light. In the completely correlated case, i.e., the coherent case, the ensemble average factorises out in the form

$$\Gamma(x_1\ x_2,\ y_1\ y_2) = \nu^*\ (x_1\ x_2)\ \nu(y_1\ y_2)$$

but in general it does not and the ensemble average is a sum of many such amplitudes. The Van Cittert-Zernicke theorem states that if one starts with a primary radiating souce where any two points are uncorrelated, at a sufficiently long distance from the source, coherence between the two points sets in automatically; the degree of correlation increases with increasing distance. The magnitude of the correlation is dependent upon the solid angles subtended at the points far away from the source, by the primary radiating source.

The theorem was later examined by Wolf who clarified the setting in of coherence as follows: The correlation function $\zeta(x_1\ x_2,\ y_1\ y_2)$ is really a wave function which satisfies wave equation separately in $x_1\ x_2$ and $y_1\ y_2$. The initial configuration of τ is a singular one corresponding to uncorrelated δ function; as the wave spreads, the singularity becomes more and more diffuse, so that the onset of coherence is related to the spreading of wave function τ. This becomes more transparent when we recast the argument of Wolf functions denoting ray densities.

The way to define the Wolf function in this case would be to write

$$\Gamma(x_1 + (1/2)y_1,\ x_2 + (1/2)y_2,\ x_1 - (1/2)y_1,\ x_2 - (1/2)y_2)$$

and take a Fourier transform with respect to y.

$$W(x_1\ x_2\ p_1\ p_2) = \int \Gamma(x_1 + 1/2\ y_1,\ x_2 + 1/2\ y_2,\ x_1 - 1/2\ y_1,\ x_2 - 1/2\ y_2)\ e^{ipy}\ dy\ .$$

The primary radiating source is now completely independent of p, but depends on x and y corresponding to change of intensity i.e., it is radiating in all direction. In this ray picture, the Van Cittert-Zernicke theorem translates as follows: Whatever be the illumination of the light source, at distances sufficiently far away,, the rays of light are almost parallel. In effect, the onset of coherence of waves at far distances from light source has been translated into purely geometric correlation of rays.

We now consider the Wolf function

$$W(x, p) = \int dy\ e^{ipy}\ \Gamma\ (x + 1/2\ y,\ x - 1/2\ y)$$

which is real but non positive. Earlier positive definite W(x, p) led essentially to results of geometric optics; to obtain extra predictions we now invoke the "reality" of W(x, p) which means that W(x, p) could be negative. We shall see that this leads to genuinely wave optical phenomena.

We know of such phenomena in elementary optics like interference, diffraction, double refraction and optical activity. Here we shall be content with demonstrating interference using ray optics. In a typical interference set up, we have the following arrangement:

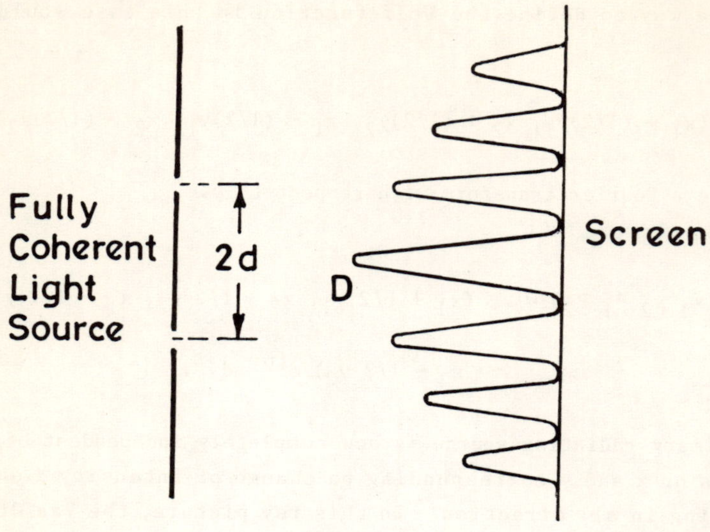

A fully coherent light source illuminates the two slits placed at a distance 2d apart and produces an interference pattern of maximum fringes on a screen placed at a distance D from the slits. The separation $\Delta = \lambda D/d$ where λ is the wavelength of incident light. We also know from elementary optics that if the optical path difference between two waves is an integral multiple of wavelength, then we have a bright fringe; if it is a half integral multiple of λ, then we have a dark fringe and so on. The phase $\alpha = 2\pi/\lambda \times$ path.

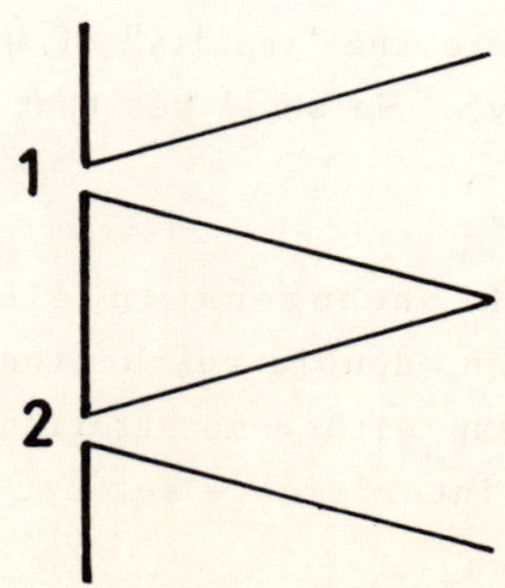

The point now is that the same interference pattern can be explained by ray picture ! Firstly, for fixed d and λ , moving the screen (increasing D) enlarges the fringe pattern linearly, so already a ray aspect of light is suggested; there is a scaling of Δ with D and the process is similar to the behaviour of a shadow in screen.

It looks like we have an apparent contradiction here, a ray aspect is suggested by scaling, but light added to light should produce darkness i.e., dark fringes which does not seem to make sense with rays. The non-positivity of W(x, p) resolves the apparent contradiction.

Consider light rays from two slits as pictured above; we shall explain the interference pattern using the fact that W(x, p) can be less than zero. The amplitudes entering on the right hand side of the Wolf function are now a sum of two parts for each of the two slits:

$$A = A_1 + A_2 ;$$

$$\Gamma = \langle A^* A\rangle = \langle A_1^* A_1\rangle + \langle A_2^* A_2\rangle + \langle A_1^* A_2 + A_2^* A_1\rangle$$

The first term corresponds to ray intensity at the slit 1, the second term the ray intensity at slit 2; but the third term is the interference term between the rays of light from slits 1 and 2; and is apparentlly emanating from the middle of the two slits ! The relative phase between A_1 and A_2 is not contained in the terms due to either of the two slits, but enters only in the third term. Strictly speaking all this is true in the Fraunhofer limit i.e., when the screen is far away from the slits only. So the rays from the two slits are redistributed. We have

$$\langle A_1 + A_2\rangle = \langle A_1^* A_1\rangle + \langle A_2^* A_2\rangle + \langle A_1^* A_2 + A_2^* A_1\rangle$$

$$= 1 + 1 + e^{i\alpha} + e^{-i\alpha} = 2(1 + \cos \alpha)$$

where 1 and 2 refer to normalised light intensity from slits 1 and 2, and α is the geometric phase difference between A_1 and A_2 and can be related to path difference between the two light rays by multiplying it with $2\pi/\lambda$. Thus we essentially recover interference using ray optics. The interference pattern can be completely eliminated by adding a statistical phase .

$$1 + 1 + \langle e^{i\alpha + i\eta} e^{-i\alpha - i\eta} \rangle = 2(1 + \cos(\alpha + \eta)$$

If $\eta = \pi/2$, dark and light fringes are interchanged, but if η is not held fixed but allowed to be random, then the fast fluctuations between dark and light fringes will completely wipe out the interference pattern. The minimum visibility of the interference pattern corresponds to uniform greyness; and maximum visibility would correspond to $2(1 + \cos\ \alpha)$.

The case of diffraction can be treated similarly; instead of two slits in a diffraction grating we have many of them and interference between any two light rays from adjacent slits takes place, and a diffraction pattern can be seen, so also for diffraaction by a narrow rectangular slit.

In conclusion, we may observe that any quantum field theoretical reformulation of classical theories would introduce new, additional effects that would not be expected in purely classical theoories. The additional effects are evidence of quantum behaviour. For example, in the case of coherent state representation, we can have weighted averages of diagonal coherent states and we can realize ensemble averages of amplitudes A. These, like W(x, p), are real, but not necessarily positive definite. As long as they are greater than zero, they can be imaged by a classical system; but as they develop non-positive values, their behaviour deviates from classical expectations. For instance, a purely classical distribution in counting statistics is a Poisson distribution, and weighted average of classical distribution is a weighted average of Poisson distribution, where variance is larger than that of a Poisson distribution. So there is no classical situation where the variance of a distribution is less than the Poisson; but such sub-Poisson distribution can arise in quantum theory.

There is much more that we can marvel and delight regarding the richness of light but this discussion will have to wait for another opportunity. Allow me to remind you that often people invoke uncertainty principle when asked what happens to a one-photon state which propogate and encounters a partially silvered mirror, or the quantum state for low level illumination of a Michelson interferometer. These and other such elementary questions do have a precise answer in terms of quantum field theory. The treatment in the latter framework is no different from the classical treatment given here - thanks due to the Fundamental Theory of Optics sometimes called the Diagonal Coherent Representation theorem.

Lattice Actions Improved by Block-Spin Renormalisation Group and Maximally Symmetric Lattices

Arunabha Guha*
Department of Physics and Astronomy
University of Rochester
Rochester, NY 14627, USA

In this talk I will discuss two different schemes of improving lattice actions based on Block Spin Renormalisation Group (BSRG) and maximally symmetric lattices.

1. THE FIRST SCHEME; BSRG FOR O(∞) MODEL

In recent studies of lattice systems, critical behavior plays a very important role because it is only in the approach to critical points that a universal behavior is recovered and we observe the scaling properties of the corresponding continuum field theory. We naturally then ask how to accelerate this approach to continuum behavior since in a numerical Monte-Carlo simulation (MC), we would like to observe the scaling behavior already on small lattices where the correlation length ξ cannot be much larger than lattice spacing a. In a series of papers[1]], Symanzik has shown how to modify the lattice action in such a way that the $O(a^2/\xi^2)$ corrections to scaling be cancelled and replaced by $O(a^4/\xi^4)$. This method is of perturbative essence. An alternative idea, based on real space renormalization group and block-spin methods, is to determine the renormalized trajectory (RT) in coupling constant space under some block spin

*Supported by DOE contract number DE-AC02-76ER13065.

operation[2]. Along such a RT, there is a one-parameter renormalization group and scaling should be observed in some optimal way since points on the RT are directly connected to the euclidian invariant continuum limit (fixed point) by exact renormalization group transformation. Also, in the vicinity of such a RT, Monte Carlo renormalization group should be an accurate and powerful method[3].

Although the idea is conceptually simple, it is difficult to locate the RT precisely. So one has to introduce some approximate trajectory which is supposedly not far from the true RT. For example, in asymptotically free theories, the trajectory can be determined by perturbative calculation in the weak coupling region. In this talk, I will discuss the usefulness of RT by using another approximation which is the block-spin transformations in the large N limit of O(N) non-linear spin model. This work was done in collaboration with M. Okawa and J. B. Zuber[4]. Extending a previous work of Hirsch and Shenker[5], we showed that these block-spin transformations are given by a simple set of equations. We determined numerically the fixed trajectory and then proposed to use this large N RT as an approximate trajectory for finite N models. Of course, this cannot be justified a priori but Monte-Carlo results show that using this approximate trajectory O(3) and O(4) models improve considerably. I hope that our analysis will illustrate clearly the mechanism of improvement and help extend these ideas to other interesting models.

I will sketch briefly our method since the details can be found in our paper. The classical O(N) Heisenberg model is described by the partition function

$$Z = \int \prod_m d\vec{S}_m \delta(\vec{S}_m^2 - 1)\exp(-H) \quad , \quad H = - N \sum_{m,n} \rho_{mn}(\vec{S}_m \cdot \vec{S}_n - 1) \qquad (1.1)$$

where $\vec{S}_m$ is a N-component unit vector at lattice site m.

We divide the d-dimensional hypercubic lattice into cubes of L sites per side and define an average block-spin t_a (of unit length) within a given block 'a'. They interact through a new block Hamiltonian $H_L(t)$. Since it is difficult, in general, to find a closed expression for H_L, we work in the large N limit where we can use the property of "factorisation"[6]. This implies that all O(N) invariant correlation functions of block-spin $\vec{t}$ can be related to the O(N) invariant two-point functions for the original spin $\vec{S}$. Thus, in the O(N) symmetric phase, it is not unreasonable to assume for H_L a form bilinear in spins and calculate the new coupling constants as

$$H_L(t) = -\sum_{m,n} \rho_L(m-n)\vec{t}_m \cdot \vec{t}_n \quad , \quad \rho_L(\vec{n}) = -\int_{-\pi}^{\pi} \frac{d^d k}{(2\pi)^d} \frac{\exp(i\vec{k}\cdot\vec{n})}{2F_L(\vec{k})} \quad (1.2)$$

where $F_L(k)$, the fourier transform of $\langle \vec{t}_m \cdot \vec{t}_n \rangle$, can be explicitly written in terms of ρ_{mn} in (1.1).

Let us now make a few observations: <u>**(i)**</u> The renormalisation group (RG) transformations $\rho \to \rho_L$ are explicit. <u>**(ii)**</u> They form a (semi-) group; a L×L blocking followed by a L'×L' blocking amounts to a LL'×LL' blocking. <u>**(iii)**</u> Our block-spin transformations do not affect the long distance behavior, as expected. <u>**(iv)**</u> In d=2, we have numerically evaluated the various block-spin couplings for blocks of size 2×2, 4×4, 8×8, etc., for various choices of ρ_{mn}. Some of the results are shown in figure (1) where we have plotted the couplings K_2

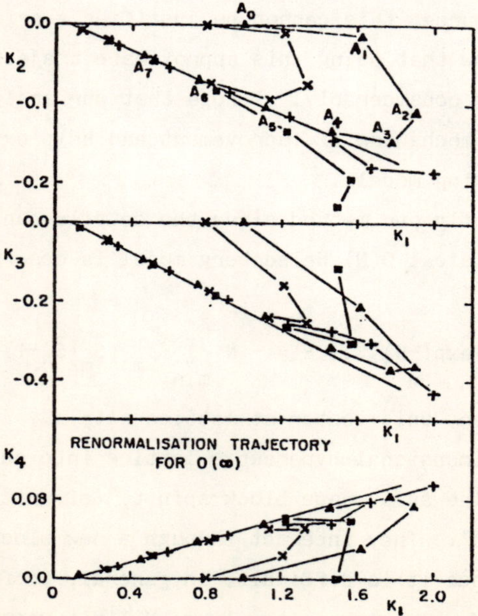

Figure 1. Trajectories starting with $K_1 = .8$, $K_i = 0$ for $i > 1$ (crosses); $K_1 = 1.0$, $K_i = 0$ for $i > 1$ (triangles); $K_1 = 1.5$, $K_2 = -.2308$, $K_3 = -.1154$, $K_i = 0$ for $i > 3$ (squares, corresponds to ST action); $K_1 = 2.0$, $K_2 = -.186$, $K_3 = -.430$, $K_4 = .095$ (pluses).

(interaction between spins $\sqrt{2}a$ apart), K_3 (for spins 2a apart) and K_4 (for spins 3a apart) against K_1 (nearest neighbor interaction) for successive blocking. For example, starting with the point A_0 ($K_1 = 1.0$, $K_i = 0$ for $i \neq 1$), we evaluate $K_i'(i=1,2...)$ for blocks of size 2×2 (point A_1), 4×4 (point A_2) etc. up to 128×128 (point A_7). This process is repeated for different choices of initial interaction. After a few blockings, all the trajectories seem to approach a limiting curve usually referred to as the renormalized trajectory (RT). **(v)** We have observed that starting from Shenker-Tobochnik action (ST) ($K_1:K_2;K_3K_4 = 1: -2/13:-1/13:0$)[3], one approaches the RT much faster than starting from standard action ($K_i = 0$ for $i > 1$) or from "tree improved" action (TIA) ($K_1:K_2:K_3:K_4 = 4/3:0:-1/12:0$)[7]. **(vi)** It is easy to determine the shape of the RT at small and large couplings. For small K_i, we can show that the RT approaches the origin along $K_2 \sim K_1^2$, $K_3 \sim -K_1^2$, $K_4 \sim K_1^3$. For large K_i, the model is asymptotic to the Gaussian model[8] which projects along the line $K_1:K_2:K_3:K_4 = 1:-.0948:-.2187:.0492$. **(vii)** Since we intend to use the action given by the RT for intermediate values of the couplings, it is convenient to take the curvature near the origin into account and to fit the RT by

$$(K_1-0.5):(K_2+.045):(K_3+.1025):(K_4-.021) = 1.0:-.1011:-.2412:.056 \quad (1.3)$$

(viii) Let us emphasize at this point that although we have assumed that effective Hamiltonians are bilinear in spins, there has been <u>no truncation</u> such as assuming that they have only a finite number of coupling constants. However, for numerical simulations, we kept only four coupling constants (K_1,K_2,K_3 and K_4) and neglected all longer range interactions which were observed to be rather small compared to K_1 for Hamiltonians near the RT. From now on, we shall refer to these four parameter actions (satisfying equation (1.3)) as "Renormalisation

Group Improved Models" (RGIM). <u>(ix)</u> Finally, let us discuss the significance of choosing a lattice action near the RT. An intuitive way of understanding this is as follows. Let us consider the points A_0 and A_5 in figure (1). Let us denote the Hamiltonians associated with these two points as $H_o(\{K\},a)$ and $H_5(\{K'\},2^5a)$ respectively. Since the point A_5 is obtained from A_0 by 32×32 blocking, they describe the same physics; the important point is that simulating $H_5(\{K'\},2^5a)$ is equivalent to simulating $H_0(\{K\},a)$ which has a much smaller lattice spacing. Thus, even for finite {K'}, we would expect H_5 to show an earlier scaling, i.e. at lower values of ξ/a.

2. MONTE-CARLO SIMULATION OF O(3) AND O(4) RGIM's

In the continuum limit, Heisenberg model reduces to the O(N) non-linear sigma model (NLSM) defined by euclidian action

$$H \rightarrow \text{Action} = (\beta/2) \int (\partial_\mu \vec{S})^2 d^d x \qquad (2.1)$$

which is known to be asymptotically free in d=2. To verify this, we have to show that physical quantitites such as ξ and susceptibility χ obey the appropriate scaling relations (for $\beta \rightarrow \infty$)[9]. In figures (2) and (3) we show the result for mass-gap of O(4) and O(3) models. In our O(4) model scaling starts at a much smaller value of $\xi \simeq 1.1 - 1.2$ compared to standard model where scaling seems to start at $\xi \simeq 3$.[10] For O(3) scaling starts at $\xi \simeq 1.8 - 2.0$. For comparison,

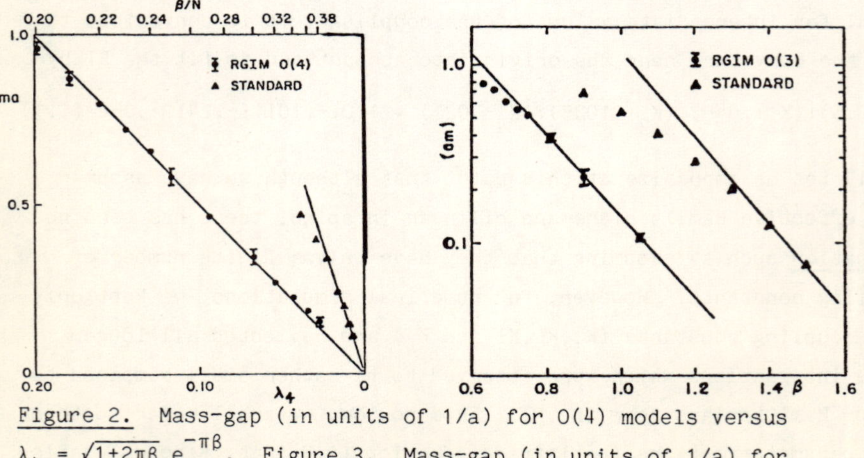

<u>Figure 2.</u> Mass-gap (in units of 1/a) for O(4) models versus $\lambda_4 = \sqrt{1+2\pi\beta}\, e^{-\pi\beta}$. <u>Figure 3.</u> Mass-gap (in units of 1/a) for O(3) models. Data for standard model taken from reference [10].

scaling starts at $\xi \simeq 4$ for TIA, $\xi \simeq 1.5 - 2.0$ for one-loop improved Symanzik action and $\xi \simeq 3$ for ST. It is gratifying to see that our model based on RT for $O(\infty)$ compares quite favorably to these other models even for N=3.

We also evaluated magnetic susceptibility scaling defect $\delta\chi$ which should approach a constant value in scaling region. The results for O(4) and O(3) are shown in figures (4) and (5). They are also consistent with the observation that scaling sets in early in RGIM.

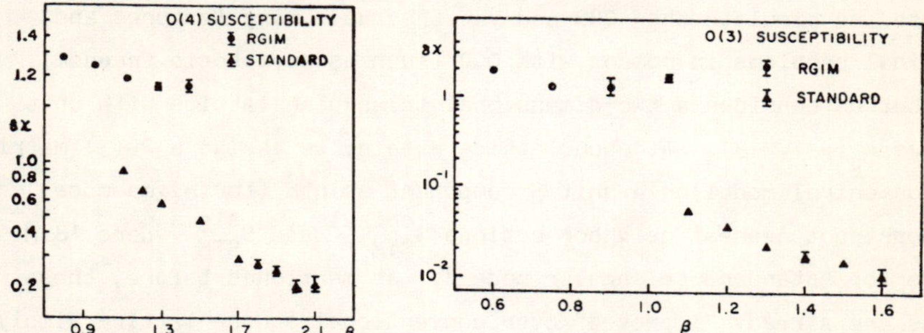

Figure 4. and 5. Magnetic susceptibility scaling defects for O(4) and O(3) models. Data for standard model taken from reference[10].

3. ACTIONS DEFINED ON MAXIMALLY SYMMETRIC LATTICES

In the last two sections, I talked about improving lattice actions by BSRG. However, a note of caution is useful at this point. In interesting but complicated theories such as Quantum Chromodynamics (QCD) it may turn out that to represent the system near the renormalised trajectory one needs a fairly complicated action (say, having different types of plaquettes in the case of gauge theory) for which Monte-Carlo updating time increases significantly. In this part of the talk, I will discuss an alternative way to usher in "scaling" faster without significantly increasing the Monte-Carlo updating time. This was done in collaboration with M. Okawa[11] and R. Basu[12]. What we did was to use a lattice with as many nearest neighbors as possible and thus having the maximal discrete rotational symmetry. Since the nearest neighbor action defined on such a lattice already has more discrete rotational invariance than the conventionsl simple hypercubic lattices, we expect the Lorentz invariance (actually in this case the euclidian rotational invariance) to set in faster. With spin models in two dimensions we did observe earlier scaling while updating time increased only slightly. The same phenomenon has also been observed in four dimensional gauge theory[13].

We shall present numerical results for two dimensional SU(3) chiral and O(3) non-linear sigma model on a triangular lattice. The reason we have chosen these models is that they are much simpler to analyse and simulate than QCD and yet they have many features and technical problems in common with QCD (such as asymptotic freedom).

Let us consider a two-dimensional triangular lattice with unit vectors e_μ (μ=1,2,3). At each lattice site n, we assign a SU(3) matrix U_n (for chiral model)or a unit 3-component spin $\vec{S}$ (for sigma model) and construct nearest neighbor actions S^{chiral}_{STM} and S^{sigma}_{STM} where 'STM' stands for "standard triangular model". As mentioned before, these models are already "improved" over conventional square lattice simply because of more rotational symmetry. We can improve it even further by repeating the block-spin renormalisation group transformations to derive a RGIM but instead, we chose to implement Symanzik's program of perturbative improvement which goes as follows.

It is well-known that in chiral and non-linear spin models, invariant Green's functions are infra-red finite to all orders in perturbation theory while non-invariant Green's functions are plagued by infrared divergences[14]. So we write down an "improved triangular model" (ITM)

$$\begin{aligned} S^{chiral}_{ITM} &= \sum_{n\mu} \{[(\frac{4\beta}{3} +.01626)\mathrm{Tr}\ U_n U^+_{n+e_\mu} -(\frac{\beta}{12} +.00407)\mathrm{Tr}\ U_n U_{n+2e_\mu}]+c\cdot c\} \\ S^{sigma}_{ITM} &= \sum_{n\mu} \{(\frac{4\beta}{3\sqrt{3}} +.02788)\vec{S}_n\cdot\vec{S}_{n+e_\mu} -(\frac{\beta}{12\sqrt{3}} +.00697)\vec{S}_n\cdot\vec{S}_{n+2e_\mu}\} \end{aligned} \tag{3.1}$$

where the co-efficients are obtained by eliminating $O(a^2)$ terms in the invariant Green's functions $\langle \mathrm{Tr}(U_o U^+_m + h.c.)\rangle$ and $\langle \vec{S}_o\cdot\vec{S}_m\rangle$ upto one-loop level.

Figures 6 and 7 show the results for mass-gap for SU(3) and O(3) models respectively. Scaling seems to start at

$\xi/a \simeq 1.7$ (chiral STM) $\simeq 2.5$ (standard square lattice)

$\simeq 1.35$ (chiral (ITM)

$\simeq 4$ (sigma STM) $\simeq 14$ (standard square lattice)

$\simeq 2.5 - 2.8$ (sigma ITM)

Solid lines are the asymptotic scaling forms given by renormalisation group equation. The improvement is quite obvious. For the same accuracy, the gain in computing power achieved is a factor of 2 for SU(3) and a factor of 12 for O(3) which is impressive given the slight increase in Monte-Carlo updating time.

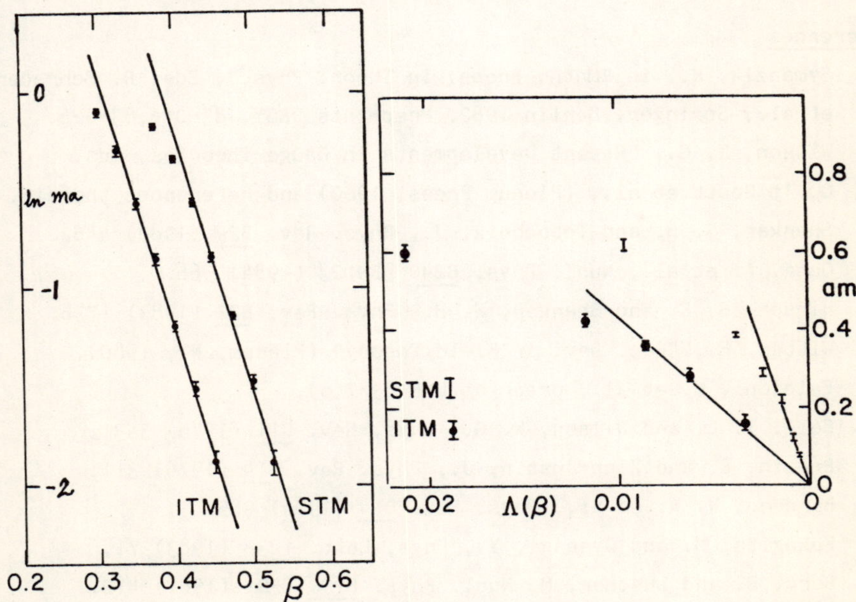

Fig. 6. Mass-gap for STM and ITM SU(3) models. Fig. 7. Mass-gap for STM and ITM O(3) models for different values of $\Lambda(\beta) = 2\pi\beta\exp(-2\pi\beta)$.

For O(3) model, we also measured the magnetic susceptibility and the four-point dimensionless coupling (at zero momentum) λ given by[7]

$$\lambda = \frac{1}{\chi^2\xi^2}\int d^2x d^2y d^2z \langle\vec{S}(o)\cdot\vec{S}(x)\vec{S}(y)\cdot\vec{S}(z)\rangle_C \qquad (3.2)$$

In the continuum, λ approaches λ^* estimated to be $-(6.66\pm.06)$.[7] Figs. 8 and 9 show the numerical results which also indicate early scaling.

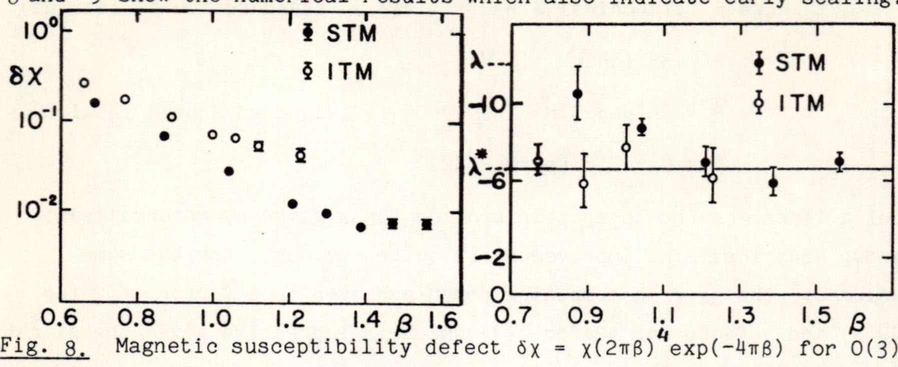

Fig. 8. Magnetic susceptibility defect $\delta\chi = \chi(2\pi\beta)^4\exp(-4\pi\beta)$ for O(3) models. Fig. 9. Four point dimensionless coupling λ for O(3) models.

References

1. Symanzik, K., in "Math. Probs. in Theor. Phys.", Eds. R. Schrader et al., Springer, Berlin 1982. Preprints DESY, 83-016,83-026.
2. Wilson, K. G., "Recent Developments in Gauge Theories, Eds. G. 't Hooft et al., (Plenum Press, 1980) and references therein.
3. Shenker, S. H. and Tobochnik, J., Phys. Rev. B22 (1980) 4462.
4. Guha, A. et al., Nucl. Phys. B240 [FS12] (1984) 566.
5. Hirsch, J. E. and Shenker, S. H., Phys. Rev. B27 (1983) 1736.
6. Witten, E., "Rec. Dev. in Field Theory" (Plenum, NY, 1980).
7. Falcioni, M. et al., preprint LNF-83/7(p).
8. Bell, T. L. and Wilson, K. G., Phys. Rev. D11 (1975) 3431.
9. Brezin, E. and Zinn-Justin, J., Phys. Rev. B14 (1976) 311. Bardeen, W. A. et al., Phys. Rev. D14 (1976) 985.
10. Fukugita, M. and Oyanagi, Y., Phys. Lett. 123B (1983) 71. Berg, B. and Luscher, M. Nucl. Phys. B190 (FS) (1981) 412.
11. Guha, A. and M. Okawa, University of Rochester preprint UR-900.
12. Basu, R. and Guha, A., University of Rochester preprint UR-903.
13. W. Celmaster, Phys. Rev. Lett. 52 (1984) 403.
14. Mckane, A. and Stone, M., Nucl. Phys. B163 (1980) 169. David, F., Nucl. Phys. B209 (1982) 433. Elitzur, S., Nucl. Phys. B212 (1983) 501.

POSSIBLE EVIDENCE FOR PURELY POWERLAW CORRECTION TO SCALING

B.P. Mahapatra

Post Graduate Department of Physics
Sambalpur University, Burla 768 019

ABSTRACT

We present phenomenological evidence for purely powerlaw correction to scaling by obtaining excellent description of the world data on deep inelastic (lepton-hadron) scattering and the Drell-Yan process in terms of a set of parton densities with asymptotic behaviour $q(x,Q^2) \to q_i(x) + (m^2_i/Q^2)\,\tilde{q}_i(x) + \ldots$ when Q^2 is large. The parton densities $q(x,Q^2)$ are constrained to satisfy the general requirements of positivity, Regge behaviour, quark counting rule, sum rules, vector dominance, correct real photon limit ($Q^2 \to 0$) and exact Bjorken scaling. There is no need for a K-factor to describe the Drell-Yan data on the nucleon. It is further shown that the above power law departure from scaling can be generated in QCD by non-perturbative running coupling constant $\alpha_s(Q^2)$ given by $\alpha_s(Q^2) = \alpha_o m^2_o/(Q^2 + m^2_o)$. In contrast, perturbative QCD requires the scale parameter Λ to vary with Q^2 to describe the DIS data and a K-factor to describe the DY process.

Perturbative quantum chromodynamics (PQCD) predicts logarithmic violation of scaling. Some definite results in PQCD are (i) trivial scaling in the Bjorken limit[1], i.e., as $\Lambda^2/Q^2 \to 0$, $F_2(x,Q^2) \to$ const. $\delta(x)$; (ii) the moments $M_n(Q^2)$ vanish asymptotically, i.e., $M_n(Q^2) \to 0$, $n \geqslant 2$, as $Q^2 \to \infty$; (iii) the scale parameter Λ is process and Q^2-independent once a renormalisation scheme is fixed; (iv) the presence of large higher order corrections (HOC) in the Drell-Yan (DY) process leading to a K-factor[2] $\sim$ 1.6-2.0.

How far the data on DIS and DY processes support the above characteristic features of PQCD ? The recent high Q^2 data[3)] on structure functions show little evidence for trivial scaling. Instead these appear to approach non-trivial finite limits at all-x. Similarly, the moments $M_n(Q^2)$ are extremely well represented as straight lines[4)] in $1/Q^2$ with nonvanishing asymptotic ($Q^2 \to \infty$) limit, i.e., $M_n(Q^2) \to A_n + B_n/Q^2$, $A_n \neq 0$. It is also established[5)] that the DIS data can not be described over the full range of Q^2, $5\ \text{GeV}^2 < Q^2 < 200\ \text{GeV}^2$ with a constant scale parameter Λ. The high Q^2 data imply[5)] $\Lambda \sim 0.2$-0.3 GeV while the low Q^2 data can be fitted[6)] in PQCD provided $\Lambda \simeq 0.5$ GeV. It is widely believed[5)] that there could be substantial nonperturbative contribution to scaling violation at low $W^2 < 10\ \text{GeV}^2$.

While the above experimental features are in contrast with PQCD, they are completely consistent with the powerlaw correction to exact scaling given by (when $Q^2 \gg m_0{}^2$)

$$F_2(x,Q^2) \to F_2(x) + (m_0{}^2/Q^2)\tilde{F}_2(x) + \ldots \tag{1}$$

A set of Q^2 dependent parton densities $q(x,Q^2)$ ($q = u_v, d_v$, S or G) constrained to satisfy (1) and other general requirements imposed by positivity, Regge behaviour, quark counting rule (QCR), quark parton sum rules, vector dominance at low Q^2 and correct behaviour in the real photon limit ($Q^2 \to 0$) provide[7)] excellent description of the world data on DIS. These parton densities are given by:

$$xu_v(x,Q^2) = 2\sqrt{x}(1-x)^{\beta_u(Q^2)}/B(1/2,\ 1+\beta_u(Q^2)) \tag{2}$$

$$xd_v(x,Q^2) = \sqrt{x}(1-x)^{\beta_d(Q^2)}/B(1/2,1+\beta_d(Q^2)) \tag{3}$$

$$xS(x,Q^2) = AQ^2(1-x)^{\beta_s}/(Q^2+m_\rho^2) \tag{4}$$

$$xG(x,Q^2) = B(Q^2)(1-x)^{\beta_G} \tag{5}$$

where $\beta_u(Q^2) = \beta_u Q^2/(Q^2+m_1{}^2)$, $\beta_d(Q^2) = \beta_d Q^2/(Q^2+m_2^2)$,

$\beta_d = \beta_u + 1$, $\beta_u \simeq 3$, $\beta_s = 7$, $\beta_G = 5$; B(x,y) is the Euler beta function; m_ρ = ρ-meson (mass). There are only two free parameters, m_1 and m_2; all other parameters are fixed by the constraints of QCR, sum rules and vector dominance.

We obtain very good fits (Figs. 1-2) to the SLAC-MIT data[8] on $F_2^{ep,ed}$; EMC data[3] on $F_2^{\mu p}$, CDHS data[3] on $F_2^{\nu N}$ and the moments[9] $M(n,Q^2)$ of F_2 for $2 < n < 10$. A total number of 446 data points yield[7] total $\chi^2 = 632$ whereas two popular parton models[6] based on PQCD yield $\chi^2 \simeq 1300$ (sixteen parameters) and $\chi^2 \simeq 2830$ (more than 30 parameters). The values of m_i^2 obtained from our fits are : $m_1^2 = 2.1 + 0.23$ GeV^2 and $m_2^2 = 0.75 + 0.26$ GeV^2.

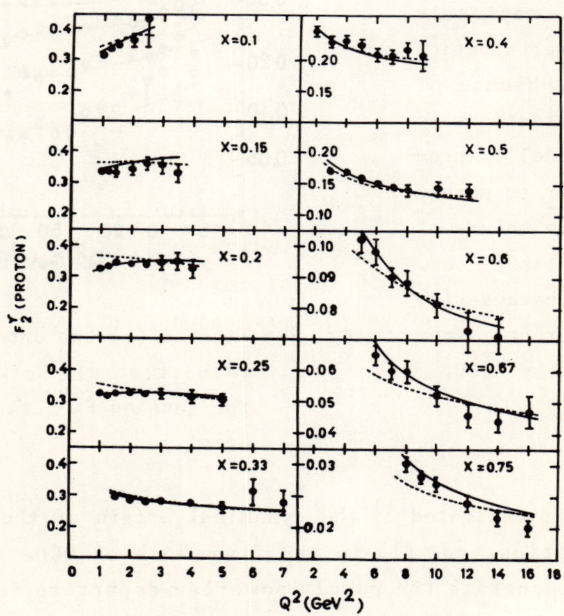

Fig. 1 : Fits (solid curves) to the SLAC-MIT data (Ref. 8) on F_2^{ep} compared with those of the PQCD parton model (dashed curves) of Buras and Gaemers (Ref. 6).

The absolute predictions[10] for the DY cross section $m^3 d\sigma/dmdy|_{y=0}$ using eqs. (2)-(5) for the parton densities describes the data (Fig. 3) quite well without the need for a K-factor. The PQCD based parton models[6], on the otherhand, fail to describe these data and require additional normalisation correction (K-factor). It is well known[2] that the large (~ 60-90%) higher order corrections contributing to the K-factor in PQCD vitiates the rationale for applying perturbation theory. The absence of such large corrections in the present model is probably a pointer to purely nonperturbative nature of scaling violation, eq. (1) which is characteristically different from that predicted by PQCD.

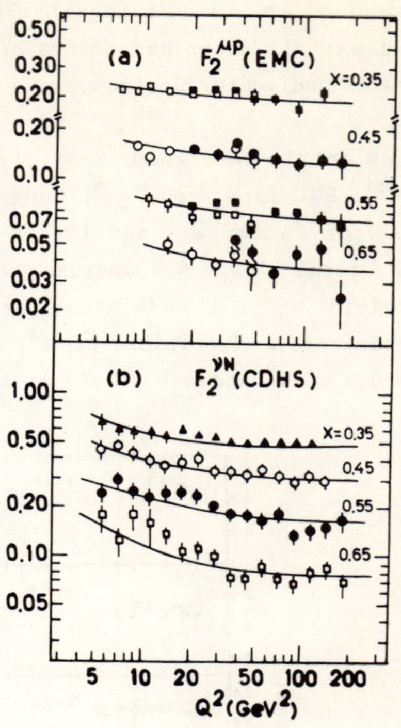

Fig. 2 : Fits to (a) the data on $F_2^{\mu p}$ (Auburt et al. Ref. 3) and (b) the data on $F_2^{\nu N}$ (Abramowicz et al.)

We have investigated[11] the dynamical origin of the pattern of scaling violation, eq. (1) in the framework of QCD. A possible mechanism to generate the purely powerlaw departure from scaling could be through a nonperturbative running coupling constant $\alpha_s(Q^2)$ with the asymptotic behaviour : $\alpha_s(Q^2) \to 1/Q^2$ as $Q^2 \to \infty$. The simplest ansatz for $\alpha_s(Q^2)$ consistent with this requirement, given by

$$\alpha_s(Q^2) = \frac{\alpha_o m_o^2}{m_o^2 + Q^2} , \quad Q^2 \geqslant m_o^2 \tag{6}$$

generates the required scaling violation, eq. (1) through the evolution equations when $\alpha_s \ll 1$. We have confronted the prediction of (6) for scaling violations of nonsinglet distributions (or, equivalently, for general structure functions $F(x,Q^2)$ for large $x > 0.35$) and obtained very good agreement[11] for the large $x > 0.35$ data on $F_2^{ep,ed}$, $F_2^{\nu N}$, $xF_3^{\nu N}$, $F_2^{\mu p}$ over the full range of Q^2, $5 < Q^2 < 200$ GeV^2. The two parameters α_o, m_o in (6) are the only free parameters which get fixed at $m_o = 2.47 + 0.02$ GeV and $\alpha_o = 0.8 + 0.05$. In contrast PQCD fails[5] to describe these data over the full range of Q^2.

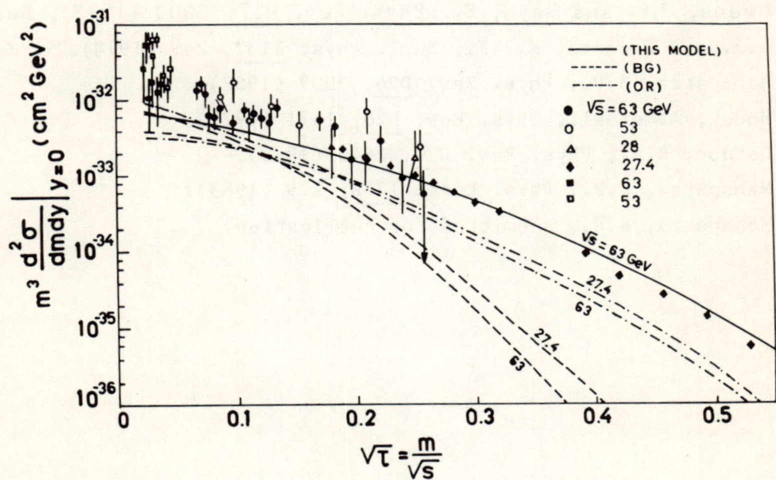

Fig. 3 : The prediction of the present model and those of the PQCD parton models (Ref. 6) compared with the data on proton induced dimuons.

In conclusion, we have presented phenomenological evidence that scaling violations observed in DIS and DY processes do not require the logarithmic violations of scaling predicted by PQCD. These data are much better described by the purely powerlaw correction to exact scaling in a very economical way. The discriminative test between the two models is possible at HERA when data will be available for $Q^2 > 200$-300 GeV^2.

REFERENCES

1. Georgi, H. and Politzer, H.D., Phys. Rev. D9, 416 (1974).
2. Parisi, G., Phys. Lett. 90B, 295 (1980); Altarelli, G., Ellis, R.K. and Martinelli, G., Nucl. Phys. B157, 461 (1979).
3. Aubert, J.J. et al., Phys. Lett. 105B, 315 (1981); Abramowicz, H. et al., Z. Phys. C17, 283 (1983).
4. Mahapatra, B.P., Phys. Lett. 97B, 299 (1980).
5. Eisele, F., Dortmund report no. DO-EXP-82/5 (unpublished); Wahlen, H., Proc. SLAC-Summer Institute in Particle Physics, 1981, Edited by Mosher, A., p. 683.
6. Owens, J.F. and Reya, E., Phys. Rev. D17, 3003 (1978); Buras, A.J. and Gaemers, K.J.F., Nucl. Phys. B132, 249 (1978).
7. Mahapatra, B.P., Phys. Rev. D26, 3002 (1982).
8. Bodek, A. et al., Phys. Rev. D20, 1471 (1979).
9. Gordon, B.A., Phys. Rev. D20, 2645 (1979).
10. Mahapatra, B.P., Phys. Lett. 126B, 279 (1983).
11. Mahapatra, B.P., submitted for publication.

CLASSICAL STABILITY OF FIELD THEORETIC SOLITONS

Ajit Kumar

Department of Physics
Indian Institute of Technology
Hauz Khas, New Delhi-110016.

ABSTRACT

Classical stability of field-theoretic solitons is reviewed. It is shown that topological solitons are absolutely stable in Lyapunov sense. Stability of Skyrmions is reviewed in detail.

The existence of soliton solutions to classical equations of motion in many field theories has led to relevent changes in quantum field theory (QFT). As a result non-perturbative methods of calculations in QFT have been developed[1-4]. The central idea is to look for classical solutions to non-linear field equations and construct quantum states around that. In this connection the stability of classical solitons acquires special interest.

Stability of Non-Topological Solitons

Studies on Lyapunov stability of non-topological solitons (NTS)[5-11] in models of field theories of fields with spin or without (including models with spontaneously broken symmetries) have established that NTS are unstable. However, they can have conditional stability related to internal symmetry of the theory.

In models with charged scalar field ϕ the internal symmetry group is U(1). The corresponding conserved quantity is the charge Q. It has been shown[5-8] that stationary soliton solutions of the type $\phi_o = u(\vec{x}).\exp(-i\omega t)$; $u^* = u$ ω = const. in these models are stable only against perturbations for which $Q[\phi] = Q[\phi_o]$, ϕ being the perturbed soliton. The sufficient condition for stability is given by the so-called Q-theorem: $(dQ_o/d\omega) < 0$, $Q_o \equiv Q[\phi_o]$.This stability is called Q-stability.

The effect of electromagnetic interaction on the stability of these solitons has also been studied[15] using perturbation expansion in the electromagnetic coupling constant e. It again turns out that stability can only be Q-stability, sufficient condition for which is given by $(\partial^2 Q_o/\partial\omega\, \partial\beta) < 0$; $\beta \equiv e^2$. It is the generalization of the above mentioned Q-theorem.

Similar analysis done[19] for solitons in models with higher groups of internal symmetry[12,13] shows that here too the solitons are stable only if the projection of iso-spin on a fixed axis is conserved.

Stability of Topological Solitons

It has been established[14,16] that solitons in topological models can exist only among static solutions. In most of such models it is possible to estimate the Hamiltonian from below through the topological charge Q^t:$H \geq \alpha . Q^t$; α = const. In the case when the lower bound is achieved for the static part of the Hamiltonian H_s the corresponding soliton solutions are absolutely stable since the Lyapunov functional $V = \int d^3x\, H_s$ automatically satisfies all the requirement of the stability theorem[20].

As an example consider ϕ^4-theory in 1+1 dimensions with the Lagrangian

$$L = \frac{1}{2}\, \partial_\mu\phi\, \partial^\mu\phi - U(\phi); \mu = 0,1;\ U(\phi) = \left(\frac{\lambda}{4}\right)[\phi^2 - \frac{\mu^2}{\lambda}]^2; \quad (1)$$

where μ = const. and λ is the non-linear self-coupling constant.

The static solutions satisfy the equation

$(d\phi/dx) = \pm \sqrt{2U(\phi)}$ and H_s can be decomposed as

$$H_s = \int_{-\infty}^{+\infty} [(d\phi/dx) + \sqrt{2U}]^2 \, dx + \int_{-\mu/\sqrt{\lambda}}^{\mu/\sqrt{\lambda}} \sqrt{2U(\eta)} d\eta, \qquad (2)$$

where $\phi = \pm \mu/\sqrt{\lambda}$ are the vacuum solutions.
For static solutions the squared term is zero and so $H_s = Q^t$. So the lower bound of H_s is achieved and the corresponding soliton solutions are absolutely stable.

Similarly stability of Nielsen-Olesen vortices and Hooft-Polyakov monopoles can be proved. However, in these cases it can be done only for a particular choice of the coupling constant λ.

Classical Skyrmions and their Stability

In this model the main object is the chiral field $\chi(\vec{\phi})$ with values in SU(2)[17]. It is parametrized with an iso-vector field ϕ^a, a = 1,2,3 by $\chi(\vec{\phi}) = \exp[i\sigma_n \Theta(\vec{x})]$, where $\sigma_n = \vec{\sigma}.\vec{n}$, $\vec{n} = \vec{\phi}/|\vec{\phi}|$; $|\vec{\phi}| = \sin\Theta$ and $\vec{\sigma}$ are Pauli-matrices. The Lagrangian is

$$L = (1/4\lambda^2)\mathrm{Tr}(L_\mu L^\mu) + (\varepsilon^2/16)\mathrm{Tr}\{[L_\mu, L_\nu][L_\mu, L_\nu]\}, \qquad (3)$$

where $\mu,\nu = 0,1,2,3$ and $L_\mu = \partial_\mu \chi\, \chi^{-1}$ is the left chiral current.

The topological charge is given by the degree of mapping $S^3 \to S^3$ and takes only integer values. The Hamiltonian in the model can be estimated as $H \geq (2\pi^2\varepsilon/\sqrt{2}\lambda)|Q^t|$ where $|Q^t| = \int |\mathrm{Tr}\{\varepsilon_{ikj} L_i [L_k, L_j]\}| d\vec{x}$, (i,j,k = 1,2,3). For static fields $L_o = 0$ and putting $\vec{n} = (\sin\beta \cos\gamma, \sin\beta \sin\gamma, \cos\beta)$ we get

$$H_s = \int d\vec{x}\{(1/2\lambda^2)[(\nabla\theta)^2 + \sin^2\theta\{(\nabla\beta)^2 + \sin^2\beta(\nabla\gamma)^2\}]$$
$$+ \varepsilon^2 \sin^2\theta([\nabla\theta\ \nabla\beta]^2 + \sin^2\beta[\nabla\theta\ \nabla\gamma]^2 + \sin^2\theta\sin^2\beta[\nabla\beta\ \nabla\gamma]^2\}$$
$$(4)$$

To prove the achievement of the lower bound we follow the work of Rybakov[18)].

Theorem. Lower bound of H_s with $Q^t = 1$ is achieved in the class of functions $\vec{\phi}(\vec{r})$ for which $\vec{n} = \vec{r}|r$ and $\theta = \theta(r)$.

Proof. Minimizing the functional (4) in β and γ with $Q^t = (-1/2\pi^2)\int d\vec{x}\{\sin^2\theta \sin\beta(\nabla\gamma[\nabla\theta\ \nabla\beta])\} = 1$ we get that the extremals must satisfy the condition $(\nabla\theta\ \nabla\beta)=(\nabla\theta\ \nabla\gamma) = (\nabla\beta\ \nabla\gamma) = 0$. From this and (4) it follows that inf H_s is achieved when $(\nabla\beta)^2 = \sin^2\beta(\nabla\gamma)^2 = f(\theta,|\nabla\theta|)$ where f is an arbitrary function. Hence $H_s \geq F[\theta] = \int dx\ G((\nabla\theta)^2, \sin^2\theta)$ where $G \geq 0$. From the boundedness of F it follows that $(\partial\theta/\partial r) \not\equiv 0$ and hence the inequality $(\nabla\theta)^2 \geq (\partial\theta/\partial r)^2$ yields that inf F is achieved in the class of spherically symmetric function $\theta = \theta(r)$. Then writing the above condition for extremals and the equation $(\nabla\beta)^2 = \sin^2\beta(\nabla\gamma)^2$ in the spherical coordinates r,ψ,α we get that $\beta = c_1\ \psi + c_2$, $\gamma = c_3\alpha + c_4$ and $c_1^2.\sin^2\psi = c_3^2 \sin^2(c_1\psi + c_2)$. From these we get (upto reflection) $\beta = \psi$ and $\gamma = \alpha$. It means that $\vec{n} = \vec{r}/r$.

Now introducing $x = r/\varepsilon\lambda$ H_s can be written as

$$H_s = (4\pi\varepsilon/\lambda)I;\ I = \int_0^\infty dx\{(\theta')^2(\frac{x^2}{2} + 2\sin^2\theta) + \sin^2\theta + \sin^4\theta\ /x^2\}\ , \qquad (5)$$

The prime denotes differentiation with respect to the argument.

Now we construct the minimizing sequence $\{\theta_n(x)\}$ with $\theta_n(0) = \pi$ and $\theta_n(\infty) = 0$ (so that $Q^t = 1$). Assume that $|\theta_n| \leq M < \infty$ and then in accordance with the asymptotic behaviour of the solutions to Euler-Lagrange equation there exists some $\ell_1 > 0$ such that for $x > \ell_1$ $|\theta_n'(x)| < M(\ell_1/x)^3$. In that case $I[\theta_n]$ can be written as

$$I[\theta_n] = (g^n,\ h^n) + \|\rho^n\|^2 + \|\eta^n\|^2$$

where $g^n = (\theta_n')^2(1+x^2)^2$; $h^n = (x^2+4\sin^2\theta_n)/2(1+x^2)^2$, $\rho^n = \sin\theta_n$; $\eta^n = \sin^2\theta_n/x$. Now from the boundedness of $I[\theta_n]$ it follows that $\int_0^\infty x^2(\theta_n')^2\,dx \leq c^2 < \infty$, c = const. and hence for $\ell > 0$ according to Schartz's inequality

$$|\theta_n(\ell)| \leq \int_\ell^\infty |\theta_n'|dx \leq \left(\int_\ell^\infty \frac{dx}{x^2}\right)^{1/2} \left(\int_\ell^\infty x^2(\theta_n')^2\,dx\right)^{1/2} < c\ell^{-1/2},$$

i.e. for $x \geq \ell > 0$ the inequality $|\theta_n(x)| < cx^{-1/2}$ holds. Hence the sequence $\{\theta_n\}$ is bounded in the Hilbert space $H^1(0,\infty)$ with the norm $\|\theta_n\| = (\int_0^\infty x^2(\theta_n')^2\,dx + |\theta_n(\ell)|^2)^{\frac{1}{2}}$ and, therefore, according to Tikhonov-Alaoglu theorem one can choose a weakly converging in H^1 subsequence from $\{\theta_n\}$. Let $\omega - \lim_{n\to\infty} \theta_n(x) \equiv \theta(x)$.

From the definition of H^1 it follows that $\{\theta_n(x)\}$ is bounded in the Hilbert space $W_2^1(a,b)$, for $x \in [a,b]$; $0 < a < b\ \infty$ with the norm $\|\theta_n\| = (\int_a^b [(\theta_n')^2 + \theta_n^2]dx)^{1/2}$ and hence in $W_2^1(a,b)$ weakly converges to $\theta(x)$. Since the embedding $W_2^1(a,b) \to c(a,b)$ is fully continuous the sequence $\{\theta_n\}$ converges strongly to $\theta(x)$ in $c(a,b)$. Hence the sequences $\{h^n\}$, $\{\rho^n\}$ and $\{\eta^n\}$ converge (strongly in $c(a,b)$ and weakly in $L_2(0,\infty)$) to $h(x,\theta)$, $\rho(\theta)$ and $\eta(x,\theta)$ respectively;$\{g^n\}$ converges weakly in $L_2(0,\infty)$ to $g(x,\theta')$.

Now in L_2 we have $\lim_{n\to\infty}(x_n,y_n) \geq (x,y)$ with $x = \omega - \lim x_n$ and $y = \omega - \lim y_n$; x_n, $y_n \in L_2$. Hence $I[\theta] \leq \lim_{n\to\infty} I[\theta_n] = \inf I$. From this and the definition of the lower bound we get $\inf I[\theta_n] = I[\theta\ ]$. The theorem is proved.

Thus we have proved that classical Skyrmions are absolutely stable.

REFERENCES

1. Rajaraman, R., Phys. Rep. 21C, 227(1975).
2. Jackiw, R., Rev. Mod. Phys., 49, 681(1977).
3. Christ, N.H. and Lee, T.D., Phys. Rev. D12,1606(1975).
4. Faddeev, L.D. and Korepin, Phys. Rep., 42, 1(1978).
5. Friedberg, R., Lee, T.D. and Sirlin, A., Phys.Rev. D13, 2739 (1976).
6. Kumar Ajit, Nisičhenko, V.P. and Rybakov, Y.P., Int. J. Theo. Phys. 18, 6(1979).
7. Kumar Ajit, Izvestiya Vuzov "FIZIKA", 9, 29(1979).
8. Kumar Ajit, Physica Scr.,22, 6(1979).
9. Zastavenko,L.G., Appl. Math. and Mech., 29, 430(1965).
10. Vakhitov, N.G. and Kolokolov, A.A., Izv. Vuz. "Radio-Phys., 16, 1020(1973).
11. Makhankov, V.G., Phys. Rep. 35C, 1(1978).
12. Friedberg R., Lee,T.D. and Sirlin A., Nucl. Phys. B115, 1(1976).
13. Friedberg, R., Lee, T.D. and Sirlin A., Nucl. Phys. B115, 32(1976).
14. Coleman, S. in "New Phenomena in Subnuclear Physics", edited by Zichichi, A. (Plenum New York, 1977)Part B.
15. Kumar Ajit, Phys. Letts. 93A, 8(1983).
16. Jackiw, R., "Semiclassical Analysis of QFT", Latin American School of Physics (1976).
17. Skyrme, T.H.R., Proc. Roy. Soc. 262A, 237(1961).
18. Rybakov, Yu. P. in "Problems of Theory of Gravitation and Elementary Particles", edited by Staniukovich, K.P., (Atomizdat, Moscow), 10(1979).
19. Kumar Ajit, "Stability of Solitons in Friedberg-Lee-Sirlin Model", 1983 (unpublished).
20. Kumar Ajit, Commun. JINR (Dubna), E2-81-85 (1981).

SOME OBSERVATIONS ON FEYNMAN PATH INTEGRALS IN NONRELATIVISTIC QUANTUM MECHANICS

S.V. Lawande
Theoretical Physics Division
Bhabha Atomic Research Centre, Bombay 400 085

1. INTRODUCTION

Path integrals were first introduced by Feynman[1)] as an alternative Lagrangian formulation of non-relativistic quantum mechanics (NRQM). Since then there has been a large number of applications of this formulation in many areas of physics. A large number of path integral theories have been presented in this workshop in the areas of particle and condensed matter physics. The basic theme of this seminar is the summability of paths in NRQM.

The Feynman propagator is the quantum mechanical transition amplitude for a particle to be found at the position x" at time t′ given that the particle was at position x′ at an earlier time t′ and has the form

$$K(\vec{x}'', t''; \vec{x}', t') = \int \exp\{(i/\hbar)\, S[\vec{x}(t)]\}\, D\vec{x}(t) \tag{1}$$

where $S[\vec{x}(t)]$ is the classical action function

$$S[\vec{x}(t)] = \int_{t'}^{t''} L\, dt, \quad L = (m/2)\, \dot{\vec{x}}^2 - V(\vec{x}) \tag{2}$$

The symbol Dx(t) indicates "sum over all paths". An operational meaning of the summation of paths is as follows[2]. Partition the time interval T = t"-t´ into n subintervals $t_{j+1} - t_j$. Assuming for simplicity that all subintervals are of equal length, we have

$$t' = t_0 < t_1 < \dots < t_{N-1} < t_N = t'; \ \vec{x}_j = \vec{x}(t_j) \tag{3}$$

$$S_N = \varepsilon \sum_{j=1}^{N} L \left[\frac{(\vec{x}_j + \vec{x}_{j-1})}{2}, \frac{(\vec{x}_j - \vec{x}_{j-1})}{\varepsilon} \right] \tag{4}$$

and the path differential measure

$$D\vec{x}(t) \rightarrow \frac{1}{\Delta} \prod_{j=1}^{N-1} \frac{dx_j}{\Delta} \tag{5}$$

where $\Delta = (m/2\pi i \hbar\varepsilon)^{3/2}$. This choice of Δ yields the correct propagator for a free particle and the aanalogy to Wiener integrals is evident here. The propagator (1) is then difined as

$$K(\vec{x}", t"; \vec{x}', t') = \lim_{\substack{N \to \infty \\ (\varepsilon \to 0,\ N\varepsilon = t"-t')}} \Delta^{-N} \int \dots \int \exp\left[\frac{i}{\hbar} S_N\right] \prod_{j=1}^{N-1} d\vec{x}_j . \tag{6}$$

2. EXACT PROPAGATORS

Now the task is to evaluate the multiple Riemaan integrals in (6) and then take the $N \rightarrow \infty$ limit. This is possible only in few cases and only in those situations 1) where the integrals involve Gaussians or 2) where a recursive formula is available. The well-known cases of a free particle[2] or the harmonic oscillator with either a constant[2] or a time-dependent frequency[3] fall under the case 1. In general quadratic actions involving potentials of the form[4]

$$V(x) = a(t) + b(t)\, x + c(t) \tag{7}$$

and their three-dimensional analogues can be path integrated. The answer depends on the classical path ($\delta^2 S = 0$) and is given by the well-known Van Vleck - Pauli formula

$$K(x'', t''; x', t') = \left[1/2\pi i\hbar \left| (\delta^2 S_{cl}/\delta x'' \delta x') \right|\right]^{1/2} \exp\ (i/\hbar) S_{cl} \tag{8}$$

Heuristic understanding of why this is so is provided by the expansion of the action $S[x(t)]$ around "classical paths" in which $\delta^2 S = 0$ and terms beyond δS vanish identically for quadratic actions.

A nontrivial case where Gaussian path integrals are involved is the nonlocal action of the type

$$S = \int_0^T (m/2)\, x^2\ 1/2 \int_0^T\!\!\int_0^T G(s,t)x(t)x(s)dsdt + \int_0^T f(t)x(t)dt \tag{9}$$

for which the Lagrangian is difficult to write but the classical equation of motion is

$$mx_c = \int_0^T G(t,s)\, x_c(s)ds \tag{10}$$

and the propagator has the form

$$K(x, T; x_0, 0) = F(T) \exp\ (i/\hbar)\, S_{cl} \tag{11}$$

The exponent is as in the formula (8) but the normalization factor F(T) is quite different and has been obtained in an exact form[5]. Furthermore we have derived exact analytical propagator even if a more general term of the type $F(\int_0^T f(t)\, x(t)\, dt)$ is included in the action. However in this case the propagator can only be expressed as an integral.

We might add that the limiting procedures inherent in the definition (6) may be avoided by employing prodistribution theory[6]. This method essentially maps the infinite dimensional Feynman integral onto a finite dimensional one and Gaussian path integrals including the case of nonlocal action (9) can be conveniently handled by this technique[7].

Recursive formula for evaluation multiple Rieman integrals is available[8] if the potential is of the form[3]

$$V(x) = \Omega^2 (t)\, x^2 + g/x^2 \;, \; 0 < x < \infty \tag{12}$$

where g is a constant such that $g > - \hbar^2/8$ and $\Omega^2 (t)$ is an arbitrary function of time. In particular for $V(x) = g/x^2$, $0 < x < \infty$ the propagator takes the form

$$K(x'',t''; x',t') = \frac{\sqrt{x'x''}}{i\hbar\ T} \exp\left[\frac{i}{2\hbar} \frac{(x''^2+x'^2)}{(t''-L')}\right] I_\gamma\left(\frac{x'x''}{i\hbar T}\right) \tag{13}$$

where I_γ is a Bessel function with $\gamma = (1/2)\{1 + 8(g/\hbar^2)\}^{1/2}$. This corresponds to the radial propagator for a free particle in three dimensions with an angular momentum (γ - 1/2). In fact in polar coordinates more general potentials of the form

$$V(r) = \Omega^2 r^2 + a^2/r^2(1/\cos^2 a\theta + 1/\sin^2 a\theta)\ 0 < \theta < \pi/2a \tag{14}$$

where Ω, a, f and g are constant ($f > - \hbar^2/8$, $g > -h^2/8$ can be exactly treated[9] because of the possibility of expansion of the propagator in terms of some well-known polynomials of the angular coordinate θ. The radial propagator may then be evaluated because it involves potentials which are essentially of the form (12).

3. TIME DEPENDENT POTENTIALS

Lagrangians which depend explicitly on time pose further problems in path integration. A great deal of simplification arises if the classical Lagrangian admits an invariant of motion. In particular, if the invariant is quadratic in momentum p, we have some interesting results[10]. It is known that the existence of such an invariant requires that the potential V(x,t) be of the form[11]

$$V(x,t) = (\frac{\ddot{\rho}}{\rho}\vec{\alpha} - \ddot{\vec{\alpha}})\cdot\vec{x} - \frac{\ddot{\rho}}{2\rho}\vec{x}.\vec{x} + \frac{1}{\rho^2}F(\frac{\vec{x}-\vec{\alpha}}{\rho}) \tag{15}$$

where $\rho(t)$, $\alpha(t)$ and $F((x-\alpha)/\rho)$ are arbitrary functions of time. The associated invariant has the expression

$$I(x,p,t) = \tfrac{1}{2}\left[\rho(\vec{p}-\dot{\vec{\alpha}}) - \dot{\rho}(\vec{x}-\vec{\alpha})\right]^2 + F(\frac{(\vec{x}-\vec{\alpha})}{\rho}). \tag{16}$$

If we make use of the form (15) in the definition (1) of the propagator, then a phase factor can be removed a priori. The remaining propagator can be reduced by a change of variables

$$\vec{\xi} = (\vec{x} - \vec{\alpha})/\rho \ , \ \tau(t) = \int^{t} \rho^{-2}(s)ds \tag{17}$$

and the associated transformation of the path differential measure. We then obtain the neat result[12)]

$$K(\vec{x}'', t''; \vec{x}', t') = (\rho'\rho'')^{-\frac{1}{2}} \exp\{\frac{i}{h}\left[x(t'')-x(t')\right]\}\overline{K}_0 (\vec{\xi}''.\tau''; \vec{\xi}',\tau') \tag{18}$$

where

$$K_0(\vec{\xi}'',\tau'',\vec{\xi}',\tau') = \exp\left[\frac{i}{\hbar}\int_{\tau''}^{\tau''} L_0 \, d\tau\right] D\vec{\xi}(\tau) \tag{19}$$

$$L_0 \equiv L_0(\vec{\xi}, \frac{d\vec{\xi}}{d\tau}) = \tfrac{1}{2}(\frac{d\vec{\xi}}{d\tau})^2 - F(\vec{\xi}) \tag{20}$$

$$\chi = \frac{\dot{\rho}}{2\rho}\vec{x}.\vec{x} + (\dot{\vec{\alpha}} - \dot{\rho}\frac{\vec{\alpha}}{\rho}).\vec{x} - \tfrac{1}{2}\int^{t}\rho^2\left[\frac{d}{dt}(\frac{\vec{\alpha}}{\rho})\right]^2 dt. \tag{21}$$

Thus the propagator for the original time-dependent problem is related to the propagator for an associated time-independent problem. In particular when $F(\vec{\xi}) = 0$, K_0 corresponds to the free particle propagator and most of the text book examples can be explicitly worked out on the 'back of an envelope'. It is also possible to show easily that the propagator admits an expansion in terms of the eigenfunctions of the invariant operator.

REFERENCES

1. Feynman, R.P., Rev. Mod. Phys. 20, 367 (1948).
2. Feynman, R.P., and Hibbs, A.R., ´Quantum Mechanics and Path Integrals´ (McGraw Hill, New York, 1966).
3. Khandekar, D.C. and Lawande, S.V., J. Math. Phys. 18, 712 (1975).
4. Khandekar, D.C. and Lawande, S.V., J. Math. Phys. 20, 1870 (1979).
5. Khandekar, D.C., Lawande, S.V. and Bhagwat, K.V., J. Phys. A16, 4209 (1983).
6. DeWitt-Morette, C., Maheshwari, A. and Nelson, B., Phys. Rept. 50, 255 (1979).
7. Khandekar, D.C., Datta Sunil and Lawande, S.V., Phys. Rev. A (to appear).
8. Peak, D. and Inomata, A., J.. Math. Phys. 10, 1422 (1969).
9. Khandekar, D.C. and Lawande, S.V., J. Math. Phys. 18, 712 (1977).
10. Lawande, S.V. and Dhara, A.K., Phys. Lett. 97A, 353 (1983).
11. Lewis, H.R.Jr. and Leach, P.G.L., J. Math. Phys. 23, 2371 (1982).
12. Dhara, A.K. and Lawande, S.V., J. Phys. A17, 2423 (1984).

EXACT PROPAGATOR FOR GENERAL TWO-TIME QUADRATIC ACTIONS

D. C. Khandekar
Theoretical Physics Division
Bhabha Atomic Centre
Bombay, 400 085.

1. INTRODUCTION

In the context of a path integral theory, non-local quadratic actions arise in many physical applications. The general form of such actions is

$$S = (m/2)\int_0^T \dot{x}^2 \, dt - \int_0^T dt \int_0^T ds \, G(t,s)(x(t) - x(s))^2 \qquad (1)$$

where $G(t,s)$ (without any loss of generality) is a symmetric function of t and s. Physically $G(t,s)$ is a simple phenomenological way of characterizing memory effects which arise when the given system is in interaction with a larger system (e.g. a heat both or a reservoir).

Historically, Feynman[1]) was the first to introduce a non-local action of the type (1) in his path integral theory of the polaron problem. A simple form of (1) was also used by Bezak[2]) for treating an electron gas in a random potential. It has also been exploited in the calculation of the density of electron states in disordered systems[3-4]). In an altogether different context, the action has been found useful for discussing the propagation of waves in random media[5]).

2. DERIVATION OF THE PROPAGATOR

According to the conventional polygonal approach the propagator is to be evaluated as:

$$K(x'', t''| x', t') = \lim_{N\to\infty} K_N(x'', t''| x', t') \tag{2}$$

where K_N is defined as

$$K_N = A_N \int_{-\infty}^{\infty}\dots\int_{-\infty}^{\infty} \prod_{k=1}^{N-1} dx_k \exp[(i/\hbar)S_N] \tag{3}$$

Here A_N, the normalization factor required to define the path - differential measure, is given by $(m/2\pi i\hbar\varepsilon)^{N/2}$ and S_N, is the discretized form of the action defined over the partition of the time interval [t', t"] into N subintervals each of length ε. With the notation, $x_k = x(t_k)$, $x_0 = x'$, $x_N = x''$ we can write S_N as,

$$S_N = [m/2\varepsilon] \sum_{k=1}^{N} (x_k - x_{k-1})^2 - \varepsilon^2 \sum_{j=1}^{N} \sum_{k=1}^{N} G_{jk}(x_j - x_k)^2 \;] \tag{4}$$

When the expressions for A_N and S_N are inserted in (3) the expression for K_N in Eq.(3) takes the form

$$K_N(x,T|x_0,0) = (\alpha/i\pi)^{N/2} \exp\{i\alpha(x_0^2+x_N^2+)\} \int_{-\infty}^{\infty} dX \exp\{i\alpha[\widetilde{X} P X - 2 \widetilde{X} Y]\} \tag{5}$$

where $\alpha = (2\hbar\varepsilon/m)^{-1}$ and $\varepsilon = T/N$. X and Y are column vectors having (N-1) components. The components of X are $x_1\ x_2 \dots x_{N-1}$ while Y has the components $y_1 = x_0$, $y_j = 0, 2 \leqslant j \leqslant N-2$ and $y_{N-1} = x_N$. Further $\tilde{X}$ and $\tilde{Y}$ denote respectively the row vectors corresponding to X and Y. P is an (N-1)-dimensional square matrix with the following structure

$$P_{ij} = P_{ji}, \quad P_{j,j+1} = g_{j,j+1} - 1, \; P_{ij} = g_{ij} \quad (i \neq j, \; j \neq 1)$$

$$P_{jj} = 2 + g_{jj} - \Omega_j^2, \; g_{i,j} = (4 \, \varepsilon^3/m) \, G_{i,j}, \quad \Omega_j^2 = \sum_{i=1}^{N} g_{ij} \qquad (6)$$

The gaussian integral in (5) may be easily evaluated, yielding the result

$$K_N = D_N \exp(-i \, \chi_N) \qquad (7)$$

where χ_N and D_N are given by

$$\chi_N = \alpha \, [x_0^2 + x_N^2 - \tilde{Y} \, P^{-1} \, Y]\} \qquad (8)$$

$$D_N = (\alpha/i\pi \det P)^{1/2} \qquad (9)$$

We need next to evaluate the exponent χ_N and the normalization factor D_N in the limit, $N \to \infty$ with $N \varepsilon = T$. These limiting expressions have been derived by Khandekar et. al[6]). When these are inserted in Eq.(7), we obtain the desired expression for the propagator associated with the action (1).

$$K \, (x, \, T; \, x_0, \, 0) = (m/2\pi i h \varepsilon Q_1 Q_2)^{1/2} \exp((i/h) S_{cl}) \qquad (10)$$

where S_{cl} is the action along the classical path and is given by

$$S_{cl} = (1/2) m (x - x_0)^2 \; \dot{v}(T)/(v(T) - v(0)) \qquad (11)$$

Here $v(T)$ denotes the value of $v(t)$, the non constant solution of the differential equation

$$m\ddot{v} + 4\int_{c}^{T} G(t,s)(v(t) - v(s)) ds = 0 \qquad (12)$$

at $t = T$. The quantities Q_1 and Q_2 are obtained by solving two equations. $Q_1(T)$ is the value of $Q(t)$ which satisfies the equation

$$m\ddot{Q}(t) + 4\, Q(t)\int_0^T G(t,s)ds = 0, \quad Q(0) = 0, \ Q(0) = 1 \qquad (13)$$

at t =T. Finally Q_2 is given by

$$Q_2 = \exp(\int_0^T dt \int_{-1}^{1} d\mu\, R(t,t,\mu)) \qquad (14)$$

where R(t,s,μ) satisfies the equation

$$\ddot{R}(t,s) + R(t,s)\int_0^T G(t,s)ds = G(t,s) + \mu\int_0^T G(t,t'')R(t'',s)dt'' \qquad (15)$$

subject to the conditions R(0,s) = R(T,s) = 0 for all s in [0,1]. The explicit expression for K will depend on the nature of the kernal G(t,s). For a particular case when G(t,s) = constant = $\Omega^2/4T$ the propagator expression (10) simplifies to

$$K(x,\ t;\ x_0,\ 0) = (m/2\pi ihT)^{1/2}(\Omega T/2\sin(\Omega T/2))\exp[(im\Omega/4h)\cot(\Omega T/2)\times (x - x_0)^2] \qquad (16)$$

3. APPLICATION

We shall now apply the propagator expression obtained in Eq.(16) to find the density of electronic states (DOS) in a positionally disordered solid. The central problem here is to evaluate the average propagator $K(x,\ T;\ x_0,\ 0)$ associated with the electron motion in the field of randomly distributed scatterers. The DOS is given by the fourier transform of K(x,T; x, 0) over T. Several other approaches to this problem[7-8]) yield only "the most probable propagator" and therefore do not describe DOS correctly over entire range of energy. Using the path integral approach it has been shown in Ref.3 that the problem of finding the <u>average Green's function</u> and hence DOS reduces to path integrating an action functional containing a non local term. Under certain assumptions on the strength of the scattering potential[2]) the average propagator is described by Eq.(16). The parameter Ω depends on the range and strength of the scattering potential. Therefore to obtain the DOS for this model problem we have to merely obtain the fourier transform of the diagonal part of K of Eq.(16). Using the relation between K of Eq.(16) with the propagator associated with the harmonic oscillator (HOP) of frequency Ω[9]) and the expansion of the HOP in terms of eigenfunctions of Schrödinger equation it is possible to show that DOS n(E) is given by[10])

$$n(E) = \int_{-\infty}^{\infty} \exp(iET/h) K(x, T; x, 0) dT$$

$$= (m\,\Omega^2/2\pi h)\,\sqrt{\pi/2}\,\sum_{n=c}^{M} 1/(E - (n+1/2)h\Omega)^{3/2} \qquad (17)$$

and zero otherwise. Here M is the largest integer such that $E - (n+1/2)h\Omega > 0$. This result is different in two ways than the earlier results. The DOS obtained by other approaches predict a finite exponentially decaying DOS as $E \to -\infty$. In contrast Eq.(17) implies that $n(E) = 0$ if $E < 1/2\, h\Omega$. Further the behaviour of n(E) is not exponential. We have also found that similar conclusions hold even if a more realistic electron ion potential is used. The propagator in this case was evaluated in first cumulant approximation with respect to a trial action (1)and $G(t,s) = \Omega^2/4T$.

4. REFERENCES

1. Feynman R.P., Phys.Rev. 97, 660 (1955)
2. Bezak V., Proc. R. Soc. 315A, 1224 (1970)
3. Edwards S.F. and Gulyavev V.B., Proc. Phys. Soc. 83, 495 (1964)
4. Sa Yakanit V., Phys. Rev. 19B, 2266 (1979)
5. Dashen R., J. Math. Phys. 20, 894 (1979)
6. Khandekar D.C., Bhagwat K.V. and Lawande S.V., J. Phys. 16A, 4209 (1983)
7. Kane E.O., Phys. Rev.131, 79(1963)
8. Halperin B.I. and Lax M., Phys. Rev. 148, 722 (1966)
9. Dhara A.K., Khandekar D.C. and Lawande S.V., J. Math. Phys.24, 844 (1983)
10. Khandekar D.C.,Singh Vijay A., Bhagwat K.V. and Lawande S.V. (To appear in Phys. Rev.)

ANTIFERROMAGNETIC EXCITATION ON PERCOLATION CLUSTERS

Deepak Kumar

Department of Physics, University of Roorkee
Roorkee, U.P., 247667, India

ABSTRACT

The long wave length excitations of a diluted antiferromagnet near percolation threshold, p_c, are studied. Within the hydrodynamic theory, the excitation frequency depends on two parameters A and χ_t. A is the stiffness associated with the variation in the staggered magnetisation and χ_t is the transverse susceptibility. The behaviour of A near p_c is known. We develop a mean field theory to calculate χ_t and find that it diverges like $|\ln (p - p_c)|$ as the concentration p approaches p_c.

1. INTRODUCTION

The study of linear excitations in diluted systems close to their percolation threshold is a challenging problem, because the infinite cluster over which such excitations occur is extremely disordered. In fact, such clusters are very nice physical examples of random fractals with non-integral dimensionality. Recently, significant progress has been made in studying a class of processes whose dynamic equations involve the finite difference analog of the Laplician operator[1]. These processes include ferromagnetic spin waves, diffusion, random resistor networks etc. The present study of antiferromagnetic excitations is partly motivated by the fact that these excitations are different from the class described above, and exhibit a new type of critical behaviour at the percolation threshold.

This is best understood by considering the energy functional for the long wavelength excitations of an antiferromagnet. If $\vec{m}_a(\vec{r})$ and $\vec{m}_b(\vec{r})$ denote the sublattice magnetisations, the energy functional is given by [2)]

$$E = \sum_{\alpha=x,y} \int d\vec{r} \left\{ \frac{A}{2M_s^2} \left[\vec{\nabla} (m_a^\alpha - m_b^\alpha) \right|^2 + \frac{H_E}{2M_s} (m_a^\alpha + m_b^\alpha)^2 \right\} \qquad (1.1)$$

where M_s is the magnitude of the sublattice magnetisation which is along z-axis for both the sublattices, A and H_E are phenomenological constants, which may also be derived from the Heisenberg model. From (1.1), the frequency of the excitations can be easily seen to be [2)]

$$(\omega/\gamma)^2 = (2A\, H_E/M_s)\, q^2 \qquad (1.2)$$

where γ is the gyromagnetic ratio. For the diluted system, both A and H_E show critical behaviour as the percolation threshold p_c is approached. A, which is associated with the energy associated with the spatial variation of the staggered magnetisation varies like $(p-p_c)^t$, where p denotes the concentration and t is the index with which the conductivity of a diluted resister network vanishes at p_c. H_E, on the other hand, is related to the transverse susceptibility of the antiferromagnet in the ordered phase by the relation, $\chi_t = (M_s/H_E) \cdot \chi_t$ is also known to diverge [2,3)] as $p \to p_c$ like $(p-p_c)^{-\tau}$, where τ is a new index. The divergence of χ_t is physically due to fluctuations in the occupation of A and B sublattices [2)], as a result of which any small region of the lattice has some unbalanced moment which can give large response to the transverse field.

2. TRANSVERSE SUSCEPTIBILITY AT T = 0

The above discussion shows that at long wavelengths, the dispersion of antiferromagnetic excitations depends upon χ_t. To calculate χ_t, we consider the following Hamiltonian

$$H = \sum_{\langle ij\rangle} J_{ij}\, \vec{S}_i, \vec{S}_j - \Gamma \sum_i (S_i^z)^2 - g\mu_B \sum_i H_i^x S_i^x \qquad (2.1)$$

where <ij> denotes sum over nearest neighbours, and J_{ij}'s are random variables taking values J and 0 with probabilities p and (1-p) respectively. At T = 0, for small fields H_i^x, one can write

$$\vec{S}_i = S\left[\varepsilon_i(1-\theta_i^2/2)\,\hat{z} + \theta_i\,\hat{x}\right] \tag{2.2}$$

where θ_i's are small deviations from the Neel state and $\varepsilon_i = \pm 1$ if i is on A sublattice and $\varepsilon_i = -1$ if i is on B - sublattice. Substituting (2.2) into Eq.(2.1) yields the energy to be

$$E = E_o + \frac{1}{2}S^2 \sum_{ij} B_{ij}\,\theta_i\,\theta_j - g\mu_B S \sum_i H_i^x\,\theta_i \tag{2.3}$$

from which the transverse susceptibility is found to be

$$\hat{\chi}_t = \chi_t/(g\mu_B)^2 = \langle \sum_{ij} (B^{-1})_{ij} \rangle \tag{2.4}$$

where the brackets denote the configuration averaging over J_{ij}'s and

$$B_{ij} = (\sum_k J_{ik} + \Gamma)\,\delta_{ij} + J_{ij} \tag{2.5}$$

To calculate (2.4), we define a partition function [4)]

$$Z\{h_i\} = \int \prod_i d\theta_i \exp\left[-\frac{1}{2}\sum_{ij} B_{ij}\,\theta_i\,\theta_j + \sum_i h_i\,\theta_i\right] \tag{2.6}$$

in terms of which,

$$\hat{\chi}_t = \frac{\partial^2}{\partial h^2} \langle \log Z\,\{h_i = h\}\rangle \tag{2.7}$$

The configuration average of log Z is now done using the replica track [4)], for which purpose we calculate $\langle Z^n \rangle$. This is given to be

$$\langle Z^n \rangle = \int \sum_{i,\alpha} d\theta_i^\alpha \exp\left[\sum_i \vec{h}_i\,\vec{\theta}_i - \frac{\Gamma}{2}\sum_i \vec{\theta}_i^2 + \sum_{\langle ij\rangle} \log\{1+ve^{-\frac{J}{2}(\vec{\theta}_i+\vec{\theta}_j)^2}\}\right] \tag{2.8}$$

where θ_i^α denote n-replica variables at each site and $v = p/(1-p)$. It turns out that the variables $\Psi_i^\alpha = \varepsilon_i\theta_i^\alpha$ are more convenient to work with, as then the ensuing analysis becomes close to that of diluted resistor networks. In terms of Ψ_i^α's, we write

$$\ln\{1+ve^{-\frac{J}{2}(\vec{\Psi}_i-\vec{\Psi}_i)^2}\} = \frac{1}{z}\sum_{\vec{p}} B_{\vec{p}}\, e^{i\vec{p}.(\vec{\Psi}_i-\vec{\Psi}_j)}$$

$$= \frac{1}{z}\sum_{\vec{p}} B_{\vec{p}}\, Q_{\vec{p}}(\vec{\Psi}_i)\, Q_{-\vec{p}}(\vec{\Psi}_j) \tag{2.9}$$

where Z is the coordination number of the lattice and

$$B_{\vec{p}} = z \sum_{\ell=1}^{\infty} (-1)^{\ell+1} \frac{v^{\ell}}{\ell} \exp(-p^2/2J\ell) \tag{2.10}$$

Substituting Eq.(2.9) into Eq.(2.8) yields

$$< Z^n > = \int \prod_i d\vec{\Psi}_i \exp\Big[\sum_i (\varepsilon_i \vec{h}_i . \vec{\Psi}_i - \frac{\Gamma}{2} \vec{\Psi}_i^2) + \sum_{ij} A_{ij} \sum_p B_{\vec{p}} Q_{\vec{p}}(\vec{\Psi}_i)\, Q_{-\vec{p}}(\vec{\Psi}_i) \Big] \tag{2.11}$$

where $A_{ij} = 1$ if i and j are nearest neighbours and $A_{ij} = 0$ otherwise. A Gaussian transformation is now made to remove the intersite coupling in Eq.(2.11). As a result

$$<Z^n> = \int \prod_{i,\vec{p}} ds_i(\vec{p}) \exp\Big[-\frac{1}{2} \sum_{\vec{p}} B_{\vec{p}} \sum_{ij} A_{ij}^{-1} s_i(\vec{p}) s_j(-\vec{p}) + \sum_i u_i\{s_i(\vec{p})\}\Big] \tag{2.12}$$

where

$$\exp|u_i\{s_i(\vec{p})\}| = \int d\vec{\Psi} \exp\Big[\varepsilon_i \vec{h}_i . \vec{\Psi} - \frac{\Gamma}{2} \vec{\Psi}^2 + \sum_{\vec{p}} B_{\vec{p}}\, s_i(p) e^{-i\, \vec{p}.\vec{\Psi}} \Big] \tag{2.13}$$

Eq.(2.12) is in a form which is most amenable to do further field theoretic calculations. Before proceeding further, we recall how susceptibility may be expressed in terms of new field variables. If we have a uniform field $\vec{h}_i = \vec{h}$, it is a staggered field for $\vec{\Psi}_i$, which therefore has different values for the two sublattices, and

$$\hat{\chi}_t = \frac{\partial}{\partial h} \sum_i \varepsilon_i <\Psi_i^{\alpha}> = \frac{N}{2} \frac{\partial}{\partial h} (<\Psi_A^{\alpha}> - <\Psi_B^{\alpha}>)$$

$$= i \frac{\partial^2}{\partial p_{\alpha}\, \partial h} (<s_B(\vec{p})> - <s_A(\vec{p})>)\Big|_{p=0}$$

$$= \int \frac{d^n w}{(271)^n} w^{\alpha} \frac{\partial}{\partial h} (<s_B(\vec{w})> - <s_A(\vec{w})>) \tag{2.14}$$

where

$$s_i(\vec{w}) = \int e^{i\, \vec{p}.\vec{w}} s_i(\vec{p})\, d\vec{p} \tag{2.15}$$

3. MEAN FIELD APPROXIMATION

The effective free energy functional needed for the evaluation of $<Z^n>$ involves an infinite number of fields at each site. To understand the behaviour of the system, it is judicious to employ the simplest approximation, namely the mean field approximation, namely the mean field approximation, which is already quite nontrivial for this system. In the mean field approximation, Eq.(2.12) is evaluated using the steepest descent method, which leads to the following self-consistent equation for the average value $<s_i(\vec{p})>$

$$\sum_j A^{-1}_{ij} <s_j(-\vec{p})> = \frac{\int d\vec{\Psi}\left[\exp -i\ \vec{p}.\vec{\Psi} + f_i(\vec{\Psi})\right]}{\int d\vec{\Psi}\ \exp(f_i(\vec{\Psi}))} \tag{3.1}$$

where

$$f_i(\vec{\Psi}) = \varepsilon_i\ \vec{h}_i.\vec{\Psi} - \frac{\Gamma}{2}\vec{\Psi}^2 + \sum_p B_{\vec{p}} < s_i(\vec{p}) > e^{-i\ \vec{p}.\vec{\Psi}} \tag{3.2}$$

On taking the fourier transform with respect to $\vec{p}$ variable, (3.1) becomes

$$< s_i(w) > = \sum_j A'_{ij} \frac{1}{Z_{ij}} \exp\left[\varepsilon_j\ \vec{h}_j.\vec{w} - \frac{\Gamma\vec{w}^2}{2} + \sum_p B_{\vec{p}} < s_j(\vec{p}) > e^{-i\ \vec{p}.\vec{w}}\right] \tag{3.3}$$

where

$$Z_{ij} = \int d\vec{\Psi}\ \ \exp\ (f_j(\vec{\Psi})) \tag{3.4}$$

To understand the implications of the mean field equation (3.3), one needs to make further simplifications. Consider the limit $J\to\infty$,in which $B_{\vec{p}}$ (= Z $\ell n(1+v) = x$) becomes independent of p and Eq.(3.3) is considerably simplified physically when $J\to\infty$, the clusters act rigidly as single entities, and the problem is left with geometrical features only. To see this take $\vec{h}_i = \varepsilon_i h$, to get

$$<s(\vec{w})> = \frac{1}{Z_1} \exp\left|\ \vec{h}.\vec{w} - \frac{\Gamma\vec{w}^2}{2} + x < s(\vec{w}) >\right| \tag{3.5}$$

Eq.(3.4) is essentially same as what one gets in the percolation problem and has the following solution [4)]

$$< s(\vec{w}) > = \sum_{m=1}^{\infty} p_m\ e^{m\left|\vec{h}.\vec{w} - \Gamma\vec{w}^2/2\right|} \tag{3.6}$$

with

$$p_m = \frac{(m\ x)^{m-1}}{m\,!}\ e^{-m\,x} \tag{3.7}$$

p_m can be interpreted as the probability that a given site belongs to a cluster of m-sites. The quantity $p_\infty = 1 - \langle s(0) \rangle$, which is the probability that a given site belongs to the infinite cluster, is the order parameter. It is easily seen that

$$p_\infty = 0 \quad \text{if } x \leq 1$$
$$= 2\,|1 - x| = 2\,|r| \quad \text{if } x > 1 \tag{3.8}$$

The percolation transition occurs at $x = 1$, which corresponds to $p_c = 1 - \exp(-1/z)$.

Having established the connection with the percolation problem, let us consider the situation when the external field is uniform. The values of $\langle s_i(\vec{w})\rangle$ now depend on whether i is on A or B sublattice. We have the following coupled equations

$$\langle s_A(\vec{w})\rangle = \frac{1}{Z_1} \exp\left[-\vec{h}.\vec{w} - \Gamma\vec{w}^2/2 + x \langle s_B(\vec{w})\rangle \right] \tag{3.9a}$$

$$\langle s_B(\vec{w})\rangle = \frac{1}{Z_1} \exp\left[\vec{h}.\vec{w} - \Gamma\vec{w}^2/2 + x \langle s_A(\vec{w})\rangle \right] \tag{3.9b}$$

The physical significance of these equations becomes clear, when one examines the percolation problem in a staggered field. It is known that the s-state Potts model is related to the percolation problem in the $s \to 1$ limit [5]. Using this connection, it can be established that the mean field equations for the Potts model in a staggered field are related to Eqs.(3.9) in the $s \to 1$ limit. Furthermore, a cluster expansion of s = 1 Potts model shows that [6]

$$2\Psi_A = 1 - \sum_{n_A, n_B} n_A\, K(n_A, n_B) e^{-\vec{h}.\vec{w}(n_A - n_B)} \tag{3.10a}$$

$$2\Psi_B = 1 - \sum_{n_A, n_B} n_B\, K(n_A, n_B) e^{-\vec{h}.\vec{w}(n_A - n_B)} \tag{3.10b}$$

where $\Psi_{A(B)} = 1 - \langle s_{A(B)}(\vec{w})\rangle$ and $K(n_A, n_B)$ denote the number of clusters per site containing n_A A-sites and n_B B-sites. (Here we have set $\Gamma = 0$ for clarity of interpretation.) Eqs.(3.10) show that the staggered field couples to the fluctuations of $(n_A - n_B)$ in the clusters, which are precisely the fluctuations responsible for the divergence of $\chi_t \cdot \hat{\chi}_t$ can be calculated in the $J = \infty$ limit and is seen to be

$$\hat{\chi}_t = \int \frac{d^n w}{(2\pi)^n} \frac{w_\alpha^2 < s(\vec{w}) >}{1+x <s(\vec{w})>} \xrightarrow[n\to 0]{} \int du \frac{S_1(u)}{1+x\, S_1(u)}$$

$$\simeq \frac{1}{2\Gamma} \sum_m \frac{1}{m} p_m \simeq \frac{1}{2\Gamma} (A + B\,|r|^{2-\alpha_p}) \tag{3.11}$$

where α_p is the usual critical index ($\alpha_p = -1$ in m.f.t) and

$$S_1(u) = \sum_m p_m e^{-mu} \tag{3.12}$$

Note that in this limit χ_t is inversely proportional to the anisotropy constant Γ. Physically this response comes from finite clusters with unbalanced moments. In the absence of anisotropy, such clusters, because of their net moments, would turn in the direction of the field without any resistance, thereby giving infinite response.

Let us now consider the case of finite J. On very general ground [6], it can be shown that for $J >> \Gamma$

$$\hat{\chi}_t = \frac{1}{2\Gamma} \frac{<(n_A - n_B)^2>}{<n_A + n_B>} + \frac{1}{2J} \sum_\lambda \frac{1}{\lambda} (\sum_i \varepsilon_i \phi_\lambda(i))^2 \tag{3.13}$$

where λ and $\phi_\lambda(i)$ denote respectively the eigen values and the eigen vectors of the matrix B_{ij}. Thus we can obtain an answer for χ_t by constructing an expansion of the mean field equations in powers of J^{-1} and retaining the first two terms. Thus we write

$$B_{\vec{p}} = x - \frac{p^2}{J} \sum_i (-1)^{\ell+1} \frac{v^\ell}{\ell^2} = x - \frac{p^2}{b} \tag{3.14}$$

Substituting (3.14) into mean field equations for uniform field, we find

$$<s_{B(A)}(\vec{w})> = \frac{1}{Z_1} \exp\left[\pm \vec{h}.\vec{w} - \Gamma\vec{w}^2/2 + x <s_{A(B)}(\vec{w})> + \frac{1}{b} \nabla_u^2 <s_{A(B)}(\vec{w})>\right] \tag{3.15}$$

where + holds for B and - for A. Using 1/b as an expansion parameter these equations can be solved to obtain the susceptibility. A rather lengthy expression is obtained, but the divergence comes from a term which in the $n \to 0$ limit has the form

$$\hat{\chi}_t \sim \int_0^\infty \frac{u\, S_1(u)\, S_1''(u)\, du}{1 - x\, S_1(u)} \tag{3.16}$$

for $|r| = |1-x| << 1$, this expression can be analytically evaluated to yield

$$\hat{\chi}_t \sim |(\ell n\ |r|)| \tag{3.17}$$

Thus in the mean field theory $\tau = 0$, which agrees remarkably with the heuristic relation

$$\dot{\tau} \geq t - \beta_p - (d-2)\,\nu_p$$

due to Harris and Kirkpatrick[2].

REFERENCE

1 Alexander, S. in "Percolation Structure and Processes" Eds.G. Dentscher, R. Zallen and J. Adler, Annals of Israel Physical Society, Vol 5(1984).

2 Breed, D.J., Gilijamse, K., Sterkenberg, J.W.E. and Miedema, A.R., Physica 68 303 (1973).

3 Harris, A.B. and Kirkpatrick, S. Phys. Rev. B16, 542(1977).

4 Stephens, M.J., Phys. Rev. B17, 4444 (1978).

5 Fortuin, C.H. and Kasteleyn, P.W., Physica 57, 536(1972).

6 Kumar, D. and Harris, A.B. to be published.

CHANNELING RADIATION APPLICATIONS IN DEFECT STUDIES IN SOLIDS

Anand P. Pathak and Prasanna K. John Balagari
School of Physics, University of Hyderabad
Central University Post Office
Hyderabad 500134, India

ABSTRACT

Using the Nielsen type power law interatomic potential and continuum approximation for planar channeling of relativistically fast positrons along Si (110),(100) and (111) planes, the channeling radiation frequencies have been calculated and compared with previous calculations and experimental results. The anharmonic terms in the planar potential are used to calculate the line widths. The effects of defects on channeling radiation are then discussed and dislocations case considered in some detail.

1. INTRODUCTION

During last five years or so, the interest on the radiation emitted by the charged particles channeled along major crystallographic planes or axes, has grown rapidly. After initial theoretical calculations by Kumakhov[1)] predicting the existence of such radiation and consequent experimental observation,[2)] several theoretical and experimental investigations have been reported.[3,4)] The basic phenomena is conceptually simple enough and is essentially an application of electrodynamics, namely any charged particle must radiate whenever it undergoes acceleration or deacceleration. Thus the transverse oscillatory motion of a channeled charged particle emits radiation with its characteristic frequency, according to the classical electrodynamics. Quantum-mechanically, the emitted frequency corresponds to difference of energies of quantized energy levels of transverse motion. Although this difference is generally of the order of eV, the relativistic transformation and Doppler effect gives rise to a factor $2\gamma^2$ in the forward direction ($\gamma=1/\sqrt{1-v^2/c^2}$ where v is the channeled particle velocity and c is light velocity) so that the observed photon energy is in the hard X-ray or γ-ray region. The polarization properties, monochromaticity and other properties of this channeling radiation make it a unique source of radiation which can be extremely useful for certain applications.

2. SPECTRAL FEATURES OF CHANNELING RADIATION.

In order to keep the results in analytically closed form with defects applications in mind, a simple power law type interatomic potential used by Nielsen[5)] has been used to calculate the planar potential in continuum approximation[6)]. The interatomic potential is,

$$V(R) \quad = \quad \frac{Z_1 Z_2 e^2}{R}\,\frac{a_o}{\pi R} \equiv \frac{Z_1 Z_2 e^2}{R}\,\frac{p}{R} \qquad (1)$$

where $p=a_o/\pi$ is a constant, Z_1 is charge number of the incident ion, Z_2 is the charge number of host atom, R is ion-atom distance, and a_o is the Bohr radius. The corresponding planar potential due to one plane at a distance y from the plane is

$$V(y) \quad = \quad N_p \int_0^\infty \frac{Z_1 Z_2 e^2 p \; 2\pi r \, dr}{(y+a)^2+r^2}$$

$$= \quad \pi Z_1 Z_2 e^2 N_p \; p \ln \frac{1}{4N_p(y+a)^2} \qquad (2)$$

where the Thomas Fermi screening distance a has been added to y because the channeled positive particles (positrons in this case) do not come closer than a. Here N_p is planar density of atoms and suitable cut off similar to that used by Erginsoy[7)] has been used in evaluating the integral. For positrons, the potential seen is that due to two planes surrounding the particle trajectory, namely

$$V_2(x) \quad = \quad V_o \ln \frac{1}{4N_p[(1+a)^2-x^2]} \qquad (3)$$

where $V_o = 2\pi Z_1 Z_2 e^2 pN_p$ and x is the distance from the mid point of the two planes and 2l is the interplanar separation. This can be expanded in ascending powers of x as

$$V_2(x) \quad = \quad V(o)+k_1 \frac{x^2}{2} + k_2 \frac{x^4}{4} + \ldots . \qquad (4)$$

where $V(o) \quad = - V_o \ln [4N_p(1+a)^2]$,

$$k_1 \quad = \quad 2V_o/(1+a)^2, \quad k_2 \; = \; 2V_o/(1+a)^4$$

The characteristic frequency in transverse space, in the harmonic oscillator potential (i.e.second term in (4)) is given by

$\omega_o=\sqrt{k_1/m}$. For relativistic case m is replaced by γm and a factor $2\gamma^2$ is multiplied to get the frequency observed in lab.[3,4)] So that

$$\omega_L \quad = \quad 2\gamma^{3/2}\sqrt{k_1/m} \qquad (5)$$

gives the observed radiation peak frequency in the forward direction in harmonic approximation. Using the first order perturbation theory[4)] the energyshift, due to the quartic term, in the nth energy level of harmonic oscillator is given by

$$E_n^{(1)} \quad = \quad \hbar\omega_o \frac{\varepsilon}{4} (2n^2+2n+1) \qquad (6)$$

so that energy levels are given by

$$E_n = (n+\frac{1}{2})\hbar\omega_o + \frac{\varepsilon}{4}\hbar\omega_o(2n^2+2n+1) \qquad (7)$$

The transition energies are given by

$$\hbar\omega_{n,n-1} = 2\gamma^2\hbar\omega_o(1+n\varepsilon) \qquad (8)$$

where

$$\varepsilon = 0.75\ \hbar k_2/k_1 m\omega_o\gamma$$

The calculated frequencies and possible widths are given in Table 1 alongwith the experimental results and previous calculations.

Table 1

Plane	γ	Peak energies and widths (keV)			
		Experimental[2]	Calculations[4]	Present calculations	widths
(110)	111	42.5 ± 0.5	27.7-43.38	36.25-42.74	15.18%
(100)	111	46.7 ± 0.5	35.64-45.99	40.72-47.47	14.21%
(111)	111	35.3 ± 0.5	26.63-43.03	33.65-40.26	16.4%

Thus we see that present results with Nielsen's potential compare well with experimental results and earlier calculations.

3. APPLICATIONS IN DEFECT STUDIES

The effects of a low concentration of lattice defects (so that complicated defect agglomerations or loops are not nucleated) on the transverse motion of channeled particles have been discussed in the literature[7]. For example the presence of dislocations tend to induce some distortions in the axial and planar channels which decreases as one moves away from the dislocation core. The particles are completely dechanneled if they approach within a region enclosed by dechanneling cylinder. Outside this dechanneling region, the state of particle motion is influenced resulting in energy spectrum of transmitted or backscattered particles being severely modified. Similarly the spectral features of channeling radiation will be modified. Although the detailed calculations to estimate the spectrum are not yet complete, a simple model calculation does illustrate the usefulness of channeling radiation to study the defects in solids.

Consider the <u>well channeled particles in undistorted</u> regions of the crystal along planar channels which will not produce any channeling radiation because of absence of transverse motion. When these particles, move on to distorted channel, they start oscillating with a finite amplitude which has been calculated in an approximation which amounts to assuming the curvature of a given channel to be constant (arc of circle approximation). The amplitude is given by [8)]

$$x_{amp} = 0.49\ (1+a)^3\ E\ b/4\pi^3 Z_1 Z_2 e^2 N_P C a^2 r_o^2 \tag{9}$$

and the corresponding period of oscillation being

$$T = [2m(1+a)((1+a)^2 - x_{amp}^2)/\pi Z_1 Z_2 e^2 N_P C a^2]^{\frac{1}{2}} F(x_{amp}/(1+a)) \tag{10}$$

where b is the Burgers vector and r_o is distance of the channel from the dislocation core. So that one expects to observe the effects of dislocation as additional enhancement near the frequency given by $\omega_{obs} = 4\pi\ \gamma^2/T$

4. CONCLUSIONS

Although the above analytical results are obtained only in model calculations, they do underline the need of accurate experiments and more detailed calculations. Recent experimental results[9)] on channeling radiation from silicon crystals with and without platelets can only be qualitatively understood mostly because the structure of the platelets having carbon and nitrogen atoms is fairly complicated. The platelets look like dislocations along (100) planes but their concentration is so high that the separation between two dislocation cores is smaller than the diameter of dechanneling cylinder.[7)] This is why no channeling radiation is observed along (100) Si with platelets. Along other directions, these platelets present themselves as stacking faults which do not produce any distortions but rather present themselves as obstructions. The analysis for this case is presently underway. Meanwhile, the channeling radiation experiments with smaller concentration of one type of defects are needed to establish the effects more quantitatively so that a comparison with analytical calculations may be possible.

REFERENCES

1. Kumakhov, M.A., Phys. Lett. A57, 17 (1976); Soviet Physics JETP 45, 781 (1977).

2. Alguard, M.J., Swent, R.L., Pantell, R.H., Berman, B.L., Bloom, S.D. and Datz, S., Phys. Rev. Lett. 42, 1148 (1979).

3. Andersen, J.U. and Laegsgaard, E., 44, 1079 (1980).

4. Rath, B. and Pathak, A.P., Rad. Eff. 63, 227 (1982).

5. Nielsen, K.O., Electromagnetically enriched isotopes and Mass Spectroscopy, p.68.

6. Lindhard, J., K. Dan Vidensk. Selsk. Mat. Fys. Medd. 34, No.14 (1965).

7. Quere, Y., Ann. Physique 5, 105 (1970); J. Nucl. Matter 53, 262 (1974); Pathak, A.P., Rad. Eff. 61, 1 (1982).

8. Pathak, A.P., Phys. Rev. B13, 4688 (1976); ibid, Phys. Rev. B (in press, Dec. 1984).

9. Park, M., Pantell, R.H., Swent, R.L., Kephart, K.O., Berman, B.L., Datz, S. and Fearick, R.W., J. Appl. Phys. 55, 358 (1984).

THEORY OF PHONONS AND NEUTRON SCATTERING IN RANDOM BINARY ALLOYS

M. Yussouff
Department of Physics
Indian Institute of Technology Kanpur
Kanpur - 208016, India

A new approach to the theory of phonons and neutron scattering in random binary alloys that yields new results is described.

Our formalism is new and general enough to be applied to any disordered system. It is based on the Augmented space method of Mookerjee[1)]. The problem of random binary alloys can be easily handled by this method.

Calculations have been performed for actual three dimensional systems using our formalism. To illustrate the method, let us consider the example of a linear chain with no topological disorder. Atoms of masses m_A and m_B are distributed at random for the "alloy" $A_c\,B_{1-c}$, c is the concentration of A type of atoms. At each site i, one can define a random number n_i with the probability

$$p(n_i) = c\,\delta\,(n_i - 1) + (1 - c)\;\;\delta\,(n_i) \tag{1}$$

The vibrations of such a linear chain with force constants ϕ_{ij} is described by the random dynamical matrix

$$K = \sum_i (m_B - m_A) n_i \omega^2 \, |i\rangle\langle i| - \sum_{i \neq j}\sum \phi_{ij}(1) n_i n_j \, |i\rangle\langle i|$$
$$- \sum_{i \neq j}\sum \phi_{ij}(2)(n_i + n_j) \, |i\rangle\langle i| + \sum_{ij}\sum \phi_{ij}(1) n_i n_j \, |i\rangle\langle j|$$
$$+ \sum_{ij}\sum \phi_{ij}(2)(n_i + n_j) \, |i\rangle\langle j| \qquad (2)$$

Here $\phi(1) = \phi(AA) + \phi(BB) - 2\,\phi(AB)$ and $\phi(2) = \phi(AB) - \phi(BB)$.

The phonon properties can be found from the Green´s function

$$G = (m_B \ \omega^2 \ I - K)^{-1} \qquad (3)$$

which must be averaged over all possible configurations.

In the augmented space method[1)], one constructs a two-dimensional mathematical space with basis vectors $|\gamma_o{}^i\rangle$ and $|\gamma_1{}^i\rangle$ at each site such that

$$p(n_i) = (1/\pi) \, \mathrm{Im} \, \langle \gamma_o^i \, | (n_i + io - m_i)^{-1} | \, \gamma_o^j \rangle \qquad (4)$$

with the matrix M_i given by

$$M_i = \begin{pmatrix} c & \sqrt{c(1-c)} \\ \sqrt{c(1-c)} & 1-c \end{pmatrix} \qquad (5)$$

The augmented space is a product space spanned by $|i, \gamma_o{}^j\rangle = |\,i\rangle \times |$ $\times \, | \, \gamma_o{}^j\rangle$. It has been rigorously proved[1)] that the configurationally averaged Green´s function is given by

$$G_{ij} = \int \ldots \int \prod_\ell dn_\ell \, p(n_\ell) \, G_{ij}(\{n_\ell\})$$
$$= \langle i, \gamma_o \, | \, (m_B \ \omega^2 \tilde{I} - \tilde{K})^{-1} \, | j, \gamma_o \rangle \qquad (6)$$

where $\tilde{K}$ operator can be constructed from the dynamical matrix K using M_i in direct product form wherever n_i occurs in K. This $\tilde{K}$ operator acts in the Augmented space.

Once the Augmented space is formed, it is easy to evaluate $< G_{ij} >$ using the theorem of equation (6) by recursion method. This has been done for the electronic problem[2)] and its correspondence with the cluster CPA formalism has been established.

This new formalism which uses Augmented space method exhibits its superiority over other methods for phonons in disordered systems. It not only overcomes the difficulties encountered in other methods but also agrees better with the experiments. The other approaches had three major difficulties. The first difficulty was related to the task of incorporating the force constant sum rule in CPA theories[3)]. In our formalism[4)] the force constant sum rule can be incorporated exactly. The second difficulty is regarding the analytic (herglotz) properties of $< G_{ij} >$. In the usual CPA approach, one approximates G_{ij} before performing the configuration averaging. The result, in general, may not preserve the analytic properties of G_{ij} and does in fact yield unphysical localized states. Analyticity is exactly preserved in the Augmented space formalism because approximations are done <u>after</u> the configuration averaging has been done.

The final difficulty of usual CPA methods is related to the inclusion of the effective medium while doing the self-consistent calculations on a cluster. This has been done very effectively in our theory. The results are very good in one dimension[4)]. For an actual three-dimensional system $Ni_{55}\,Pd_{45}$, calculation[5)] based on our formalism shows vast improvement over other theoretical calculations and agrees better with experiments. In particular, the phonon width functions have different shapes in different directions as seen experimentally. However, there is no complete agreement with experiments. This is not a real disagreement with our theory because the experimental curves are obtained by neutran scattering and the theory used to derive these results in disordered systems, itself needs modifications as described below.

The theory of thermal neutron scattering in ordered systems is well known[6)]. The interaction Hamiltonian is contact type

$$H_{int} = \sum_j \frac{2\pi\hbar^2}{m} b_j \delta(\vec{\gamma} - \vec{r}_j - \vec{u}_j) \quad (7)$$

and the Born approximation cross-section is expanded for small displacements u_j to write it in terms of the Green's function G_{ij}. This is averaged thermally and over nuclear spins to yield, for perfect crystals

$$\frac{d^2\sigma}{d\Omega dE} = \frac{k'}{2\pi\hbar k} \sum_{ij\mu\nu} B_i B_j n(\omega) q^\mu q^\nu e^{i\vec{q}\cdot(\vec{r}_i - \vec{r}_i)} \operatorname{Im} G^{\mu\nu}_{ij}(\omega) \quad (8)$$

where scattering is from $\vec{k}$ to $\vec{k}'$ with $\vec{q} = \vec{k} - \vec{k}' + \vec{Q}$

$$\hbar\omega = \frac{\hbar^2}{2M}(k^2 - k'^2), n(\omega) = (e^{h\omega/k_B T} - 1)^{-1}, B_j = b_j e^{-\frac{1}{2}\langle(\vec{q}\cdot\vec{u}_j)^2\rangle_{th}}$$

For disordered systems one has to average over possible configurations. We employ the Augmented Space method again to perform the configuration averaging. The calculations are lengthy and the final results exhibit completely new terms in the cross-section that arise due to proper configuration averaging (which was never done in earlier calculations). Thus we get[7)]

$$\left\langle \frac{d^2\sigma}{d\Omega dE} \right\rangle = \frac{d^2\sigma_{inc}}{d\Omega dE} + \frac{d^2\sigma_{coh}}{d\Omega dE} \quad (9)$$

with

$$\frac{d^2\sigma_{inc}}{d\Omega dE} = \frac{k'}{2\pi\hbar k} \sum_\lambda \frac{\beta^{(1)}_\lambda(k,\omega)\sigma^I_\lambda}{(\omega^2 - \omega^2_{\sigma\lambda} - \sigma^R_\lambda)^2 + \sigma^{I\,2}_\lambda}$$

$$- \frac{\beta^{(2)}_\lambda\{\xi'_\lambda(2\xi'_\lambda\xi''_\lambda + \eta''_\lambda W - \sigma^I_\lambda\eta''_\lambda) + \xi''_\lambda(\xi'^2_\lambda - \xi''^2_\lambda + \eta'_\lambda W + \eta''_\lambda\sigma^I_\lambda)\}}{(\xi'^2_\lambda - \xi''^2_\lambda + \eta'_\lambda W + \eta''_\lambda\sigma^I)^2 + (2\xi'_\lambda\xi''_\lambda + \eta''_\lambda W - \sigma^I_\lambda\eta'')^2}$$

$$+ \frac{\beta^{(3)}_\lambda\{\sigma^I_\lambda(\eta'_\lambda W + \eta''_\lambda\sigma^I_\lambda - \xi'^2_\lambda + \xi''^2_\lambda) + W(\eta''_\lambda W - \sigma^I_\lambda\eta_\lambda)\}}{(\eta'_\lambda W + \eta''_\lambda\sigma^I - \xi'^2_\lambda + \xi''^2_\lambda)^2 + (\eta''_\lambda W - \sigma^I_\lambda\eta'_\lambda)^2} \quad (10)$$

and a similar expression for $\frac{d^2\sigma_{coh}}{d\Omega\, dE}$. Here $W = \omega^2 - \omega^2_{o\lambda}\ \sigma^R_\lambda$ and the dependence of the cross-section on the frequency is explicitly exhibited. The first term in equation (10) is what has so far been used to interpret the neutron scattering experiments and the second and third terms are completely new. One important feature of our new result is that it gives asymmetric phonon line shape as observed experimentally whereas earlier theories corresponding to the first term of equation (10) give Lorentian symmetric line shape. Work is under progress for further quantitative comparison between our results and the neutron scattering experiments.

REFERENCES

1. Mookerjee, A., J. Phys. C6, L205 and 1340 (1973).
2. Mookerjee, A., J. Phys. C8, 29, 1524 and 2943 (1975), Bishop, A.R. and Mookerjee, A., J. Phys. C7, 2165 (1974).
3. Mostoller, M. and Kaplan, T., Phys. Rev. B19, 4939 (1979).
4. Yussouff, M. and Mookerjee, A., J. Phys. C17, 1009 (1984).
5. Singh, R.P., Vibrational Properties of Disordered Alloys (Ph.D. Thesis, IIT Kanpur 1982).
6. Elliot, R.J., et al., Rev. Mod Phys. 46, 465 (1974).
7. Yussouff, M. and Mookerjee, A., Theory of Neutron Scattering in Disordered Alloys, ICTP preprint No. IC/84/135 (1984).

QUANTIZED HALL EFFECT AND LOCALIZATION IN TWO DIMENSIONS

Vipin Srivastava
School of Physics, University of Hyderabad
Hyderabad 500 134, INDIA

ABSTRACT

A brief introduction of the quantized Hall Effect is presented with emphasis on the underlying physical principles responsible for it. While the presence of extended states is essential for carrying the Hall current, localized states must be present to cause the observed plateaus in Hall resistance. It is argued that the number of localized states increases as the magnetic field increases.

1. INTRODUCTION

Hall Effect is one of the easier concepts to reckon with in the condensed matter physics and can be observed under very ordinary laboratory conditions. Quite unexpectedly certain surprises have lain hidden for a long time and were unravelled less than a decade ago.[1)] It became a subject of utmost interest about three years ago when the predicted new effects[1)] were verified experimentally.[2-4)] The new phenomena are due to quantum mechanical effects which show up when (a) the system is two dimensional, (b) the temperature is very low (below liquid helium temperature) and (c) the magnetic field is very high - of the order of few tens of kilo-Gauss. The Hall resistance, R_H, has been found to show plateaus over certain ranges of magnetic field in contrast with the monotonic increase as expected classically. At the plateaus the R_H takes values,

$$R_H = h/e^2\ell, \qquad \ldots. \quad (1)$$

where ℓ is an integer which takes values 1,2,3,....; h is Planck's constant and e is the electron charge. Remarkably, the plateaus are horizontal to a great precision so much so that the fine structure constant,

$$\alpha = \frac{\mu_0}{2} \cdot \frac{e^2}{h} = \frac{\mu_0}{2} (\ell R_H)^{-1} \text{ , ... (2)}$$

can be determined to an accuracy of 1 part in 10^7.

This phenomenon has attracted attention from theorists working both in condensed matter and field theory. Consequently the theoretical work has progressed in two main directions - one led by Laughlin[5] and the other led by field theorists who relate Chiral anomaly to the quantized Hall Effect (QHE).[6] The aim of this talk is to put forward qualitative explanations to certain important features exhibited in the experimental results (shown in fig.1) on the basis of the underlying physical principles. No mention shall be made about the work contained in refs. 5 and 6.

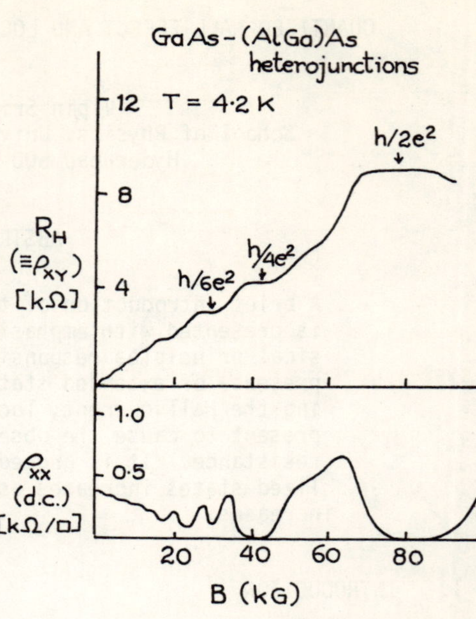

Fig.1 : Experimental results (ref.3) for Hall resistance R_H and d.c. resistivity ρ_{xx}.

The result that will emerge eventually and will prove to be of central importance is that the plateaus are caused by the presence of localized states[7] in the 'dirty' system under consideration. I will also not discuss the 'fractional' QHE[4] which is believed to be caused by electron-electron interaction and is observed at higher magnetic fields and lower temperatures.

2. EXPERIMENTAL RESULTS

Three important features have to be noticed in the experimental results on R_H and the d.c. resistivity ρ_{xx} :

(i) the zeros of ρ_{xx} coincide with the plateaus of R_H;

(ii) the plateaus at higher B correspond to lower ℓ; and

(iii) the plateaus at higher B are relatively wider.

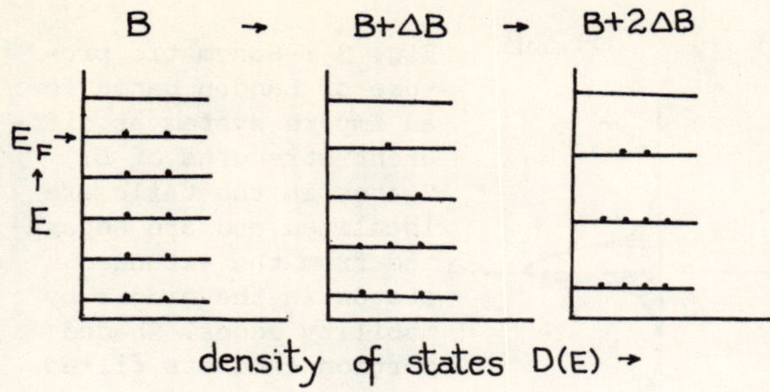

Fig.2 : Schematic picture of Landau levels and their filling at different strengths of magnetic field B.

We shall aim at understanding these features in a two dimensional dirty system containing non-interacting electrons.

3. PHYSICAL PRINCIPLES

We shall consider the system to be dirty, i.e. the one that contains scattering centres, however, to begin with we shall pretend as though we do not know anything about the localization phenomenon.[7] We shall introduce localization at an appropriate stage.

3.1 Application of Magnetic Field

On application of a magnetic field B perpendicular (z-direction) to the two dimensional system (x-y plane) the free electrons are locked into the cyclotron orbits of radii given by,

$$r_\ell = \left[\frac{2\hbar}{eB} (\ell - 1/2) \right]^{1/2} . \qquad \text{.... (3)}$$

They move with constant frequency,

$$\omega_c = eB/m. \qquad \text{.... (4)}$$

The system is now described by a harmonic oscillator Hamiltonian and the continuous energy spectrum breaks down to be represented by,

$$\epsilon_\ell = (\ell - 1/2)\hbar\omega_c \ , \qquad \text{..... (5)}$$

with each Landau level having the degeneracy,

$$s = eB/h \ . \qquad \text{.... (6)}$$

The electrons fill these levels in accordance with Pauli's exclusion principle; the highest filled level is the Fermi level to be denoted

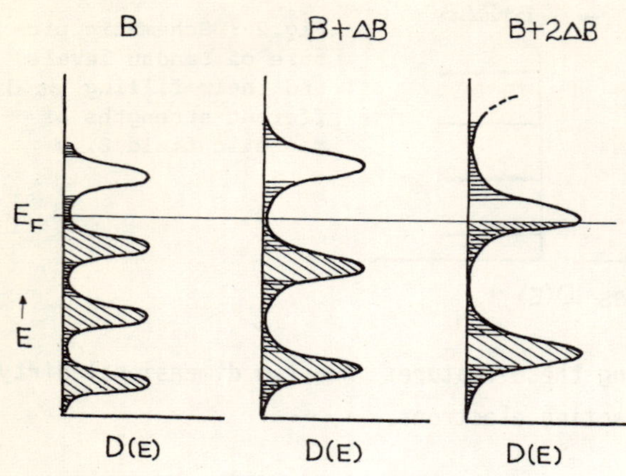

Fig. 3 : Schematic picture of Landau bands for an impure system at different strengths of B. States in the tails are localized and are separated from the extended states in the middle by mobility edges. Shaded portion exhibits filled states.

by E_F.

As the magnetic field is increased, ω_c increases, thereby increasing the separation between the levels; the degeneracy s also increases, which demands that the electrons should rearrange themselves among the discrete energy levels. This should be viewed as a 'spectral flow' in which the levels split appropriately, move upwards and downwards and then merge with others to stabilize into a new set of Landau levels with greater s. Note that stable situations will occur whenever s takes integer values (excluding the spin considerations), i.e. whenever B is an integral multiple of h/e (see eqn.6). Also, as ω_c increases, r_ℓ decreases and stabilizes at values when the cyclotron orbit traps an integral number of flux quanta of size h/e. This is understood as 'flux quantization.' Figure 2 shows schematically the stable situations reached in the course of increasing B. A careful examination will reveal that (a) the E_F is more or less pinned, it may move a little bit ($\sim\hbar\omega_c/2$) upwards or downwards; and (b) the topmost level may be completely filled or may remain partially filled after the electrons have been rearranged at a given stage. The fact (b) becomes crucial when an electric field $\mathcal{E}$ is applied perpendicular to B.

3.2 Application of Electric Field

On application of an electric field $\mathcal{E}$ in say y-direction the electrons start flowing in x-direction such that the centres of the cyclotron orbits move along equipotential surfaces. On the way they experience elastic scattering due to disorder if the topmost level is partially filled. The scattering makes the electrons move into the unfilled orbits of the same level and a contribution to ρ_{xx} is found; ρ_{xy} ($=R_H$) also increases. Interesting situation arises when the topmost level is completely filled. Then the electrons do not feel the presence of the scattering centres because there are no empty orbits in the same level where they can go after the scattering; the next higher level is too high to be reached. Consequently there is no contribution to the d.c. resistivity ρ_{xx}. The movement in such a situation will be super one-dimensional with long range 'phase-rigidity' like a super current. Thus, as B increases one expects to observe occasional dips in ρ_{xx} whenever the topmost level is completely filled; ρ_{xy}($=R_H$), however, should remain unaffected and show a constant increase. This is not what one finds in experiment. We are instigated to include disorder-effects properly.

3.3 Effects of Disorder

It is well known that in presence of disorder the levels are broadened. So we should consider bands in place of sharp levels. The states in the tails of these Landau sub-bands are localized[7] and are separated from the extended states in the middle by 'mobility edges'. This considerably alters the picture (see fig.3). Suppose the E_F lies in a 'mobility gap' and all the extended states below it are completely filled so that $\rho_{xx}=0$. Consider a small increase in B. The set of sub-bands will move upwards with an increased separation among them and an increased degeneracy. Now after the rearrangements of electrons if few are left after completely filling the topmost extended states, they do not have to go to the next higher unfilled sub-band, instead they find (localized) states in the vicinity and move into them. Since these localized electrons do not participate in conduction, transport properties remain unchanged in this process. Consequently no change will occur in R_H and ρ_{xx} shall remain zero. Note that the status quo will maintain till E_F sweeps through the mobility gap as B increases. Thus R_H remains unchanged as long as ρ_{xx} remains **zero** and a plateau is formed. This explains the observation (i) in the experimental results. The observation (ii) is easily understood by noting that the set of sub-bands moves upwards as B increases so that E_F makes a relative movement to lower sub-bands (designated by lower values of ℓ). Also note that the mobility gap increases as B increases. This, coupled with the fact that a plateau occurs as long as E_F remains in the mobility gap, indicates that the plateaus at higher B shall be wider as stated in the observation (iii). It is important in this connection to realize that the whole mobility gap must be filled with localized states even as it becomes wider with increasing B - this is essential in order to have a single plateau corresponding to a given value of ℓ. This clearly indicates that the number of localized states is increasing as B increases. Can we account for it in simple qualitative terms ?

Consider following three mechanisms that either help delocalization or create more localized states.

(1) The effective temperature of the system goes up in presence of electric field $\mathcal{E}$. This tends to decrease the localizing tendencies.

(2) The electrons occupying lower sub-bands screen the long range potential so that the remaining electrons can experience potential fluctuations only over limited ranges which may not always prove to be strong enough to cause localization. Thus localization in higher bands may be lesser than in lower bands. This effect will not be important at very high B when only lower few bands are filled.

(3) Semiclassical considerations show that the wave function remains non-zero only along equipotential surfaces over a range of the order of r_ℓ about the surface. For large but finite B there will be a non-vanishing overlap integral η between two adjacent regions,[8)]

$$\eta \sim \exp\,(-d^2 eB/4\hbar) \quad , \qquad ..... \quad (7)$$

where d is the separation between two closest points of the adjacent regions. As B increases the wave function shrinks more to the vicinity of the equipotential surfaces because the r_ℓ reduces. In this process η reduces and certain closed equipotential regions can delink themselves from larger and open regions, thus giving birth to new localized states. Also, new localized states may peel off from large localized states. In this manner more and more localized states are created as B increases. Note that these arguments shall hold if the potential can be taken to be slowly varying. This condition is fulfilled at high B where r_ℓ's are quite small compared to macroscopic length scales and the potential can be taken to be slowly varying over distances of the order of r_ℓ.

These mechanisms viewed in the light of the observation (iii) indicate that the last one dominates over the other two and as a balance new localized states are constantly created in the mobility gap. It is important to note that the mechanism (3) does not convert an extended state into a localized state, it only segregates localized states from extended states which are mixed up through η at lower values of B.

The analysis presented thus far raises two questions.

(1) Are there extended states present at B=0 or they are created at B>0 ?

(2) How are the localized states created in the middle of the spectrum ?

While the question (1) pertains to the recent claim made by Abrahams et al.[9] that all states become localized in two dimensional dirty systems at arbitrarily small disorders, the question (2) indicates that the familiar picture of Mott-CFO model which puts localized states only in the tails of a band at B=0 should be examined with care - how should we generalize the Mott-CFO model to include a situation like the present one where a single band is <u>split</u> into several sub-bands ?

One may feel that the problems at B=0 and B>0 are two independent ones and should not be linked. Then the above two questions become unnecessary. On the other hand one may want to treat the two situations as contiguous. We propose to take the latter attitude.

While we are examining these questions in detail it will not be out of place to make certain tentative remarks here.

As regards question (1) we observe that the evidences produced here which satisfactorily explain the experimental findings indicate that new localized states are created by a mechanism of segregation as B increases leaving a core of extended states in the limit $B\to\infty$. Thus there will be a greater number of extended states at B=0 contradicting the result of Abrahams et al.[9] Indeed, if disorder exceeds a critical value there will not be any Hall Effect. But it is indicated that the critical value of disorder is not vanishingly small.[10]

A possible way of accommodating localized states in the middle of the spectrum is by allowing the existence of degenerate extended and localized states which are separated in space. Such localized states obviously can not hold electrons permanently and may possible be represented by poles (of Greens function) lying in second or higher Riemann sheets. On application of magnetic field localized and extended states are affected differently so that the aforesaid degeneracy is lifted and the poles collapse into the first sheet and fall in the midst of the branch-cut that represents the continuum of extended states. Consequently such localized states regain their genuine localizing character and break the branch cut at many places. This argument needs to be polished.

REFERENCES

1) Ando, T., Matsumoto, Y. and Uemura, Y., J.Phys.Soc.Jpn.39,279(1975).
2) von Klitzing, K., Dorda, G. and Pepper, M., Phys.Rev. Lett.45,494 (1980)
3) Tsui, D.C., Gossard, A.C., Field, B.F., Cage, M.E. and Dziuba, R.F., Phys. Rev. Lett. 48 3 (1982).
4) Tsui, D.C., Stormer, H.L. and Gossard, A.C., Phys. Rev. Lett. 48, 1562 (1982).
5) Laughlin, R.B., Phys. Rev. B23, 5632 (1981).
6) See e.g. Friedman, M.H., Sokoloff, J.B., Widom, A. and Srivastava, Y.N., Phys. Rev. Lett. 52, 1587 (1984).
7) Anderson, P.W., Phys. Rev. 109, 1492 (1958).
8) Trugman, S.A., Phys. Rev. B27, 7539 (1983).
9) Abrahams, E., Anderson, P.W., Licciardello, D.C. and Ramakrishnan, T.V., Phys. Rev. Lett. 42, 673 (1979).
10) Srivastava, V. and Chaturvedi, M., J. Phys. C15, L215 (1982).

NEUTRINO NEUTRAL CURRENT INTERACTIONS IN ^{12}C

R. Parthasarathy and C.P. Yesudian

Institute of Mathematical Sciences
Madras 600 113, India

ABSTRACT

The neutrino neutral current excitations of $^{12}C^*$ (1^+; T=1; 15.1 MeV) (ISO - vector) and $^{12}C^*$ (1^+; T=0; 12.71 MeV) (ISO - Scalar) are investigated within the context of $SU(2)_L \times U(1)$ electroweak theory of Salam and Weinberg. An effective neutrino-nucleus neutral current interaction Hamiltonian is constructed in the Impulse Approximation (I.A). The variation of the total cross-section for the Iso-vector transition with the Weinberg angle is studied. The recoil polarization is evaluated. The effect of the (strong) Iso-spin mixing on the $^{12}C^*$ (1^+; 12.71 MeV) excitation is examined.

The unified electroweak theory of Weinberg[1)] and Salam[2)] predict the neutral current weak interaction. This new interaction can be used to probe the nucleus complimenting the charged interaction probes. We have investigated the excitation of $^{12}C^*$(1^+; T=1; 15.1 MeV) and $^{12}C^*$(1^+; T=0; 12.71 MeV) by neutrino neutral current (NNC). They correspond to Iso-vector and Iso-scalar transitions respectively. Starting from the neutrino-quark NC interaction Hamiltonian given by[1,2)], we have[3)] constructed an effective neutrino-nucleus NC interaction Hamiltonian using non-relativistic reduction and IA. For Iso-vector excitation, this is given by

$$H_{eff} = \frac{G}{2\sqrt{2}} (1-\vec{\sigma}_\ell . \hat{\nu}') \sum_{n=1}^{A} \Big[\{c_1 I_\ell . I_n + ic_2 \hat{\nu} . (\vec{\sigma}_n \times \hat{\nu}')$$

$$- \frac{g_v}{M} \hat{\nu} . \vec{p}_n\} + \vec{\sigma}_\ell . \{D_1 \vec{\sigma}_n + D_2 \hat{\nu} (\hat{\nu} . \vec{\sigma}_n) + D_3 \hat{\nu}' (\hat{\nu} . \vec{\sigma}_n)$$

$$+ D_4\hat{\nu}\,(\vec{\sigma}_n.\hat{\nu}') + i\,D_5 I_n(\hat{\nu}\times\hat{\nu}') - \frac{ig_V}{M}\,(\vec{P}_n \times \hat{\nu})$$

$$- \frac{g_A}{M}\,\hat{\nu}\,(\vec{\sigma}_n.\vec{p}_n)\}]\,(1 - \vec{\sigma}_\ell.\hat{\nu}) \tag{1}$$

where G is the Fermi Coupling constant and the subscripts l and n refer to lepton and nucleon respectively. C's and D's are effective coupling constants defined in Ref. 3. The matrix element for the excitation is

$$Q = \langle u_{\nu'}\,|\,\Omega\,|\,u_\nu\rangle \tag{2}$$

with

$$\Omega = \frac{G}{2\sqrt{2}}\,(1-\vec{\sigma}_\ell.\hat{\nu}')\big[c_1 M_1 + ic_2\vec{\nu}.(\vec{M}_2 \times \hat{\nu}') - \frac{g_V}{M}(\vec{\nu}.\vec{M}_3)$$

$$+ \vec{\sigma}_\ell.\{D_1\vec{M}_2 + D_2\hat{\nu}\,(\hat{\nu}.\vec{M}_2) + D_3\hat{\nu}'\,(\hat{\nu}.\vec{M}_2)$$

$$+ D_4\hat{\nu}\,(\hat{\nu}'.\vec{M}_2) + i\,D_5 M_1(\hat{\nu}' \times \hat{\nu}) - i\,\frac{g_V}{M}(\vec{M}_3 \times \hat{\nu})$$

$$- \frac{g_A}{M}\,M_4\,\hat{\nu}\}]\,(1 - \vec{\sigma}_\ell.\hat{\nu}) \tag{3}$$

where M_i's are nuclear matrix elements of the form

$$\langle J_f\,M_f\,|\sum_{n=1}^{A}\exp(i\vec{q}.\vec{r}_n)O_i|\,J_I M_I\rangle$$

with $O_1 = 1_n$; $O_2 = \vec{\sigma}_n$; $O_3 = \vec{p}_n$ and $O_4 = \vec{\sigma}_n \cdot \vec{p}_n$.

The effective NNC Hamiltonian for $\bar{\nu}$ excitation and corresponding ones for Iso-scalar transitions have been obtained in a similar way.

For the Iso-vector transition, we have used the general 1p-shell wave functions of Cohen-Kurath which describe the charged current reactions well. We present the σ_T for Iso-vector transition in Table 1.

Table 1

Total cross-sections for the excitation of $^{12}C^*(1^+;\ T=1;\ 15.1\ \text{MeV})$ in $10^{-41}\ \text{cm}^2$

$E_{\nu(\bar{\nu})}$ (MeV)	σ_T^{ν}	$\sigma_T^{\bar{\nu}}$
50	1.20	0.92
100	2.78	1.83
150	2.55	1.84
200	2.87	2.07
250	3.05	2.24
300	3.04	2.35

We define $R = (\sigma_T^{\nu} - \sigma_T^{\bar{\nu}})/(\sigma_T^{\nu} + \sigma_T^{\bar{\nu}})$. R is insensitive to nuclear models but depends very much on $\text{Sin}^2\theta_W$. Hence R can be used to measure the Weinberg angle in a nuclear reaction. For $\text{Sin}^2\theta_W = 0.22 \pm 0.04$, $R = 0.16 \begin{array}{l} +\ 0.02 \\ -\ 0.03 \end{array}$ for 250 MeV.

The average recoil nuclear polarization of $^{12}C^*(1^+;\ T=1;\ 15.1\ \text{MeV})$ is calculated following Devanathan et al[4]. This is found to be nuclear model insensitive and also not much dependent on Weinberg angle.

The Iso-scalar excitation $^{12}C^*(1^+;\ T=0;\ 12.71\ \text{MeV})$ is described in a particle-hole model, of Gillet and Vinh-Mau[5]. However, the Iso-spin mixing of 12.71 MeV and 15.1 MeV can make drastic changes in the cross-section. Accordingly we write

$$|^{12}C^*(1^+;\ 12.71\ \text{MeV})\rangle = \sqrt{1-\delta^2}\,|^{12}C^*(1^+;T=0;12.71\ \text{MeV})\rangle + \delta|^{12}C^*(1^+;\ T=1;\ 15.1\ \text{MeV})\rangle \quad (4)$$

where δ is the Iso-spin mixing parameter. A model dependent measurement of δ has been made by Cecil et al[6] and they quote $\delta = 0.054 \pm 0.011$ or 0.194 ± 0.011. One immediate consequence of this mixing is non-vanishing R for 12.71 MeV level. In the absence of mixing (pure Iso-scalar) we have $\sigma_T^{\nu} = \sigma_T^{\bar{\nu}}$ and hence R = 0. The mixing brings in 15.11 MeV (Iso-vector) level for which $\sigma_T^{\nu} \neq \sigma_T^{\bar{\nu}}$. In Table 2, we give $\sigma_T^{\nu}, \sigma_T^{\bar{\nu}}$, and R for representative values of δ. We find that R is directly related to δ and so a measurement of R will give information on δ, unambiguously.

Table 2

Total cross-sections and R for 12.71 MeV level of ^{12}C.
Cross-sections are in 10^{-42} cm^2. $\sin^2 {}_w = 0.23$.

$E_{\nu(\bar{\nu})}$ (MeV)	$\delta = 0.05$			$\delta = 0.2$		
	σ_T^{ν}	$\sigma_T^{\bar{\nu}}$	R	σ_T^{ν}	$\sigma_T^{\bar{\nu}}$	R
60	0.009	0.08	-0.79	0.489	0.616	-0.11
100	0.022	0.148	-0.74	0.827	1.04	-0.12
150	0.034	0.135	-0.60	0.816	0.987	-0.09
200	0.047	0.161	-0.55	0.921	1.101	-0.09
250	0.056	0.166	-0.50	1.00	1.162	-0.08

To summarize, we have investigated the NNC reactions in ^{12}C. The Iso-vector and Iso-scalar total cross-sections steeply rise with E_{ν} upto ~ 150 MeV and then becomes flat. the ratio R for Iso-vector transition is sensitive to the Weinberg angle. R is also a measure of parity violating effect of NNC interaction. The Iso-spin mixing in the 12.71 MeV level has been investigated. R for this level is a direct measure of the Iso-spin mixing parameter δ.

REFERENCES

1. Weinberg, S., Phys. Rev. Lett. 199, 1264 (1967).
2. Salam, A., in Elementary Particle Physics, Ed. N. Svartholm (Stockholm, Sweden, 1968) p. 367.
3. Parthasarathy, R. and Prem Kumar Yesudian, C., Ann. Phys. (N.Y.) - to appear.
4. Devanathan, V., Parthasarathy, R. and Subramanian, P.R., Ann. Phys. (N.Y.) 73, 291 (1972).
5. Gillet, V. and Vinh Mau, N., Nucl. Phys. 54, 321 (1964).
6. Cecil, F.E., et al., Phys. Rev. C9, 798 (1974).

Electron Scattering and Pion Photoproduction from Light Nuclei

Il-Tong Cheon

Department of Physics, Yonsei University, Seoul, Korea

Electron scattering is one of the best method to test the nuclear wave functions because the processes can be described by the well known electromagnetic interaction. And the distortion of the electron wave function is only due to the Coulomb force. Therefore, it is very easy to treat the processes.

In this talk, I will discuss the electromagnetic form factors of light nuclei and photoproduction of pions.

When the cross section for electron scattering is expressed as [1]

$$d\sigma/d\Omega = (d\sigma/d\Omega)_{Mott}\left((q_\nu^4/q^4)F_L^2(q^2) + (q_\nu^2/2q^2+tg^2\tfrac{1}{2}\theta)F_T^2(q^2)\right) , \tag{1}$$

the longitudinal and transverse form factors are given by

$$F_L^2(q^2) = \frac{4\pi}{Z^2}\frac{1}{(2I_i+1)}\sum_j |\langle I_f||T_J^L(q)||I_i\rangle|^2 \tag{2}$$

$$F_T^2(q^2) = \frac{4\pi}{Z^2}\frac{1}{(2I_i+1)}\sum_J |\langle I_f||T_J^T(q)||I_i\rangle|^2 \tag{3}$$

The iso-scalar and iso-vector nucleon form factors can be expressed by [2]

$$F_1^S = \frac{1}{2}\left((1-B_\omega-B_\phi) + B_\omega(1+\lambda_\omega^2q^2)^{-2}+B_\phi(1+\lambda_\phi^2q^2)^{-2}\right) , \tag{4}$$

$$F_1^V = \frac{1}{2}\left((1-B_\rho) + B_\rho(1+\lambda_\rho^2q^2)^{-2}\right) , \tag{5}$$

$$F_2^S = \frac{1}{2}\left(C_\omega(1+\lambda_\omega^2q^2)^{-2} + C_\phi(1+\lambda_\phi^2q^2)^{-2}\right) , \tag{6}$$

$$F_2^V = \frac{1}{2}C_\rho(1+\lambda_\rho^2q^2)^{-2} . \tag{7}$$

Here the λ's are the Compton wave lengths of mesons with the masses M_ρ = 770 MeV/c^2, M_ω = 783 MeV/c^2 and M_ϕ = 1020 MeV/c^2.

Then, the nucleon electric and magnetic form factors are given as

$$G_E^{\pm} = (F_1^S \pm F_1^V) - (\hbar q/2Mc)^2 (F_2^S \pm F_2^V) \tag{8}$$

$$G_M^{\pm} = (F_1^S \pm F_1^V) + (F_2^S \pm F_2^V) \tag{9}$$

where ± signs denote the proton and neutron. Figure 1 shows the results obtained with two sets of parameter values, (I) $B_\omega = 0.08$, $B_\phi = 1$, $B_\rho = 0.6$, $C_\omega = 1$, $C_\phi = -1.12$, $C_\rho = 3.706$ and (II) $B_\omega = 0.08$, $B_\phi = 1$, $B_\rho = 0.8$, $C_\omega = 0.08$, $C_\phi = -0.2$, $C_\rho = 3.706$. The nucleon form factors are possibly deformed inside the nucleus [3,4]. Since dominant contributions to the transverse form factors are from the magnetic form factors of nucleons, we may apply the free nucleon electric form factor and a phenomenological magnetic form factor,

$$G_M^p(q)/\mu_p = G_M^n(q)/\mu_n$$

$$= \frac{0.2}{1+0.002q^2} + \frac{0.8}{1+0.42q^2} \tag{10}$$

which is also shown in Fig.1.

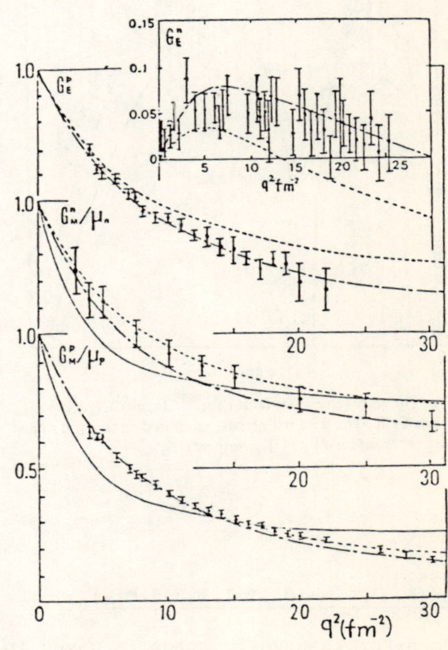

Fig.1. Nucleon form factos. The dotted and dot-dashed curves are obtained with the parameter sets I and II, respectively. The solid curves are obtained by eq.(10).

These modifications are based on the distortion of meson fields inside the nucleus. For instance, the pion field is distorted through the Δ gets an effective mass as a result [5].

$^{9}Be(e,e)^{9}Be(\frac{3}{2}^{-}, T=\frac{1}{2}; g.s.)$: Figures 2 and 3 show the theoretical predictions obtained by the Cohen-Kurath reduced matrix elements (6). The dotted and solid curves are, respectively, calculated by the free and bound nucleon form factors. Theses results may be evidences for distortion of the nucleon e.m. form factors in the nucleus. The discrepancies in the region $q \geq 2$ fm^{-1} are due to neglect of neson exchange current effects.

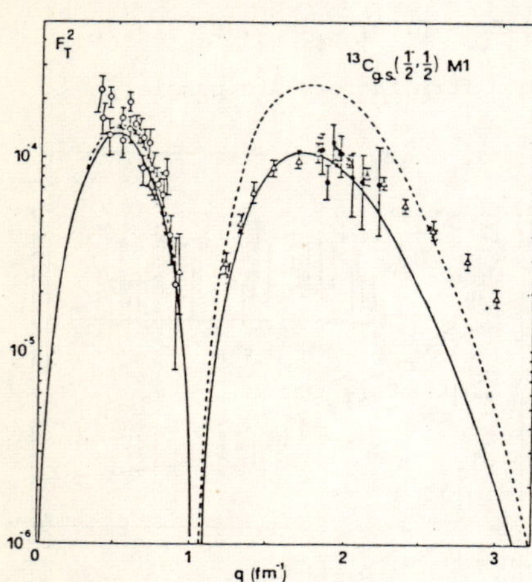

Fig. 2. The transverse form factor for $^{13}C(e,e)^{13}C(g.s.)$. Same as fig.3. The data with triangles, closed circles and open circles are from refs. [7 - 9], respectively.

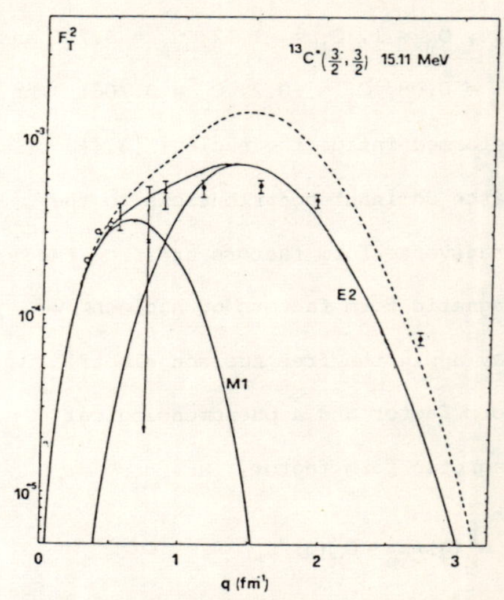

Fig.3. The transverse form factor, $F_T^2(q)$, for the process $^{13}C(e,e)^{13}C^*(\frac{3}{2}^-, \frac{3}{2}; 15.11$ MeV). The dotted curve shows the result obtained with G_E, G_M^p and G_M^n given in eqs.(8) and (9), while the solid curves are calculated with $\tilde{G}_E$ and $\tilde{G}_M$.

The experimental data with closed circles, crosses and open circles were taken from refs. [10-12], respectively.

$^{14}N(e,e')^{14}N^*(0^+, T=1; 2.313$ MeV) : The result of our calculation with the Cohen-Kurath wave function is given in Fig.4 by a dotted curve, which shows a large discrepancy compared with the experimental data. The dashed curve is obtained by normalizing the the dotted curve by an overall factor ¼. Even with this factor, we can not remove the discrepancy in the region q<1 fm^{-1}. When the Cohen-Kurath reduced matrix elements, $X^n_{\frac{1}{2}\frac{1}{2}1} = -X^p_{\frac{1}{2}\frac{1}{2}1} = -1.0922$, $X^n_{\frac{1}{2}\frac{3}{2}1} = -X^p_{\frac{1}{2}\frac{3}{2}1} = -0.051846$, $X^n_{\frac{3}{2}\frac{1}{2}1} = -X^p_{\frac{3}{2}\frac{1}{2}1} = 0.16448$ and $X^n_{\frac{3}{2}\frac{3}{2}1} = -X^p_{\frac{3}{2}\frac{3}{2}1} = 0.026753$, are modified as $X^{n}_{\frac{1}{2}\frac{1}{2}1} = -0.51267$, $X^{n}_{\frac{1}{2}\frac{3}{2}1} = -0.02295$, $X^{n}_{\frac{3}{2}\frac{1}{2}1} = 0.08983$ and $X^{n}_{\frac{3}{2}\frac{3}{2}1} = -0.01196$, then a better agreement can be obtained. The result is shown in Fig.4 by a solid curve. Validity of the modified reduced matrix elements should be examined for some other processes. We will give a discussion on the process, $^{14}N(\gamma,\pi^{+})^{14}C$(g.s.), later on.

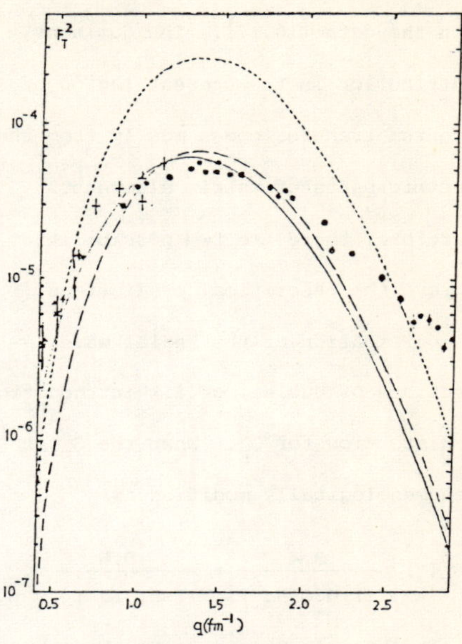

Fig.4. The transverse form factor for $^{14}N(e,e')^{14}N(0^{+},T=1;2.313$ MeV). a=1.67 fm was used. The experimental data are from ref. [13,14].

<u>$^{11}B(e,e)^{11}B$(g.s.)</u> : The reuslts calculated for the transverse form factors with the bound nucleon form factors are shown in Fig.5. Good agreements were obtained for these transitions. The dotted curve in Fig.5 was calculated with the free nucleon form factors. These results denote validity for the distortion of nucleon form factors in the nucleus. Figure 6 shows the results for the longitudinal form factors. The results obtained with the Cohen-Kurath wave function and the free nucleon form factors are shown by the dashed curves, which are in large discrepancies compared

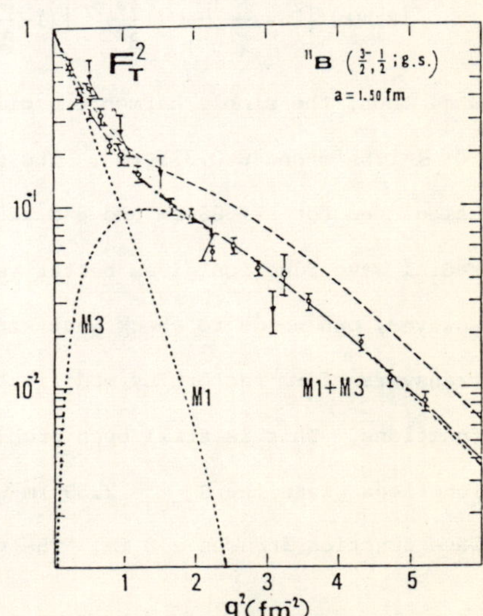

Fig.5 . The transverse form factor for $^{11}B(e,e)^{11}B$(g.s.). The data are from ref. [15] .

with the data [16,17]. The dominant contribution in the present region of momentum transfer comes mostly from the electric parts of matrix elements. Therefore, there are two possibilities to improve the theoretical predictions, i.e. i) modification of the radial wave functions of nuclei, or ii) introduction of distortion for G_E^p. When the G_E^p is phenomenologically modified as

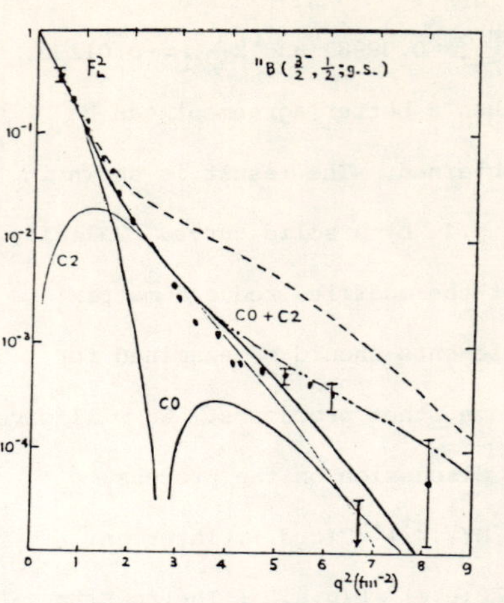

Fig. 6 . The transverse form factors.

$$G_E^p(q) = \frac{0.4}{(1+0.055q^4)} + \frac{0.6}{(1+0.31q^4)} ,$$

we can obtain better results shown by the solid curves. On the other hand, the radial part of nuclear wave function is expressed in the form

$$|R_{1l}(r)|^2 = \frac{1}{(2\pi^2)} \left(\frac{4\pi}{a^2}\right)^{3/2} \left((1-\frac{6\beta}{a^2}) + \frac{4\beta r^2}{a^4}\right) \exp\left(-\frac{r^2}{a^2}\right), \qquad (11)$$

and, then, the simple harmonic oscillator wave function can be obtained for a=1.5 fm and β=0.375 fm^2. The dot-dot-dashed curve in Fig.6 was calculated for a=1.86 fm and β=0.125 fm^2. It is obvious that the modified radial wave function gives better results for the longitudinal form factors. However, one needs to check what kind of effects are induced to the transverse form factors by modification of the radial parts of nuclear wave functions. This is still open problem. The modification of nuclear wave functions gives the R_{rms} = 2.59 fm while the simple harmonic oscillator wave function induces 2.8 fm. The formula R=1.12 $A^{1/3}$ gives 2.49 fm.

The pion photoproduction is well known as one of the best method to test the nuclear wave function. It is naturally expected that the test wave function may reproduce both processes of electron scattering and pion photoproduction. The latter process can be described by the amplitude (18)

$$T^{1}_{\gamma\pi^+} = \Big(\sum_{n=0}^{1} \sum_{\mu=-n}^{n} (-1)^{\mu}\Gamma^{\mu}_{n} K^{n}_{-\mu} \Big)\tau^- , \qquad (12)$$

$$\Gamma^{o}_{o} = 1,\ \Gamma_1 = \boldsymbol{\sigma},\ K^{o}_{o} = L \text{ and } K^1 = \mathbf{K} \qquad (13)$$

The CGLN amplitudes (19) are given by

$$K = C_1\boldsymbol{\varepsilon}+C_3\boldsymbol{\varepsilon}(\mathbf{q}\cdot\mathbf{k})+C_4\mathbf{k}(\mathbf{q}\cdot\boldsymbol{\varepsilon})+C_5(\mathbf{k}-\mathbf{q})(\mathbf{q}\cdot\boldsymbol{\varepsilon})/\big((\mathbf{k}-\mathbf{q})^2+\mu^2\big), \qquad (14)$$

$$L = C_2\mathbf{q}\cdot(\mathbf{k}\times\boldsymbol{\varepsilon}) \ . \qquad (15)$$

All notations are given in ref. (18) .

The radial part of pion wave function $\phi_\ell(qr)$ can be obtained by solving the Klein-Gordon equation with the pion-nucleus optical potential V_{op} . Let us take the local Laplacian optical potential (20) and the following form of nuclear density,

$$\rho(r)=\big(1-(3/2)\alpha C^3\pi^{3/2}\big)^{-1}(1+\alpha C^{-2}r^2)\exp(-r^2/C^2),$$

with $\alpha = 3/4$ and $C = 1.58$ fm.

To test this optical potential, Maleki (21) analyzed the process $^{13}C(\pi^+,\pi^+)^{13}C$ at $T_\pi = 47$ MeV. The result was shown in Fig.7. Since the local Laplacian optical potential seems to produce a good

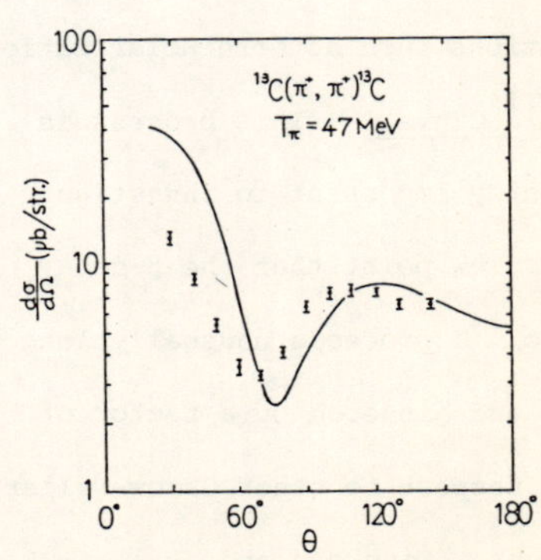

Fig.7 . Elastic scattering of positive pions from ^{13}C. Theoretical result (21) was obtained by the local Laplacian optical potential

result even for $^{13}B(\pi^+,\pi^+)^{13}B$, for which data are not available, we will use this type of optical potential to analyze the process $^{13}C(\gamma,\pi^+)^{13}B$.

The numerical calculations have been carried out for the processes $^{13}C(\gamma,\pi^+)^{13}B$(g.s.) at T_π=39.5 MeV and $^{14}N(\gamma,\pi^+)^{14}C$(g.s.) at T_π=20 MeV.

$^{13}C(\gamma,\pi)^{13}B$(g.s.) : Figure 8 show the results for the differential cross section of this process at T_π=39.5 MeV. Although the shapes of differential cross sections are not sensitive to the optical potential, the magnitudes are suppressed over all angles. This fact can be easily understood by noticing that the shapes are mainly determined by factor $Y_{\ell m}(\hat{q})$ in the matrix elements, while the magnitudes depend more or less on the values of overlap integrals of radial wave functions. We have also shown the result obtained by M. Yamazaki [22] with the SMC optical potential [23]. These results clearly indicate insensitivity of the differential cross section to the optical potential. The oscillator parameter a=1.67 fm was used here. In addition, a conclusion can be drawed that the Cohen-Kurath wave function is able to reproduce both experimental results for the electron scattering and the pion photoproduction from ^{13}C, without introducing extra complications such as core polarization.

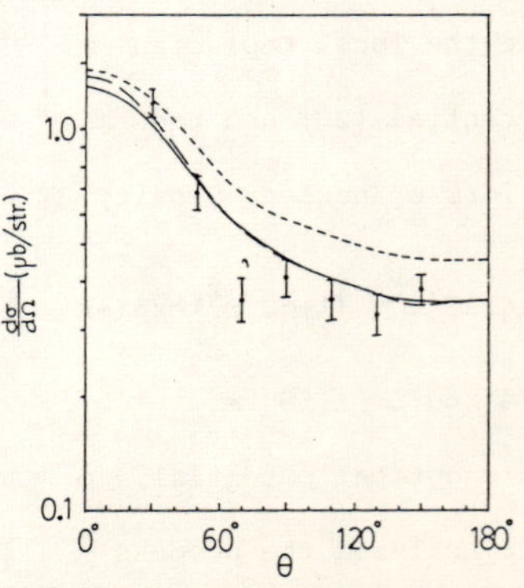

Fig. 8 The differential cross section for photoproduction of π^+ from ^{13}C at E_γ = 193 MeV. The solid curve is the result obtained with the local Laplacian optical potential. The dotted curve shows the result of plane wave calculation. These two curves were obtained with pion partial waves up to $l=2$. The dashed curve [22] was calculated with the SMC optical potential [23] for $l \leq 5$. The data are from ref. [24].

$^{14}N(\gamma,\pi^+)^{14}C$(g.s.) : This process is particularly important to investigate from the view point that the β-decay of ^{14}C to ^{14}N proceeds unusually long lifetime and hindered by a factor of 10^5 with respect to other Gamow-Teller transitions. Figureau and

Mukhopadhyay [25] analyzed the isospin analog reaction $\gamma + {}^{14}N \rightarrow \pi^+ + {}^{14}C$(g.s.) near threshold and they found that the cross section was strongly sensitive to nuclear structure. However, the wave functions used in their calculations were not examined. The process $\gamma + {}^{14}N \rightarrow \pi^- + {}^{14}O$(g.s.) was calculated by Singham and Tabakin [26] using the Cohen-Kurath wave functions. Importance of distortion of pion wave function was mainly stressed in their paper. Experimental data of electron scattering for the (0^+, T=1;2.313 MeV) state of ^{14}N can be used as a constraint on possible ^{14}C and ^{14}O ground-state wave functions. As shown above, the theoretical prediction for the process $^{14}N(e,e')^{14}N^*$(2.313 MeV) by the Cohen-Kurath wave function is as larger as by a factor 4 compared with the experimental data. It is very interesting to examine whether the modified reduced matrix elements giving a reasonable result to the electron scattering can reproduce the experimental data of $^{14}N(\gamma,\pi^+)^{14}C$(g.s.). The results are shown in Fig.9. Unfortunately, the experimental data is not available up to data.

One may also estimate the ft-value of β-decay. Our estimations of log $ft_{1/2}$ are 4.60 and 5.23 with the pure Cohen-Kurath wave function and the modified reduced matrix element, respectively. The experimental

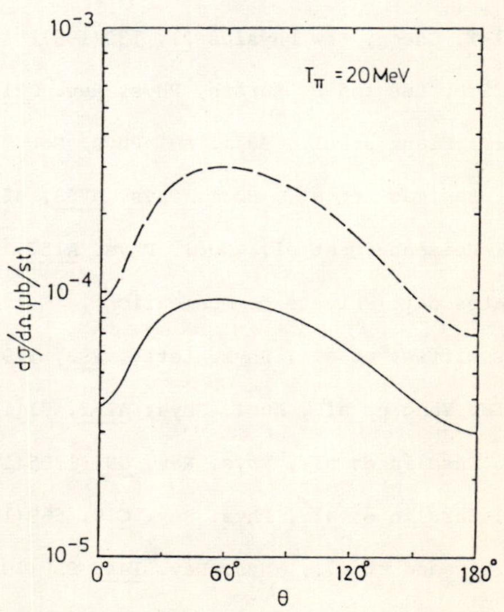

Fig.9 . $^{14}N(\gamma,\pi^+)^{14}C$(g.s.). The dashed and solid curves show the results obtained by the bure Cohen-Kurath and modified reduced elements, respectively.

result is 9.03. This large discrepancy cannot be well understood at the present stage.

In conclusion, distortion of nucleon e.m. form factors may improve the theoretical prediction of electron scattering [27]. However, it is still not conclusive. A number of investigation on electron scattering and pion photoproduction is required to lead more conclusive results.

References

1. T. deForest, Jr. and J.D. Walecka, Adv. Phys. 15, No. 57, 1(1966).

2. F. Iachello, A.D. Jackson and A. Lande, Phys. Lett. 43B, 191(1973).

3. J.V. Noble, Phys. Rev. Lett. 46, 412(1981).

4. Il-T. Cheon, Phys. Lett. 124B, 451(1983).

5. M. Ericson et al., Phys. Lett. 45B, 19(1973).
   K. Ohta and M. Wakamatsu, Phys. Lett. 51B, 325(1974).
   Il-T. Cheon, New Physics 21, 130(1981).

6. T.S.H. Lee and D. Kurath, Phys. Rev. C21, 293(1980).

7. J.B. Flanz et al., Bull. Am. Phys. Soc. 23, 583(1978).

8. L. Lapikas et al., Nucl. Phys. A253, 324(1975).

9. J. Heisenberg et al., Nucl. Phys. A157, 435(1970).

10. Bates data(Private communication).

11. G. Wittwer et al., Phys. Lett., 29B, 579(1969).

12. C.S. Yang et al., Nucl. Phys. A162, 71(1971).

13. N. Ensslin et al., Phys. Rev. C9, 1705(1974).

14. N. Ensslin et al., Phys. Rev. C19, 569(1979).

15. R.E. Rand et al., Phys. Rev. 144, 859(1966).

16. M.C. Bouten and M. Bouten, Nucl. Phys. A229, 141(1978).

17. T. Stovall et al., Nucl. Phys. 86, 225(1966).

18. Il-T. Cheon, Z. Phys. A306, 353(1982).

19. G.F. Chew, M.L. Goldberger, F.E. Low and Y. Nambu, Phys. Rev. 106, 1345(1957).

20. E.H. Auerbach et al., Phys. Rev. 162, 1683(1967).

21. S. Maleki, Ph.D. thesis at Rensselaer Polytechnic Institute, New York 1981.

22. M. Yamazaki, Private communication.

23. K. Stricker, H. McManus and J.A. Carr. Phys. Rev. C19, 929(1979).

24. K. Shoda et al., Proc. Int. Conf. on High Energy Physics and Nuclear Structure, Versailles, 1981.

25. A. Figureau and N.C. Mukhopadhyay, Nucl. Phys. A338, 514(1980).

26. M.K. Singham and F. Tabakin, Phys. C21, 1039(1980).

27. IL-T. Cheon, S.J. Choi and M.T. Jeong, 144B, 312(1984).

SOME ASPECTS OF POLARISED SPIN SYSTEMS

V. Ravishankar

Department of Physics
University of Mysore, Mysore 570 006
India

ABSTRACT

Some of the recent developments in the study of polarised spin systems, the non-oriented spin systems and their sub-classification in particular, are reported here.

1. INTRODUCTION

The polarisation analysis of high spin systems is now an integral and indispensable tool in understanding the basic dynamics underlying numerous phenomena. Moreover, the promise of more precise and complete polarisation experiments has necessitated a detailed study of spin systems. I shall restrict myself to spin systems with sharp j value, and where, single particle measurements are made. My attention will be focussed on non-oriented systems (n.o.s) which have been receiving attention only recently.

2. SU(N) REPRESENTATION OF SPIN SYSTEMS

The attention of physicists has been so far restricted to the study of oriented systems (o.s). Let us recall that one has an o.s when the eigenstates of spin density matrix ρ are the $|\,jm>$ states w.r.t. some direction in space, called the orientation axis. These systems, as is natural, posess a cylindrical symmetry about the axis of orientation. All other systems are non-oriented.

Though the conventional spherical tensor representation for ρ is sufficiently general, and very convenient in the treatment of o.s it has been realised[1] that it is more advantageous to introduce another representation, the SU(n) representation, when n.o.s are to be discussed. The advantage stems from the fact that the expansion matrrices are basis independent. The operator content underlying them may be found in ref. 2. In this new representation, ρ has, in standard notation, the explicit form

$$\rho = \frac{1}{2j+1} \left[1 + \sum_{\alpha} h_{\alpha} H_{\alpha} + \sum_{\alpha=1}^{2j} \sum_{\beta=\alpha+1}^{2j+1} \sum_{\gamma=1}^{2} e_{\alpha\beta}^{(\gamma)} E_{\alpha\beta}^{(\gamma)} \right] \quad (1)$$

The study of n.o.s in the new representation causes little complication. In fact the expression

$$p_{\alpha} = \frac{1}{2j+1} \left[1-(\alpha-1) \left\{ \frac{2j+1}{\alpha(\alpha-1)} \right\}^{\frac{1}{2}} h_{\alpha-1}^{o} + \sum_{n=\alpha}^{2j} \left\{ \frac{2j+1}{n(n+1)} \right\}^{\frac{1}{2}} h_{n}^{o} \right] \quad (2)$$

for the probability weights p_{α} when ρ is diagonalised illustrates the facility with which both the o.s and n.o.s can be handled. Moreover, one obtains elegant bounds[1] on the elements of ρ, apart from certain simplifications introduced in solving the equations of motion[3]. I must also mention here that the SU(3) representation for photons[4] has the generality of treating both real and virtual photons, a feature not shared by, for e.g., the Stokes parameter description. This may be used, with profit, in describing reactions such as electroproduction of pions.

The criteria for a system to be oriented has been discusseed in detail[5]. The answer, in summary form is that we have an o.s iff $t_{kq} \propto Y_{kq}(\theta,\phi)$, for all k, in which case (θ,ϕ) is the axis of orientation. Explicit conditions on t_{kq} for this to be satisfied have also been obtained[5].

The experimental discovery of n.o.s is yet another problem. Unambiguous answers can be obtained[6] if the system decays by photon emission. If, and only if, the system is oriented, and if $Q(\theta_o, \phi_o)$ is the orientation axis relative to the lab, does one have the following relations to be true for the intensity I and the circular polarisation asymmetry A_c of the photons of some multipole L :

$$I(\theta,\phi_1) = I(\theta,\phi_2) \; ; \; A_c(\theta,\phi_1) = A_c(\theta,\phi_2) \tag{3}$$

for <u>all</u> pairs (θ_1,ϕ_2) related by

$$\phi_1 + \phi_2 - 2\phi_o = 2\pi \tag{4}$$

and

$$I(\theta_1,\phi_o) = I(\theta_2, \phi_o) \; ; \; A_c(\theta_1,\phi_o) = - A_c(\theta_2,\phi_o) \tag{5}$$

for <u>all</u> pairs (θ_1,ϕ_2) related by

$$\theta_1 + \theta_2 - 2\theta_o = \pi \tag{6}$$

In all other cases, the system is non-oriented. Moreover, a knowledge of $I(\theta,\phi)$ and $A_c(\theta,\phi)$ determines completely in virtue of

$$t^k_q \; \alpha \left\{ \begin{array}{ll} \int I \;(\theta,\phi)\, Y_{kq}(\theta,\phi) d\Omega & ; \text{k even} \\ \int I \;(\theta,\phi)\, A_c(\theta,\phi)\, Y_{kq}(\theta,\phi) d\Omega & ; \text{k odd} \end{array} \right. \tag{7}$$

This is a self consistent method, since a knowledge of ρ amounts to a knowledge of full matter-radiation density matrix in the final state. This has been applied[7] to study the effectiveness of various basic photoproduction amplitudes in the reaction $^{10}B(\gamma, \pi^+)^{10}\vec{B}e^*$ $(2^+; 1)$ since the residual nucleus decays by γ-emission.

3. GEOMETRICAL REPRESENTATION OF SPIN SYSTEMS WITH ARBITRARY j

I shall first consider a representation for spin-1 particles[8)] and then pass on to a general discussion for arbitrary j values.

A spin-1 system is characterised by both vector polarisation (t_1 or $\vec{p}$) and the tensor polarisation (t_2 or $p_{\alpha\beta}$). The idea is to express in the Principal Axis Aligned Frame (PAAF) defined by $p_{\alpha\beta} = 0$ if $\alpha \neq \beta$. Equivalently, in PAAF, $t_{21} = \mathrm{Im}(t_{22}) = 0$. If $\vec{p} = 0$, it is easy to see that the eigenstates of ρ, are,

$$| X_A > = 1/\sqrt{2}\,(| 1 {-1} >_A - | 11 >_A) \tag{8a}$$

$$| Y_A > = i/\sqrt{2}\,(| 1 {-1} >_A + | 11 >_A) \tag{8b}$$

$$| Z_A > = | 10 >_A \tag{8c}$$

with respective eigenvalues

$$\lambda_{X,Y,Z} = 1/3(1 - p_{ii}{}^{A}) \quad ; \quad i = x,y,z. \tag{9}$$

Eqn. (8) suggests that we designate aligned systems as Cartesian Oriented systems. These are o.s only if ρ is degenerate in at least two eigenvalues. To obtain a geometric representation, define the invariant

$$R^2 = \sum_{k=1}^{2} (t_k \cdot t_k) \lesssim 2 \tag{10}$$

and consider a sphere of radius R, centred at the lab. The point $T(\eta, \zeta)$ on the sphere, defined by the pseudo angles

$$\cos\eta = (p^A{}_{yy} - p^A{}_{zz})/R \; ; \; \tan\zeta = (p^A{}_{xx} - p^A{}_{yy})/(p^A{}_{zz} - p^A{}_{xx}) \tag{11}$$

completely characterises the tensor polarisation. In writing (11) the convention that if two eigenvalues are equal, they are so chosen (as is always possible) that $p^A{}_{xx} = p^A{}_{yy}$ is adopted. Representing the vector polarisation $\vec{p}$ by a point p on the sphere either w.r.t lab. or the PAAF, one finds that the eight parameters (R,P,T, relative orientation of PAAF and lab.) provide an equivalent description of ρ , uniquely. Moreover the conditions on o.s are now stated simply thus:

$$\vec{P} \,||\, \hat{z}_A \;\; ; \;\; T \perp \hat{z}_A \tag{12}$$

which displays how small a subset the o.s of general spin systems form. This fact, though generally unnoticed, was recognised[9)] quite long ago.

To briefly describe the geometrical representation[10)] for arbitrary j, I first note that while t_1 (by definition) is represented by a vector, t_2 is characterised by two vectors:

$$t_{2q} = C_2(Y_1(\hat{q}_1) \;\; x \;\; Y_1(\hat{q}_1))_{2q} \tag{13}$$

which may be easily proved, in the canonical PAAF, by explicit construction. Indeed, the vectors lie respectively in $z_A x_A$, $x_A y_A$ and $y_A z_A$ planes according to as $(t_{20}/t_{22}) \; \varepsilon (-\infty, \; -\sqrt{2/3})$, $(-\sqrt{2/3}, \; \sqrt{2/3})$ and $(\sqrt{2/3}, \infty)$. It thus follows that a spin-1 system is, in general a triaxial system. I remark that $\hat{q}_1$, $\hat{q}_2$ are true vectors unlike the point T mentioned earlier.

The germ of the idea being contained in (13) let us propose the ansatz:

$$t_{kq} = C_k S \; (\ldots(Y_1(\hat{q}_1) \; x \; Y_1(\hat{q}_2))_2 \; x \; \ldots)_{k-1} \;\; x \; (Y_1(\hat{q}_k))_{kq} \tag{14}$$

where S is the symmetriser in all the k vectors. Eqn. (14) is proved by invoking Biedenharn's theorem[11)], whence in the coordinate system in which $t_{kk} = 0$, $\{t_{k,k-1}, \ldots, t_{ko}\}$ satisfy precisely the same algebraic equation one has for a $t_{k-1,q}$, with some (determined) constant C'_{k-1}. The proof follows by induction. In fact (14) implies that there are k coordinate systems (corresponding to k vectors:) in which $t_{kk} = 0$, and this may be proved by explicit calculation and noting that $\cot(\beta/2)e^{-i\alpha}$ ($0 \leqslant \beta \leqslant \pi$; $0 \leqslant \alpha < 2\pi$), spans the entire complex plane.

In this new picture, the complete geometrical representation being provided by (14), choosing C_k such that the k vectors are isomodular, the spin system for a given j is characterised by j(2j+1) physical vectors. It is gratifying that no recourse to higher dimensions needs to be taken to represent high spin systems. One has o.s as an extremely special case, i.e., when all j(2j+1) vectors are coaxial. More so, the method yields a complete subclassification of n.o.s as a proper generalisation of the o.s.

I have attempted to summarise, in a nutshell, the work on density matrices being carried at Mysore. I thank the organisers, especially Prof Y.R. Waghmare who encouraged me to talk on this topic. I thank Mr. T.N. Venkataramana for fruitful discussions. Finally I am grateful to Prof G. Ramachandran for his guidance and inspiration throughout the preparation of the manuscript. I thank the CSIR (India) for financial assistance through the award of a fellowship.

REFERENCES

1. Ramachandran, G. and Murthy, M.V.N., Nucl. Phys. A323, 403 (1979).
2. Von Neumann, J., "Mathematical Foundations of Quantum Mechanics", Princeton University Press, Princeton, 314 (1955).
3. Ramachandran, G. and Murthy, M.V.N., Nucl. Phys. A337, 301 (1980).
4. Ramachandran, G., Murthy, M.V.N. and Mallesh, K.S., Pramana 15, 357 (1980).
5. Ramachandran, G. and Mallesh, K.S., Nucl. Phys. A422, 327 (1984).
6. Ramachandran, G. and Ravishankar, V., Pramana (in print).
7. Ramachandran, G. and Ravishankar, V., (to be published).
8. Ramachandran, G., Mallesh, K.S. and Ravishankar, V., J. Phys. G., Nucl. Phys. L163 (1984).

9. Blin-Stoyle, R.J. and Grace, M.A., in Handbuch der Physik, Vol. XLII, Ed. S. Flugge, 557 (1957).
10. Ramachandran, G. and Ravishankar, V., Proc. Int. Conf. on Nuclear Physics (DAE), Bombay, 70 (1984).
11. Biedenharn, L.C., Ann. Phys. 4, 104 (1958).

STRINGS IN CURVED SPACE

Spenta R. Wadia
Tata Institute of Fundamental Research, Bombay 400005, India.

ABSTRACT

Some issues related to the vacuum geometry of strings are raised and discussed. In particular, we stress the role of conformal and Weyl invariance as a consistency requirement for formulating string dynamics in curved space. The case of the compact group is discussed in some detail. The current algebra representation which is obtained is equivalent to the Frenkel-Kac construction for the k = 1 level of the Kac-Moody algebra.

I. INTRODUCTION

The recent revival of interest in the string model has much to do with the continuing search for a unified theory of all interactions, including gravity. The string tension is fixed to be m^2_{planck}, and the massless modes of the string are identified with the elementary particles of the low energy world [1]. At and above planck length stringy degrees of freedom are important. Candidate string models like the $E_8 \times E_8$ heterotic string [2] or the SO(32) theory of Green and Schwarz [3] have the massless multiplets of ten dimensional supergravity and supersymmetric Yang-Mills theory.

These candidate string theories are formulaed in 10 flat space-time dimensions and hence their vacuum is perturbative and flat. Further, the symmetries of the low energy world like gauge invariance and general co-ordinate invariance are only perturbatively realized. Though their origin can be traced to the reparametrization invariance of the string model, their precise formulation as part of the symmetries of string field theory is lacking. [4]

Towards a first attempt to understand these issues we studied the purely bosonic string in a non-trivial background metric and also coupled to a Kalb-Ramond field [5]. In principle, it is worth studying the coupling of all massless modes to the string.

II. THE CURVED SPACE STRING MODEL

This model is given by specifying a map from the 2 dimensional parameter space $\sigma = \{(\xi_1,\xi_2)\}$ into a manifold M. Denote the local co-ordinate of M by $X_a(\xi)$. The simplest reparametrization invariant action is given by [6]

$$S_1 = \frac{1}{4\lambda^2} \int_\sigma d^2\xi \sqrt{g}\, g^{\mu\nu}(\xi)\, G^{ab}(x)\, \partial_\mu X_a\, \partial_\nu X_b \tag{1}$$

$g_{\mu\nu}$ is the metric on σ and G_{ab} is a metric on M. At the stationary point of S_1 with regard to $g_{\mu\nu}$, S_1 is then the area swept out by the string in space time. We shall deal with a free string and assume that σ has no handles or holes, i.e. we restrict ourselves to the leading order in the 'string coupling constant'.

The second term of our action is the coupling of the 'charged' string to the Kalb-Ramond field $A_{ab}(X)$ [7] ($A_{ab} = -A_{ba}$)

$$S_2 = ik \int_M A^{ab} \, dX_a \wedge dX_b \quad . \tag{2}$$

Note that this coupling is gauge invariant under $A^{ab} \to A^{ab} + \partial^a \Lambda^b - \partial^b \Lambda^a$. In (2) we do admit the possibility of a 'string' singularity in A^{ab}. Using Stoke's theorem we can rewrite S_2 in terms of the gauge invariant field strength $F^{abc} = \Sigma \, \partial^a A^{bc}$

$$S_2 = ik \int_\Sigma F^{abc} \, dX_a \wedge dX_b \wedge dX_c \quad , \tag{3}$$

where the boundary of Σ is M.

Including the 2-dimensional cosmological term the total action is given by [5]

$$S = \frac{1}{4\lambda^2} \int_\sigma d^2\xi \sqrt{g}\, g^{\mu\nu}(\xi)\, G^{ab}(x)\, \partial_\mu X_a \, \partial_\nu X_b \; +$$

$$+ \; ik \int_\Sigma F^{abc}(x)\, dX_a \wedge dX_b \wedge dX_c \; + \; \mu^2 \int_\sigma d^2\xi \sqrt{g} \quad . \tag{4}$$

One of the most important properties of (3) is that it is reparametrization invariant, i.e. invariant under the transformations $\xi_i \to f_i(\xi_1,\xi_2)$ in parameter space. This symmetry must remain intact in the 2 dimensional quantum field theory defined by (4). This requirement imposes restrictions on the choice of the manifold M and the values of $k\,F_{abc}$.

The general investigation of the action (4) remains to be done. Here we will state some elementary general results and then discuss the particularly instructive case when M is a group manifold.

III. CONFORMAL INVARIANCE AND THE VIRASORO ALGEBRA

An important quantity in string theory is the 2 dimensional stress tensor

$$T_{\mu\nu} = \frac{1}{\sqrt{g}} \frac{\delta S}{\delta g^{\mu\nu}} \quad . \tag{5}$$

It generates reparametrization in σ. In the string theory its classical value is zero since it expresses the equation of motion for $g_{\mu\nu}$. Introducing complex co-ordinates $z = \xi_1 + i\xi_2$ and $\bar{z} = \xi_1 - i\xi_2$, $T_{z\bar{z}} = T_{11} + T_{22}$ is the trace of $T_{\mu\nu}$ and $T_{zz} = T_{11} - T_{22} + i2T_{12}$ and $T_{\bar{z}\bar{z}} = \bar{T}_{zz}$ are its traceless parts.

Using reparametrization invariance we locally fix the conformal gauge $g_{\mu\nu} = \delta_{\mu\nu} e^{\phi}$. The residual reparametrizations in this gauge are analytic conformal transformation $z \to f(z)$. Conformal invariance in the quantum theory requires the stress tensor to be traceless: $T_{z\bar{z}} = 0$. Conformal reparametrizations $z \to f(z)$ are generated by T_{zz} and $\bar{z} \to g(\bar{z})$ are generated by $T_{\bar{z}\bar{z}}$. The conservational law of the traceless stress tensor implies that $\partial_{\bar{z}} T_{zz} = 0$ and $\partial_z T_{\bar{z}\bar{z}} = 0$. Then we have the Laurent expansion about the origin

$$T_{zz} = \sum_{n=-\infty}^{+\infty} \frac{L_n}{z^{n+2}} \quad . \tag{6}$$

The coefficients L_n are the Virasoro generators of conformal transformations and they satisfy the Virasoro algebras

$$[L_n, L_m] = (n - m)L_{n+m} + \frac{(-\lambda)}{12} (n^3 - n)\delta n{+}m,0 \quad . \tag{7}$$

$(-\lambda)$ is called the central charge. In the string theory the quantum analogue of the classical constraint $T_{\mu\nu} = 0$ are the Virasoro conditions on the physical state space

$$L_n|\psi\rangle = 0, \quad n \geqslant 1 \tag{8a}$$

$$L_0|\psi\rangle = \alpha_0|\psi\rangle \quad . \tag{8b}$$

The Virasoro conditions (8), together with $L_{-n} = L_n^+$ imply that $\langle\psi|L_n|\psi\rangle = 0$, for all non-zero n. It is worth noting that, unlike in gauge theories, the central extension in (7), does not allow us to choose $L_{-n}|\psi\rangle = 0$, $n > 0$.

A careful treatment of reparametrization invariance and associated Ward identities implies that the conserved stress tensor in the conformal gauge has two more pieces. One coming from the Faydeev-Popov ghost fields by virtue of having gone to the conformal gauge and the other being the contribution from the conformal factor ϕ. (Here ϕ is treated classically). We simply quote the result

$$T_{zz} = T^{M}_{zz} + T^{F-P}_{zz} + T^{L}_{zz}$$

$$T^{M}_{zz} = -G^{ab}(x)\, \partial_z X_a \partial_z X_a \text{ (Matter)}$$

$$T^{F-P}_{zz} = -2\, b_{zz} \partial_z c^z + \partial_z b_{zz} c^z \quad \text{(Faddeev-Popov)}$$

b_{zz} and c^z are reparametrization ghost fields

$$T^{L}_{zz} = -\frac{\lambda}{24}\ (-\partial_z \phi \partial_z \phi + 2\partial_z \partial_z \phi) \text{ (Liouville)}$$

$$(T^{L}_{zz} \text{ is a c-number}) \quad . \tag{9}$$

It is this full T_{zz} that satisfiesthe Virasoro algebra (7). Note that the central charge $(-\lambda)$ of the Virasoro algebra in (7) is the same as the coefficient of the Liouville stress tensor. Hence $-\lambda$ is the anomaly coefficient of the Weyl transformation $\phi(\xi) \to \phi(\xi) + \alpha(\xi)$ ($\alpha(\xi)$ is an arbitrary function) which corresponds to a local scaling of distances in the 2-dimensional world sheet.

Each of the commuting contributions to the stress tensor (save the Liouville piece which is a c-number) defines its respective Virasoro generators algebra. They have a value for the central charges of each contribution. Setting $-\lambda$ to zero implies that the sum of all the Weyl anomalies from all fields cancell and the theory is invariant under arbitrary translations of the conformal factor. The above formalism is valid for strings moving in an arbitrary curved space. For strings moving in flat space time ($G^{ab} = \delta^{ab}$) it is well known that the matter contribution to $-\lambda$ is d, the number of flat dimensions and the Faddeev-Popov contribution which does not depend on the space time is -26. The total anomaly is $-\lambda = (d-26)$.

IV. STRINGS MOVING ON COMPACT GROUP MANIFOLDS

From what we have said before to compute the Weyl anomalies we need to calculate the Virasoro algebras particular to the group. Now the energy momentum tensor corresponding to 2-dimensional field theory or a group manifold follows from the action (1) when G^{ab} is a metric specific to a group. To describe it let us denote group elements by $U(\theta) = \exp(t_a \theta_a)$,

θ_a are local parameters of the group and $\mathrm{tr}(t_a t_b) = -2\delta_{ab}$. The matrix can now be written in terms of vierbeins $U^\alpha_a = \mathrm{tr}\left(\frac{\partial U}{\partial\theta^a} U^{-1} t^\alpha\right)$, $G_{ab} = U^\alpha_a U^\beta_b \delta_{\alpha\beta}$. Defining the currents $L^\alpha_\mu = U^\alpha_a \partial_\mu \theta^a$, the energy-momentum tensor in the conformal gauge is given by[8]

$$T_{\mu\nu} = \frac{1}{\kappa}\left\{ L^\alpha_\mu L^\alpha_\nu - \frac{1}{2}\delta_{\mu\nu}\left(L^\alpha_\mu L^\alpha_\mu\right)\right\} . \tag{10}$$

Note that $T_{\mu\nu}$ has no contribution from the action (2). The constant is a normalization required in the quantum theory.

Note that $T_{\mu\nu} = 0$ in the classical theory. In general in a curved space such a condition cannot be maintained as σ-models are known to generate a mass gap by quantum effects. It is here that the second term in the action (4) becomes relevant. F_{abc} turns out to be the field strength of a magnetic monopole on the group manifold. It satisfies the Maxwell equations $dF = 0$ and $*d*F = 0$ on the group and its integral over a closed 3-surface in the group is an integral multiple of 2π

$$F_{abc} = \frac{1}{N_G}\,\mathrm{tr}\left(\frac{\partial U}{\partial\theta^a} U^{-1} \frac{\partial U}{\partial\theta^b} U^{-1} \frac{\partial U}{\partial\theta^c} U^{-1}\right) \quad a,b,c,\ \text{antisymmetrized.}$$

N_G is a normalization factor, e.g. for SO(3) $N_G = 24\pi$ and for SU(2), $N_G = 48\pi$. The action

$$S_2 = \frac{ik}{N_G}\int d\theta^a \wedge d\theta^b \wedge d\theta^c\ \mathrm{tr}\left[\frac{\partial U}{\partial\theta^a} U^{-1} \frac{\partial U}{\partial\theta^b} U^{-1} \frac{\partial U}{\partial\theta^c} U^{-1}\right] \tag{11}$$

is the well-known Wess-Zumino term[9]. k is an integer required by a consistent quantum theory.

The complete action for a string valued in a group and interacting with a magnetic monopole is

$$S = \frac{1}{4\lambda^2}\int_\sigma d^2\xi \sqrt{g}\, g^{\mu\nu}\, \mathrm{tr}\, \partial_\mu U\, \partial_\nu U^{-1} + $$

$$+ \frac{ik}{N_G}\int_\mu d\theta^a \wedge d\theta^b \wedge d\theta^c\ \mathrm{tr}\left(\frac{\partial U}{\partial\theta^a} U^{-1} \frac{\partial U}{\partial\theta^b} U^{-1} \frac{\partial U}{\partial\theta^c} U^{-1}\right) + \mu^2\int_\sigma \sqrt{g} . \tag{12}$$

The energy momentum tensor in (10) follows from (12).

Now a one-loop calculation indicates that for special values of the string tension, viz. $\alpha = \frac{6\lambda^2 k}{N_g} = \pm 1.$, the **quantum** energy momentum tensor is traceless and the theory is conformal invariant. The result of the one-loop calculation in the conformal gauge $g_{\mu\nu} = e^{\phi}\delta_{\mu\nu}$ and around a background configuration U_0 is [10)]

$$
\begin{aligned}
S_{eff} = & \frac{1}{4\lambda^2}\int d^2\xi\, \partial_\mu U_0\, \partial_\mu U_0^{-1} \\
& + \frac{(26 - \dim G)}{24\pi}\int_\sigma d^2\xi \left[\partial_\mu\phi\partial_\mu\phi + \mu^2 e^{\phi}\right] + \\
& + \frac{C_v}{8\pi^2}(1-\alpha^2)\int d^2\xi\, \mathrm{tr}\, \partial_\mu U_0\, \partial_\mu U_0^{-1}\, \ell n\left(\frac{\Lambda^2}{\mu^2} e^{\phi}\right) .
\end{aligned}
\tag{13}
$$

At $\alpha^2 = 1$, the effective action is conformal invariant. At dim G = 26, it is Weyl invariant.

We will show that at $\alpha^2 = 1$, the field theory is conformal invariant independent of perturbation theory. We can calculate the exact central charge of the Virasoro algebra. The method uses the current algebra following from (12) at the point $\alpha^2 = 1$. This current algebra is a reflection of the local gauge invariance $U(z,\bar{z}) \rightarrow \Omega(z)\, U(z,\bar{z})\, \Omega^{-1}(\bar{z})$ of (12) at the conformal invariant point.

Defining the complex currents

$$
\begin{aligned}
J(z,\bar{z}) &= \frac{12k}{N_G}\partial_z UU^{-1} \\
\bar{J}(z,\bar{z}) &= \frac{12k}{N_G}\partial_{\bar{z}} UU^{-1}
\end{aligned}
\tag{14}
$$

the equations of motion are statements of analyticity $\partial_z J^a = 0 = \partial_z J^a$ and the current algebra is [11)]

$$
\left[J_z^a, J_w^b\right] = f^{abc} J_z^c\, \delta(z-w) + \frac{12k}{N_G}\partial_z \delta(z-w) \tag{15}
$$

and similarly for $\bar{J}$. So we have a GxG current algebra. The equations of motion enable the Laurant expansion

$$J^a(z) = \sum_{n=-\infty}^{+\infty} \frac{J^a_n}{z^{n+1}} \tag{16}$$

and the current algebra (15) translates into the Kac-Moody algebra

$$\left[J^a_n, J^b_m \right] = f^{abc} J^c_{n+m} - m \frac{24\pi k}{N_G} \delta^{ab} \delta_{m+n,0} \quad . \tag{17}$$

The traceless part of the energy momentum tensor (10) is

$$T_{zz} = \frac{1}{\kappa} \sum_a J^a(z) J^a(z) \tag{18}$$

and the corresponding Virasoro generator is given by

$$L_n = \frac{1}{K} : \sum_{m,n} J^a_m J^a_{n-m} : \tag{19}$$

The dots indicate normal ordering: $J^a_n J^a_{-n} : = J^a_{-n} J^a_n$. To complete the definition of the theory we need to say that $J^a(z)$ is a vector current

$$\left[L_n, J^a_m \right] = -m J^a_{n+m} \quad . \tag{20}$$

Now it is easy to check that $\kappa = -(c_V + \frac{24 k}{N_G})$ and the Virasoro algebra is satisfied provided the central charge is

$$(-\lambda)_{group} = \frac{k \dim G}{k + C_V \frac{N_G}{48\pi}} \quad . \tag{21}$$

For orthogonal groups and the exceptional groups $N_G = 24$, for SU(n), $N_G = 48$.

Now if besides the group manifold the string also lives in d flat space-time dimensions the total anomaly formula (including Faddeev-Popov ghosts) is

$$(-\lambda) = d + \frac{k \dim G}{k + C_V \frac{N_G}{48\pi}} - 26 \quad . \tag{22}$$

It turns out that for $k = 1$, $\frac{\dim G}{1 + c_V \frac{N_G}{48}} = r$ the rank of the group and hence (22) becomes

$$-\lambda = d + r = 26 \quad .$$

Hence for $k = 1$, Weyl anomalies cancel if

$$d + r = 26 \quad . \tag{23}$$

V. THE SPECTRUM

The spectrum of the massless states of the closed string moving in (d flat dimensions) x Group is not difficult to write down. Let us denote the (d-2) transverse left and right moving oscillators by α_n^i and $\tilde{\alpha}_n^i$, respectively $i = 1,\ldots,d-2$ and $n = 1,2,\ldots$ In Euclidean space left and right moving are replaced by holomorphic and anti-holomorphic. Similarly let us denote the left and right moving current oscillators, satisfying the Kac-Moody algebra by J_n^a and J_n^a, respectively. The mass operator is then given by

$$\frac{\alpha'}{4} M^2 = N + \tilde{N} - 1 \quad , \tag{24}$$

where $N = \Sigma\, \alpha_{-n}^i \alpha_n^i + \frac{1}{\kappa} \sum J_{-n}^a J_{-n}^a$ and similarly for $\tilde{N}$ and $\kappa = c_V + \frac{k24\pi}{N_G}$. For the closed string we require that the string wave function be independent of translations in 'θ', where $z = e^{\tau + i\theta}$. The generator of this translation is the conformal spin and it must vanish for physical states

$$N - \tilde{N} = 0 \tag{25}$$

The massless states for any value of k then turn out to be the following:

$$(\tilde{\alpha}^i_{-1}\alpha^j_{-1} + \tilde{\alpha}^j_{-1}\alpha^i_{-1})|0\rangle = |g_{ij}\rangle \qquad \text{(graviton)}$$

$$(\tilde{\alpha}^i_{-1}\alpha^j_{-1} - \tilde{\alpha}^j_{-1}\alpha^i_{-1})|0\rangle = |A_{ij}\rangle \qquad \text{(Kalb-Ramond)}$$

$$\tilde{\alpha}^i_{-1}\alpha^i_{-1}\,|0\rangle = |D\rangle \qquad \text{(dilaton)}$$

$$\tilde{\alpha}^i_{-1}\,J^a_{-1}\,|0\rangle = |A^a_{Li}\rangle \qquad \text{(gauge field)}$$

$$\alpha^i_{-1}\,\tilde{J}^a_{-1}\,|0\rangle = |A^a_{Ri}\rangle \qquad \text{(gauge field)}$$

$$\tilde{J}^a_{-1}\,J^b_{-1}\,|0\rangle = |\phi^{ab}\rangle \qquad \text{(Higgs field)} \qquad (26)$$

The last three states are charged states in that they carry an internal symmetry index. So our investigation of strings in curved spaces naturally, and without our knowledge, lead to the Frenkel-Kac construction of these states which have been employed in the heterotic string. Formula (23) says that for d = 10, r = 16. $E_8 x E_8$ and SO(32) are of rank 16. However to determine the dimensions of the groups and hence for uniquely specifying these two groups requires a treatment of the interacting string.

In concludion our exploration of string in curved spaces lead to an understanding of internal symmetry in string theories. However, the main problems we cited at the beginning of this article remain open and yet unsolved.

This article is based on prelininary ideas presented at the Silver Jubilee Workshop at IIT, Kanpur. The details are contained in a preprint with S. Jain and R. Shankar (Ref.5.).

ACKNOWLEDGMENTS

I thank Professor R. Ramachandran and the Physics Department of IIT, Kanpur, for the opportunity to present these ideas.

REFERENCES

1) J.H. Schwarz, Phys. Rep. 89, 223 (1982);
M.B. Green, Surveys in High Energy Physics 3, 127 (1983).

2) D.J. Gross, J.A. Harvey, E. Martinec and R. Rohm, Princeton University preprints; See also P.G.O. Freund, Santa Barbara preprint (1984).

3) M.B. Green and J.H. Schwarz, CALTECH preprint CALT-68-1182 (1984).

4) A. Sen, Fermilab preprint (1985);
S. Das and S. Wadia, Fermilab/TIFR preprint (1985). The techniques of conformal invariant field theory used in this paper are due to Polyakov, Friedan, Knizhnik and Zamalodchikov.

5) R.S. Jain, R. Shankar and S. Wadia, Tata Inst. preprint (January 1985) (to appear in Phys. Rev. D).

6) L. Brink, P. Di Vecchia and P.S. Howe, Phys. Lett. 65B, 471 (1976).

7) M. Kalb and P. Ramond, Phys. Rev. D9, 2273 (1974);
Y. Nambu, Phys. Rep. 23, 250 (1976).

8) H. Sugawara, Phys. Rev. 170, 1659)1968);
C. Sommerfield, Phys. Rev. 126, 2019 (1968).

9) E. Witten, Comm. Math. Phys. 92, 455 (1984);
A. Polyakov and P.B. Weigmann, Phys. Lett. 131B, 121

10) The calculation for flat space-time ($U_0 = 1$) was originally done by A.M. Polyakov, Phys. Lett. 103B, 211 (1981). For the σ-model in flat space ($\phi = 0$) it was done by E Witten in (9).

11) V.G. Kniznik and A.B. Zamolodchikov, Nucl. Phys. B241, 83 (1984).

12) This formula is presented in Ref.5. See also D. Nemeschansky and S. Yankielowivz, SLAC preprint. The Weyl anomaly for Lie groups have been independently derived by P. Goddard and D. Olive, Cambridge preprint. A supersymmetric generalization of this work has been done by E. Berhshoeff, S. Randjbar-Daemi, A. Salam, H. Sarmadi and E. Sezgin, ICTP, Trieste, preprint

LIST OF PARTICIPANTS

1. Abdullah Shoeb, Aligarh Muslim University, Aligarh
2. Balachandran, A.P., Syracuse University, Syracuse, NY, USA
3. Banerjee, K., IIT Kanpur
4. Bansilal, IIT Kanpur
5. Bhattacharjee, J.K., IIT Kanpur
6. Bhattacharya Sumita (Ms), SINP, Calcutta
7. Biswas, S.N., University of Delhi, Delhi
8. Chakravarty Haimanti (Ms), Bose Institute, Calcutta
9. Chaturvedi, S., University of Hyderabad, Hyderabad
10. Cheon Il-Tong, Yonsei University, Seoul, South Korea
11. Deshpande, V.K., IIT Kanpur
12. Dey Biswajit, Institute of Physics, Bhubaneswar
13. Dixit A,N., Christ Church College, Kanpur
14. Ferrell, R.A., University of Maryland, College Park, Md., USA
15. Godbole Rohini (Ms), University of Bombay, Bombay
16. Govindrajan, T.R., Loyola College, Madras
17. Guha Arunabha, University of Rochester, Rochester, USA
18. Gupta Neelima (Ms), Central University, Hyderabad
19. Gyan Mohan, IIT Kanpur
20. Hanna, K.M., Atomic Energy Establishment, Cairo, Egypt
21. Hussain Faheem, Quaid-i-Azam University, Islamabad, Pakistan
22. Jain, M.C., Hindu College, Delhi
23. Joglekar, S.D., IIT Kanpur
24. Joshi Vinay, Jammu University, Jammu
25. Kapoor, A.K., Hyderabad University, Hyderabad
26. Katyal, O.P., IIT Kanpur
27. Kaushal, R.S., Ramjas College, Delhi
28. Khan, D.C., IIT Kanpur
29. Khandekar, D.C., BARC, Bombay
30. Kumar Ajit, IIT Delhi, New Delhi
31. Kumar Deepak, Roorkee University, Roorkee
32. Lawande, S.V., BARC, Bombay
33. Lohiya Daksh, University of Delhi, Delhi
34. Mahapatra, B.P., Sambalpur University, Sambalpur

35. Maheshwari, A.N., Regional College of Education, Mysore
36. Mani, H.S., IIT Kanpur
37. Manna, S.S., SINP, Calcutta
38. Misra, R.C., VSSD College, Kanpur
39. Mohanty, G.C., Institute of Physics, Bhubaneswar
40. Mukherjee, A., IIT Kanpur
41. Nayak Sachidananda, Institute of Physics, Bhubaneswar
42. Pandit Rahul, IISc., Bangalore
43. Parthasarathy, R., Institute for Mathematical Sciences, Madras
44. Pathak Anand, Hyderabad University, Hyderabad
45. Patnaik, P.K., Sambalpur University, Sambalpur
46. Paul Samir K., Institute of Physics, Bhubaneswar
47. Rajasekaran, G., Institute for Mathematical Sciences, Madras
48. Rafat, M., Aligarh Muslim University, aligarh
49. Ramachandran, R., (Convenor), IIT Kanpur
50. Ramakrishnan, T.V., Banaras Hindu University, Varanasi
51. Ramanathan, T., IIT Kanpur
52. Rao, G.N., IIT Kanpur
53. Raval, V.M., Gujarat University, Hyderabad
54. Ravi Shankar, V., Mysore University, Mysore
55. Rizvi, S. Hasim, Aligarh Muslim University, Aligarh
56. Roy, D.K., IIT Delhi, New Delhi
57. Roy, S.N., TIFR, Bombay
58. Sakita Bunji, City College of CUNY, New York, USA
69. Sandhya, S.N. (Ms), Mysore University, Mysore
60. Saran, Dayal, IIT Kanpur
61. Saran, Pankaj, Jamia Milia Islamia, New Delhi
62. Sarin, V.N., IIT Kanpur
63. Sen, Sutapa (Ms), Christ Church College, Kanpur
64. Sharma, S.K., IIT Kanpur
65. Sharma, V.K., Meerut University, Meerut
66. Shukla, A.P., IIT Kanpur
67. Singh, C.P., SSD College, Kanpur
68. Singh, V.A., IIT Kanpur
69. Singru, R.M. (Head, Physics), IIT Kanpur
70. Sirsi Swarnamala (Ms), Mysore University, Mysore
71. Srinivasan, T.M., IIT Kanpur

72. Sriram, M.S., Allahabad Uiversity, Allahabad
73. Srivastava, R.C., IIT Kanpur
74. Srivastava, V., Hyderabad University, Hyderabad
75. Sudarshan, E.C.G., Institute for Mathematical Sciences, Madras
76. Subbarao, D., IIT Delhi, New Delhi
77. Thareja, R.K., IIT Kanpur
78. Tulsi Das, IIT Kanpur
79. Wadia, S., TIFR, Bombay
80. Waghmare, Y.R., IIT Kanpur
81. Yussouff, M., IIT Kanpur
82. Zahid Mohammad, Jamia Millia Islamia, New Delhi

List of student participants from IIT Kanpur

1. Agarwal, A.K.
2. Dabholkar, A.
3. Ghosh, B.
4. Godre, S.S.
5. Kumar Krishna
6. Maiti, S.
7. Mani, T.K.
8. Mathur, M.
9. Misra, A. (Ms)
10. Mukherji, G.
11. Raina, P.K.
12. Singh, J.P.
13. Srivastava, P.K.
14. Tripathi, P.N.
15. Trivedi, S.
16. Tyagi, A. (Ms)